The Biology of Crustacea
VOLUME 10

ECONOMIC ASPECTS: FISHERIES AND CULTURE

The Biology of Crustacea

Editor-in-Chief

Dorothy E. Bliss

Department of Invertebrates
The American Museum of Natural History
New York, New York*

*Present address: Brook Farm Road, RR5, Wakefield, Rhode Island 02879

The Biology of Crustacea

VOLUME 10
Economic Aspects: Fisheries and Culture

Edited by

ANTHONY J. PROVENZANO, JR.
Department of Oceanography
School of Sciences and Health Professions
Old Dominion University
Norfolk, Virginia

ACADEMIC PRESS, INC.   1985

Harcourt Brace Jovanovich, Publishers

Orlando   San Diego   New York   Austin
London   Montreal   Sydney   Tokyo   Toronto

ACADEMIC PRESS, INC.
Orlando, Florida 32887

United Kingdom Edition published by
ACADEMIC PRESS INC. (LONDON) LTD.
24–28 Oval Road, London NW1 7DX

Library of Congress Cataloging in Publication Data

Main entry under title:

The Biology of crustacea.

    Includes bibliographies and indexes.
    Contents: v. 1. Systematics, the fossil record, and
biogeography / edited by Lawrence G. Abele -- v. 2.
Embryology, morphology, and genetics / edited by Lawrence
G. Abele -- [etc.] -- v. 10. Economic aspects / edited by
Anthony J. Provenzano, Jr.
    1. Crustacea--Collected works. I. Bliss, Dorothy E.
[DNLM: 1. Crustacea. QX 463 B615]
QL435.B48   595.3   82-4058
ISBN 0-12-106410-7 (v. 10)

85 86 87 88   9 8 7 6 5 4 3 2 1

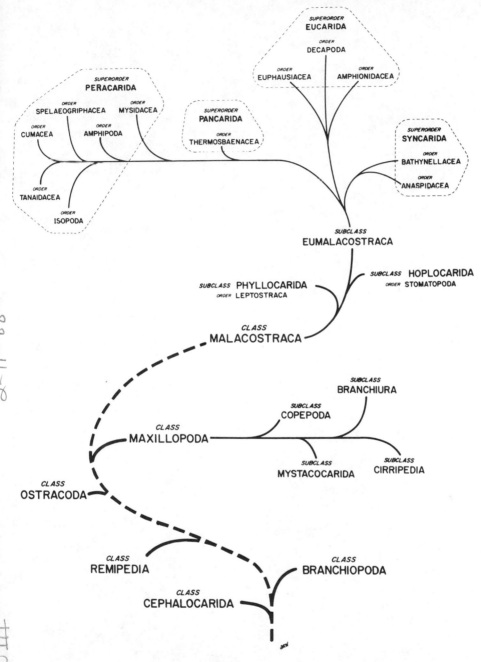

A visual representation of the Bowman and Abele classification of Crustacea (see Chapter 1, Volume 1). This is not intended to indicate phylogenetic relationships and should not be so interpreted. The dashed line at the base emphasizes uncertainties concerning the origin of the five classes and their relationship to one another.

Contents

1. Fisheries Biology of Shrimps and Shrimplike Animals

RICHARD A. NEAL AND ROBERT C. MARIS

2. The Biology and Exploitation of Crabs

PAUL A. HAEFNER, JR.

List of Contributors

Numbers in parentheses indicate the pages on which the authors' contributions begin.

J. Stanley Cobb (167), Department of Zoology, The University of Rhode Island, Kingston, Rhode Island 02881

Paul A. Haefner, Jr. (111), Biology Department, Rochester Institute of Technology, Rochester, New York 14623

Robert C. Maris (1), Department of Oceanography, Old Dominion University, Norfolk, Virginia 23508

Richard A. Neal[1] (1), International Center for Living Aquatic Resources Management, Makati, Metro Manila, Republic of the Philippines

Anthony J. Provenzano, Jr. (249, 269), Department of Oceanography, School of Sciences and Health Professions, Old Dominion University, Norfolk, Virginia 23508

Denis Wang[2] (167), Department of Zoology, The University of Rhode Island, Kingston, Rhode Island 02881

[1]Present address: Office of Agriculture, Bureau of Science and Technology, U.S. Agency for International Development, Washington, D.C. 20523.

[2]Present address: The Kildonan School, Amenia, New York 12501.

General Preface

In 1960 and 1961, a two-volume work, "The Physiology of Crustacea," edited by Talbot H. Waterman, was published by Academic Press. Thirty-two biologists contributed to it. The appearance of these volumes constituted a milestone in the history of crustacean biology. It marked the first time that editor, contributors, and publisher had collaborated to bring forth in English a treatise on crustacean physiology. Today, research workers still regard this work as an important resource in comparative physiology.

By the latter part of the 1970s, a need clearly existed for an up-to-date work on the whole range of crustacean studies. Major advances had occurred in crustacean systematics, phylogeny, biogeography, embryology, and genetics. Recent research in these fields and in those of ecology, behavior, physiology, pathobiology, comparative morphology, growth, and sex determination of crustaceans required critical evaluation and integration with earlier research. The same was true in areas of crustacean fisheries and culture.

Once more, a cooperative effort was initiated to meet the current need. This time its fulfillment required eight editors and almost 100 contributors. This new treatise, "The Biology of Crustacea," is intended for scientists doing basic or applied research on various aspects of crustacean biology. Containing vast background information and perspective, this treatise should be a valuable source for zoologists, paleontologists, ecologists, physiologists, endocrinologists, morphologists, pathologists, and fisheries biologists, and an essential reference work for institutional libraries.

In the preface to Volume 1, editor Lawrence G. Abele has commented on the excitement that currently pervades many areas of crustacean biology. One such area is that of systematics. The ferment in this field made it difficult for Bowman and Abele to prepare an arrangement of families of Recent Crustacea. Their compilation (Chapter 1, Volume 1) is, as they have stated, "a compromise and should be until more evidence is in." Their arrangement is likely to satisfy some crustacean biologists, undoubtedly not all. Indeed, Schram (Chapter 4, Volume 1) has offered a somewhat different arrangement. As generally used in this treatise, the classification of Crustacea follows that outlined by Bowman and Abele.

Selection and usage of terms have been somewhat of a problem. Ideally, in a treatise, the same terms should be used throughout. Yet biologists do not agree on certain terms. For example, the term *ostracode* is favored by systematists and paleontologists, *ostracod* by many experimentalists. A different situation exists with regard to the term *midgut gland,* which is more acceptable to many crustacean biologists than are the terms *hepatopancreas* and *digestive gland.* Accordingly authors were encouraged to use *midgut gland.* In general, however, the choice of terms and spelling was left to the editors and authors of each volume.

In nomenclature, consistency is necessary if confusion as to the identity of an animal is to be avoided. In this treatise, we have sought to use only valid scientific names. Wherever possible, synonyms of valid names appear in the taxonomic indexes.

Every manuscript was reviewed by at least one person before being accepted for publication. All authors were encouraged to submit new or revised material up to a short time prior to typesetting. Thus, very few months elapse between receipt of final changes and appearance of a volume in print. By these measures, we ensure that the treatise is accurate, readable, and up-to-date.

Dorothy E. Bliss

General Acknowledgments

In the preparation of this treatise, my indebtedness extends to many persons and has grown with each succeeding volume. First and foremost is the great debt owed to the authors. Due to their efforts to produce superior manuscripts, unique and exciting contributions lie within the covers of these volumes.

Deserving of special commendation are authors who also served as editors of individual volumes. These persons have conscientiously performed the demanding tasks associated with inviting and editing manuscripts and ensuring that the manuscripts were thoroughly reviewed. In addition, Dr. Linda H. Mantel has on innumerable occasions extended to me her advice and professional assistance well beyond the call of duty as volume editor. In large part because of the expertise and willing services of these persons, this treatise has become a reality.

Also deserving of thanks and praise are scientists who gave freely of their time and professional experience to review manuscripts. In the separate volumes, many of these persons are mentioned by name.

Lastly, thanks are due to all members of the staff of Academic Press involved in the preparation of this treatise. Their professionalism and encouragement have been indispensable.

Dorothy E. Bliss

Preface to Volume 10

Not only are the Crustacea as a whole of major ecological value, they have an impact also on the affairs of human beings in an important way. Some of the economic effects are poorly appreciated by the general public and even by the scientific community. For example, the effect of barnacles on maritime fuel and maintenance costs is probably on the order of hundreds of millions of dollars annually. The damage done to wooden structures by peracarids in the nearshore marine environment is likewise substantial. In the early planning for this volume, we had hoped to include coverage of all major economic aspects of the biology of crustaceans, but that proved to be a task beyond our current capacity. We chose instead to limit the volume to a consideration of those elements of crustacean biology associated primarily with production of human food, namely, fisheries and culture.

The role of the brine shrimp *Artemia* in crustacean culture, as in fish culture, is significant and worldwide, both for freshwater and saltwater hatchery operations. Limitations in supplies of the cysts of *Artemia,* the form in which this shrimp is most valuable, have frequently been described as bottlenecks to more rapid expansion of artificial propagation efforts for larval fish and crustaceans. Because of the importance of *Artemia* in crustacean culture and in research applications, original plans included a chapter on this important crustacean. Several recent major volumes have been published on the biology and economic impact of the brine shrimp, however, so such detailed treatment is not included here.

The shrimps and prawns support the largest and most valuable crustacean fisheries in the world. In Chapter 1, Neal and Maris review the groups comprising the commercially important shrimps and prawns and their near relatives as well as the generally used fishing methods. The major genera and the location of the major fisheries are summarized. Various aspects of the biology and the fisheries management problems of shrimp are discussed. A section on food technology with emphasis on postharvest handling is included, as is a summary of trends and research needs.

The brachyuran and anomuran crabs support large marine fisheries in several oceans. In Chapter 2, Haefner points out the role and impact of body form in the biology and especially the fisheries of crabs. He follows this with a review of the more important crab fisheries, including geographical location, scope or scale, and gear used in the fisheries, and discusses biological factors important in fisheries of crabs.

Chapter 3 by Cobb and Wang deals with lobsters and their kin. Significant differences in the anatomy and biology of these economically important groups greatly affect their respective fisheries. The impact on fisheries methods and management approaches of behavioral responses to environment, modes of reproduction, recruitment, and population dynamics are discussed. The chapter begins with biological and behavioral considerations and concludes with consideration of fisheries management problems and research needs.

With the steadily increasing demand and high prices offered for crustaceans, many fisheries are under severe pressure, and in some instances maximum yields from natural populations may already have been reached. Methods for the artificial propagation and production of crustaceans have developed rapidly in recent years. In Chapter 4, Provenzano reviews culture methods and factors important in managing systems through water quality control. Desirable characteristics of candidate species for crustacean aquaculture are discussed. The peculiar feature of molting and its effect on approaches to crustacean culture are included.

Chapter 5 deals with the large-scale culture of major decapod groups, opening with the general biological characteristics of decapods relevant to aquaculture. This is followed by sections on the culture of shrimp, prawns, anomurans, crabs, lobsters, and crayfishes.

Several of the present chapters, originally intended for inclusion in an earlier volume, were submitted substantially in advance of others, and hence their publication was delayed. I wish to thank those authors for their extraordinary patience. Dr. Philip Mundy kindly reviewed the fisheries chapters and made a number of helpful suggestions. Mr. Robert Maris assisted me in several ways, especially in the preparation of the indexes.

In a work of this type no manuscript is ever complete, and all could be indefinitely revised and updated. Nevertheless, we believe we have offered the reader a stimulating general review of the important economic aspects of crustacean biology and trust that we have encouraged the further exploration of some of the most fascinating and exciting problems in applied crustacean biology.

Anthony J. Provenzano, Jr.

Classification of the Decapoda*

Order Decapoda Latreille, 1803
  Suborder Dendrobranchiata Bate, 1888
        Family Penaeidae Rafinesque, 1815, *Penaeus, Metapenaeus, Penaeopsis,*
                        Trachypenaeopsis
            Aristeidae Wood-Mason, 1891, *Gennadus, Aristeus*
            Solenoceridae Wood-Mason and Alcock, 1891, *Solenocera,*
                Hymenopenaeus
            Sicyoniidae Ortmann, 1898, *Sicyonia*
            Sergestidae Dana, 1852, *Sergestes, Lucifer, Acetes*
  Suborder Pleocyemata Burkenroad, 1963
    Infraorder Stenopodidea Claus, 1872
        Family Stenopodidae Claus, 1872, *Stenopus*
    Infraorder Caridea Dana, 1852
        Family Procarididae Chace and Manning, 1972, *Procaris*
            Oplophoridae Dana, 1852, *Oplophorus, Acanthephyra, Systellaspis*
            Atyidae De Haan, 1849, *Atya, Caridina*
            Nematocarcinidae Smith, 1884, *Nematocarcinus*
            Stylodactylidae Bate, 1888, *Stylodactylus*
            Pasiphaeidae Dana, 1852, *Leptochela, Parapasiphae*
            Bresiliidae Calman, 1896, *Bresilia*
            Eugonatonotidae Chace, 1936, *Eugonatonotus*
            Rhynchocinetidae Ortmann, 1890 *Rhynchocinetes*
            Campylonotidae Sollaud, 1913, *Bathypalaemonella*
            Palaemonidae Rafinesque, 1815, *Palaemon, Palaemonetes,*
                        Macrobrachium
            Gnathophyllidae Dana, 1852, *Gnathophyllum*
            Psalidopodidae Wood-Mason and Alcock, 1892, *Psalidopus*
            Alpheidae Rafinesque, 1815, *Alpheus, Synalpheus, Athanas*
            Ogyrididae Hay and Shore, 1918, *Ogyrides*
            Hippolytidae Dana, 1852, *Hippolyte, Thor, Latreutes, Thoralus, Lysmata*
            Processidae Ortmann, 1896, *Processa*
            Pandalidae Haworth, 1825, *Pandalus, Parapandalus, Heterocarpus*

Prepared by Lawrence G. Abele.

Section Oxystomata H. Milne Edwards, 1834
   Family Dorippidae MacLeay, 1838, *Ethusina, Dorippe*
      Calappidae De Haan, 1833, *Calappa*
      Leucosiidae Samouelle, 1819, *Persephona, Randallia*
Section Oxyrhyncha Latreille, 1803
   Family Majidae Samouelle, 1819, *Maja, Hyas*
      Hymenosomatidae MacLeay, 1838, *Hymenosoma*
      Mimilambridae Williams, 1979, *Mimilambrus*
      Parthenopidae MacLeay, 1838, *Parthenope*
Section Cancridea Latreille, 1803
   Family Corystidae Samouelle, 1819, *Corystes*
      Atelecyclidae Ortmann, 1893, *Atelecyclus*
      Pirimelidae Alcock, 1899, *Pirimela*
      Thiidae Dana, 1852, *Thia*
      Cancridae Latreille, 1803, *Cancer*
Section Brachyrhyncha Borradaile, 1907
   Family Geryonidae Colosi, 1923, *Geryon*
      Portunidae Rafinesque, 1815, *Portunus, Carcinus, Callinectes, Scylla*
      Bythograeidae Williams, 1980, *Bythograea*
      Xanthidae MacLeay, 1838, *Rhithropanopeus, Panopeus, Xantho, Eriphia,*
                                                                                 Menippe
      Platyxanthidae Guinot, 1977, *Platyxanthus*
      Goneplacidae MacLeay, 1838, *Frevillea*
      Hexapodidae Miers, 1886, *Hexapodus*
      Belliidae, 1852, *Bellia*
      Grapsidae MacLeay, 1838, *Grapsus, Eriocheir, Pachygrapsus, Sesarma*
      Gecarcinidae MacLeay, 1838, *Gecarcinus, Cardisoma*
      Mictyridae Dana, 1851, *Mictyris*
      Pinnotheridae De Haan, 1833, *Pinnotheres, Pinnixa, Dissodactylus*
      Potamidae Ortmann, 1896, *Potamon*
      Deckeniidae Bott, 1970, *Deckenia*
      Isolapotamidae Bott, 1970, *Isolapotamon*
      Potamonautidae Bott, 1970, *Potamonautes*
      Sinopotamidae Bott, 1970, *Sinopotamon*
      Trichodactylidae H. Milne Edwards, 1853, *Trichodactylus, Valdivia*
      Pseudothelphusidae Ortmann, 1893, *Pseudothelphusa*
      Potamocarcinidae Ortmann, 1899, *Potamocarcinus*
      Gecarcinucidae Rathbun, 1904, *Gecarcinucus*
      Sundathelphusidae Bott, 1969, *Sundathelphusa*
      Parathelphusidae Alcock, 1910, *Parathelphusa*
      Ocypodidae Rafinesque, 1815, *Ocypode, Uca*
      Retroplumidae Gill, 1894, *Retropluma*
      Palicidae Rathbun, 1898, *Palicus*
      Hapalocarcinidae Calman, 1900, *Hapalocarcinus*

Contents of Previous Volumes

Volume 5: Internal Anatomy and Physiological Regulation
Edited by Linda H. Mantel

Volume 9: Integument, Pigments, and Hormonal Processes
Edited by Dorothy E. Bliss and Linda H. Mantel

Fisheries Biology of Shrimps and Shrimplike Animals

RICHARD A. NEAL and ROBERT C. MARIS

THE BIOLOGY OF CRUSTACEA, VOL. 10
Copyright © 1985 by Academic Press, Inc.

I. INTRODUCTION

A. Scope of Chapter

More nations harvest shrimps (natantian, decapod crustaceans) than near-ly any other kind of marine product, and in at least 20 countries, shrimp fishing is a substantial industry of vital economic importance. Shrimps have increased in popularity among consumers in the past few decades, so cer-tain areas have switched from subsistence fishing to more intensive opera-tions (Considine, 1982). The major harvesters, of shrimps and shrimplike animals, in 1981 were (in order) the Soviet Union, China, the United States, Indonesia, India, Thailand, Japan, Malaysia, Brazil, Norway, Mexico, Greenland, and the Philippines (FAO, 1983a).

Collectively, shrimps are one of the most economically valuable fishery products, and worldwide 1,697,910 metric tons (M.T.) were harvested in 1981 (FAO, 1983a). Even though shrimps constitute only ~1–2% of the total fishery harvest, they represent more than 5% of the total value (Wick-ins, 1976). In the United States, shrimps are the fourth most important fishery product in quantity, but first in value (Thompson, 1983) (Table I).

The terms "shrimp" and "prawn" have often been used interchangeably. Preferably, *prawns* are larger species and often have big claws, whereas *shrimps* are typically smaller (Considine, 1982). For the purposes of this review, all will generally be termed *shrimps*. Likewise, the color attributed to shrimps usually results from natural chromatophoral pigment dispersion or concentration, but the term sometimes refers to coloration resulting from cooking processes (Fitzgibbon, 1976).

Shrimp fisheries differ according to the biological aspects of the two large subgroups of shrimps. Penaeid shrimps form the basis of many tropical and subtropical fisheries, whereas caridean shrimp fisheries are generally found in cooler, temperate regions (Allen, 1966a; King, 1981). Shrimplike animals of commercial importance include the krill, or euphausiids; mantis shrimps,

TABLE I

United States Fishery Landings: 1982[a]

Group	Pounds (× 10³)[b]		%	Dollars (× 10³)		%
Menhadens	2,766,061	(1)	43.3	107,741	(5)	4.5
Salmons	607,420	(2)	9.5	391,999	(2)	16.4
Crabs	349,602	(3)	5.5	282,233	(3)	11.8
Shrimps	283,717[c]	(4)	4.4	509,118	(1)	21.3
Tunas	261,409	(5)	4.1	145,724	(4)	6.1
Total	6,382,201			2,388,749		

[a] Compiled from Thompson, 1983.
[b] Values in parentheses are rankings. One metric ton = 2,204.62 pounds.
[c] Heads on.

or stomatopods; and mysid shrimps. Listings of economically important shrimps are presented in Tables II and III.

The primary shrimp fishing gear is the otter trawl, which is most effective on smooth surfaces. Some species are harvested by traps, seines, lift nets, or cast nets, but these methods account for only a small percentage of the total shrimp harvest (Kurian and Sebastian, 1976).

The major categories of processed shrimps include (1) whole or headless, (2) cooked or uncooked, (3) peeled or unpeeled, and (4) raw or frozen (Wigley, 1973). The abdominal muscle is the most desirable edible portion, and frozen "tails" are the most common form for international trade (FAO, 1983b). In 1982, consumption in the United States was 1.52 lb (0.68 kg) edible meat per person (Thompson, 1983). Efficient utilization of by-catch and by-products of processing forms an important, but often overlooked, part of the shrimping industry (Meyers and Rutledge, 1973).

The purpose of this chapter is to describe the major shrimp fisheries, the importance of these fisheries, and the biological aspects that relate directly to man's utilization of shrimps.

B. Major Groups of Commercial Importance

The warm-water penaeid shrimps [order Decapoda, suborder Dendro-branchiata, superfamily Penaeoidea, family Penaeidae; and superfamily Sergestoidea, family Sergestidae (Bowman and Abele, 1982)] are by far the most commercially important group. The penaeids are characterized by a complex life cycle, during which eggs are broadcast freely and hatch as free-swimming planktonic naupliar larvae. Postlarvae move into estuaries,

TABLE II

Major Shrimp Fishery Landings and Producers as Compiled from Various Sections of FAO[a]

Species	1981 Production (metric tons)	Producers (% of total catch)
Penaeoidea		
Acetes japonicus (*akiami* paste shrimp)	163,100	China (95), Korea (5)
Artemesia longinaris (Argentine stiletto shrimp)	114	Argentina (100)
Haliporoides triarthrus (knife shrimp)	530	South Africa (100)
Metapenaeus joyneri (shiba shrimp)	1,819	Korea (100)
Metapenaeus spp. (*Metapenaeus* prawns)	36,808	Indonesia (61), Thailand (32), Philippines (6), Papua New Guinea (1)
Parapenaeopsis atlantica (Guinea shrimp)	0 (1980 = 122)	Gambia (100)
Parapenaeus longirostris (deepwater rose shrimp)	16,135	Spain (97), Soviet Union (3)
Penaeus aztecus (northern brown shrimp)	67,469	United States (100)
P. brevirostris (crystal shrimp)	3,051	Panama (75), Costa Rica (13), El Salvador (11), Guatemala (1)
P. californiensis (yellowleg shrimp)	500	Guatemala (88), El Salvador (10), Costa Rica (2)
P. duorarum (northern pink shrimp)	21,098	United States (85), Cuba (15)
P. japonicus (kuruma prawn)	5,421	Japan (85), Korea (15)
P. kerathurus (triple-grooved shrimp)	7,095	Italy (85), Spain (15)
P. latisulcatus (western king prawn)	1,721	Thailand (100)
P. merguiensis (banana prawn)	63,746	Indonesia (72), Thailand (27), Papua New Guinea (1)
P. monodon (giant tiger prawn)	16,572	Indonesia (98), Thailand (2)
P. notialis (northern pink shrimp)	3,573	Nigeria (56), Gabon (44)
P. orientalis (fleshy prawn)	38,031	China (99), Korea (1)
P. semisulcatus (green tiger prawn)	2,313	Thailand (90), United Arab Emirates (10)
P. setiferus (northern white shrimp)	37,407	United States (100)

TABLE II (*Continued*)

Species	1981 Production (metric tons)	Producers (% of total catch)
Penaeus spp. (penaeid shrimps)	33,976	Philippines (44), Panama (16), Colombia (8), Honduras (7), Ivory Coast (7), Cuba (4), Cameroon (3), El Salvador (3), Costa Rica (2), Guatemala (2), Peru (2), Sierra Leone (2)
Pleoticus muelleri (Argentine red shrimp)	2,616	Argentina (100)
P. robustus (royal red shrimp)	258	United States (100)
Plesiopenaeus edwardsianus (scarlet shrimp)	1,686	Spain (100)
Sergestidae (sergestid shrimps)	60,015	Malaysia (44), Philippines (33), Thailand (23)
Sicyonia brevirostris (rock shrimp)	1,935	United States (100)
Xiphopenaeus kroyeri (Atlantic seabob)	3,999	United States (100)
Xiphopenaeus, Trachypenaeus spp. (Pacific seabobs)	12,238	Panama (56), El Salvador (16), Colombia (12), Guatemala (12), Costa Rica (4)
Caridea		
Crangon crangon (common brown shrimp)	45,573	Federal Republic of Germany (32), Soviet Union (22), Spain (17), Netherlands (11), Denmark (5), Faeroe Islands (4), Belgium (3), France (3), England (2), Algeria (1)
Heterocarpus reedi (Chilean nylon shrimp)	2,942	Chile (100)
Macrobrachium rosenbergii (freshwater giant prawn)	3,742	Indonesia (100)
Macrobrachium spp. (freshwater prawns)	13,125	Brazil (71), Mexico (29)
Palaemon serratus (common prawn)	1,917	Algeria (63), France (29), Iceland (4), Spain (3), Denmark (1)
Palaemonidae (freshwater prawns)	18,577	Japan (32), Philippines (30), Thailand (21), Indonesia (16), Honduras (1)

(*continued*)

TABLE II *(Continued)*

Species	1981 Production (metric tons)	Producers (% of total catch)
Pandalus borealis (northern prawn)	100,818	Norway (41), Greenland (39), Iceland (8), Denmark (4), Faeroe Islands (3), Japan (2), Sweden (2), France (1)
Pandalus spp. (pink, pandalid shrimps)	15,793	Canada (86), United States (8), Soviet Union (4), Scotland (2)
Pandalus spp., *Pandalopsis* spp. (Pacific Ocean shrimps)	30,659	United States (100)
Natantian decapods	896,982	India (10), Brazil (5), Indonesia (4), Malaysia (4), Mexico (4), Thailand (4), Viet Nam (4), Japan (3), Australia (2), Pakistan (2), Ecuador (1), Hong Kong (1), Korea (1), 48 others (55)
Euphausiacea		
Euphausia superba (Antarctic krill)	448,266	Soviet Union (94), Japan (6)
Stomatopoda		
Squilla mantis (mantis shrimp)	4,881	Italy (100)
Squillidae (stomatopods)	449	Philippines (100)

[a] FAO (1983a). Used with permission of the Food and Agriculture Organization of the United Nations, Rome, Italy.

TABLE III

Locations of Economically Important Shrimp Fisheries Not Tabulated by FAO (1983a)[a]

Species	Fishery location
Penaeoidea	
Acetes americanus americanus	Brazil, Surinam
A. chinensis	China, Korea, Taiwan
A. erythraeus	China, Southeast Asia, India, East Africa
A. indicus	Southeast Asia, India
A. intermedius	Philippines, Taiwan
A. serrulatus	China
A. sibogae	Southeast Asia
A. vulgaris	Southeast Asia
Haliporoides diomedeae	Chile, Peru, Panama
Metapenaeopsis palmensis	Japan

TABLE III (Continued)

Species	Fishery location
Metapenaeus affinis	India
M. bennettae	Australia
M. brevicornis	India
M. dobsoni	India
M. endeavouri	Australia
M. macleayi	Australia
M. monoceros	India, East Africa, China
Parapenaeopsis stylifera	India
Penaeus brasiliensis	Caribbean Sea, Venezuela
P. esculentus	Australia
P. indicus	East Africa, India
P. kerathurus	Mediterranean Sea
P. occidentalis	Pacific Central America, South America
P. penicillatus	India, Pakistan
P. plebejus	Australia
P. schmitti	Caribbean Sea, Venezuela
P. stylirostris	Pacific Central America, South America
P. trisulcatus	Red Sea
P. vannamei	Pacific Central America
Plesiopenaeus edwardsianus	Southeast Atlantic Ocean
Sergia lucens	Japan
Trachypenaeus byrdi	Pacific Central America
T. faoe	Pacific Central America
Caridea	
Heterocarpus gibbosus	India
H. wood-masoni	India
Macrobrachium malcolmsonii	India
Palaemon adspersus	Europe
Pandalopsis dispar	Canada, United States (Alaska)
Pandalus danae	Canada, United States (Alaska)
P. goniurus	Canada, United States (Alaska)
P. hypsinotus	Canada, United States (Alaska)
P. jordani	Canada, United States (Alaska)
P. kessleri	Japan
P. montagui	Europe
P. platyceros	Canada, United States (Alaska)
Parapandalus spinipes	India
Plesionika ensis	India
P. martia	India

[a] Data taken from Cobb et al. (1973), Suseelan (1974), Omori (1974, 1978), Wickins (1976), Hayashi and Sakamoto (1978), and Considine (1982).

where they develop into the juvenile stage. Juveniles or subadults then migrate offshore to develop further and to spawn (Wigley, 1973; King, 1981). There are, however, deep-water penaeids, which spend their entire lives offshore, as well as other species, which undergo their complete life cycles in estuaries (Omori, 1974).

The caridean shrimps [order Decapoda, suborder Pleocyemata, infraorder Caridea (Bowman and Abele, 1982)] constitute a large and diverse group, which are more widely distributed but less commercially exploited than penaeids (King, 1981). Eggs are generally fewer in number than in penaeids and are incubated on the pleopods of the female. Larvae pass the naupliar stages in the egg and hatch into either a protozoea or a zoea (Omori, 1974). The Caridea includes several superfamilies of commercial importance: Crangonoidea, Palaemonoidea, and Pandaloidea.

Potentially, the commercial importance of krill, or euphausiids (order Euphausiacea, family Euphausiidae) may overshadow that of all other shrimplike animals combined. Krill are basically open-ocean pelagic animals (Omori, 1978), and their abundance over large areas of the Antarctic Ocean has encouraged extensive investigation of their potential use as a raw material in animal nutrition or as food for humans. The first commercial scale harvesting of this resource has begun in recent years (Bakus et al., 1978).

In some localities, several other shrimplike crustaceans are harvested in limited quantities for human consumption. These include mantis shrimps (order Stomatopoda) and mysid shrimps (order Mysidacea).

A more detailed account of shrimp landings in various parts of the world was given by Gulland (1970), who summarized current landings and predicted the potential for 14 regions of the world. Shrimp landings in the 14 regions have been updated and expanded to 19 regions, with krill included (FAO, 1983a) (see Fig. 1).

C. Characteristics of Importance to Fisheries

Shrimps ranging in size from a few to 100 g or more are harvested commercially. Although fishes of these sizes might be difficult to process, the shrimp's anatomical design facilitates handling and processing. Shrimps are composed of a distinct "head" (cephalothorax) and "tail" (abdomen), which can be easily separated (Green, 1949a). Major internal organs, with the exception of parts of the ovaries and straight gut, are all located in the head. The abdomen is primarily muscle tissue and, in penaeids, constitutes 60% of the entire body weight (Bayagbona et al., 1971; Cobb et al., 1973). The gut and ovarian lobes can be easily removed from the abdomen; however, in many places consumers do not object to eating this "vein" with the flesh.

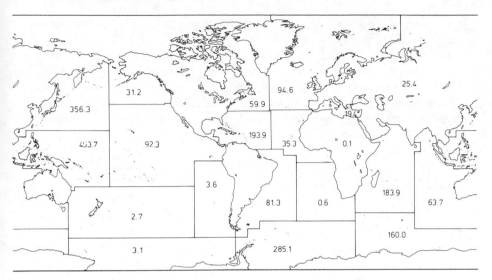

Fig. 1. Harvest of shrimps and Antarctic krill for regions of the world in thousands of metric tons for 1981. (Modified from Gulland, 1970, using figures from FAO, 1983a, with permission of the Food and Agriculture Organization of the United Nations, Rome, Italy.)

The exoskeleton forms a firm, protective covering around the tail, which prevents damage during handling and resists entry of bacteria that would cause spoilage (Green, 1949b; Jacob et al., 1962; Kurian and Sebastian, 1976). The exoskeleton, together with the fact that shrimp flesh freezes well (Green, 1949a), facilitates handling and processing. Most shrimps are marketed as frozen tails (FAO, 1983b), the processing consisting only of removing the head, washing, and freezing.

Freshly molted shrimps and shrimplike animals are easily damaged when captured, because their exoskeleton is not firm. In addition, their flesh is soft and of inferior quality (Eddie, 1977), probably because a quantity of water is taken up immediately after molting. Many of these animals are discarded by the fishermen. They are also lost to the fishery and from the population because they are invariably killed if caught in a trawl. Some freshly molted shrimps are taken at all times. However, the molting cycle is related to the lunar cycle (Brown, 1961). During certain periods of the month, high proportions of molting shrimps may appear in the catch. During this time, fishermen are advised to discontinue fishing for a few days or to move to new fishing grounds. Younger shrimps molt more frequently than older ones (Omori, 1974, 1979; Etzold and Christmas, 1977), so it is advisable for a fisherman to move to deeper waters to fish an older population, thereby reducing the proportion of molting individuals.

The almost universal acceptability and demand for shrimps as a food item is a result of their unique flavor, texture, and versatility. Consumer preference tests in the United States indicated that more than 90% of consumers find shrimps a desirable food, whereas only about 50% view fishes as a desirable food (Anonymous, 1970). Another survey in the United States indicated that more shrimps were eaten than all other shellfishes (oysters, clams, scallops, crabs, lobsters) combined (Manar, 1973). In nearly all major countries of the world, strong demand exists for shrimps, and projections based on per capita consumption indicate that the demand may double by the year 2000 (Anonymous, 1970).

II. FISHING GEAR AND METHODS

Shrimps mainly live at the bottom and walk by means of their pereopods. They also swim with pleopods and can spring from the surface when disturbed. Adult shrimps often live at sea, and juveniles live in estuaries. Light and olfactory stimuli attract shrimps. A consideration of these and other specific behavioral traits is necessary before determining which nets, traps, or other devices might be effective for shrimp collecting. Mode of operation and differences in mesh size likewise affect the size of the catch (Kurian and Sebastian, 1976).

With regard to a moving net, *Penaeus japonicus* shows the following four types of behavior: (1) jumping backward by bending the body quickly, (2) swimming forward with the pleopods, (3) sticking onto the net, and (4) crawling with the pereopods. As towing speed increases, jumping and swimming decrease, while sticking to the net increases. Crawling occurs only at the slowest tow speed (0.5 m/sec) (Ko et al., 1970).

Nets have different herding capabilities, due to the angle of tow (lower angles are better than higher ones). Maximum escape speed for *P. japonicus* is about 1.7 m/sec, with a maximum sustained distance of about 17 m.

The most common shrimp-collecting devices are nets and traps (Kurian and Sebastian, 1976). Nets include cast nets, bag nets, stake nets, drag or trawl nets (beam and otter trawls), barrier nets, seines (drift and gill nets; boat and shore seines), Chinese lift nets, and man-powered nets (scoop, dip, pushing and skimming nets). Nets are generally towed from a boat, attached between stationary objects, or operated by hand on shore. There is a wide variety of trap designs, including cover basket (plunge basket) trap, cage trap, conical cage trap (aproned cone trap), raft trap, baited float, hut box (box-type basket), net trap, and trap with a longitudinal valve. Traps are worked by hand to cover shrimps from above, located within currents to trap animals passing through, or baited to attract the catch.

The most widely used gear for harvesting both penaeid and caridean shrimps is the otter trawl (Fig. 2). A footrope, or leadline, forms the bottom of

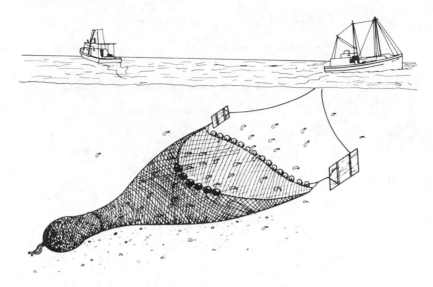

Fig. 2. Sketch of shrimp otter trawl being towed along the bottom. (Drawn by Yvonne Wilson, after Knake, 1958.)

the mouth and holds the net on the bottom. The headrope carries a number of floats for buoyancy to keep the mouth open; wings of the net are attached to heavy wooden otter boards by towing cables (Guest, 1958; Kurian and Sebastian, 1976).

The fishing methods, which evolved primarily in the Gulf of Mexico, have in recent years been widely adopted by fishermen around the world. A detailed account of this and earlier fishing methods was given by Knake (1958) and Anderson (1966). The use of an otter trawl is limited to depths suitable for motor-powered boats and areas where the bottom is smooth enough to prevent net snagging; small nets (5 m in width) may be towed from an outboard-powered boat, whereas the sizes typically used in modern commercial fishing offshore (15–20 m in width) can be towed only with larger vessels. The addition of a tickler chain and a decrease in the number of floats (keeping the net at a lower depth) produced a 70% increase in shrimp catch (Deshpande, 1960). Otter trawl efficiencies ranged from only 33 to 50% in mark–recapture studies of *Penaeus aztecus* (Loesch *et al.*, 1976).

The double-rig shrimp trawl is used extensively in the Gulf of Mexico (Bullis and Floyd, 1972). Basically, two trawls are towed using a single pair of otter boards. The two trawls are joined at the headrope and footrope to a "neutral door" connected to a third bridle leg (see Fig. 3). Generally, the

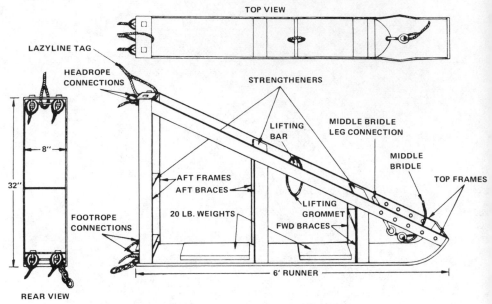

Fig. 3. Double-rig shrimp trawl sled. (From Bullis and Floyd, 1972.)

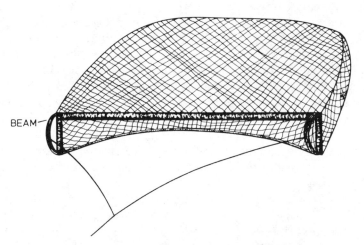

Fig. 4. Beam trawl. (Drawn by Yvonne Wilson.)

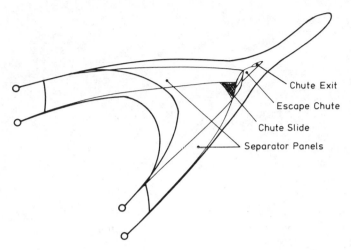

Chute Exit

Escape Chute

Chute Slide

Separator Panels

Fig. 5. V-type vertical separator panel design concept. (From Watson and McVea, 1977.)

same total headrope length is maintained for each twin-rig component as was used in single trawl units. No comparative tests have been conducted; however, certain advantages to the double-rig trawl have been noted: an increased fishing efficiency of up to 25%; light weight and ease of handling; two, 35 ft (10.7 m) trawls, as opposed to a single 70 ft (21.3 m) trawl; the net can be spread with a single pair of seven-foot (2.1 m) doors, as opposed to 10–12 ft (3.0–3.7 m) doors; sharper turns are possible; and towing can be made at significantly less than maximum horsepower.

The beam trawl is also widely used in shrimp harvesting (Fig. 4). Construction is similar to the larger otter trawl, but the wings are held open by a rigid horizontal beam of wood or aluminum. The beam limits the size of the net and makes handling difficult (Browning, 1980). In beam trawl tests, Deshpande (1960) determined that 21% of shrimps in the area escaped capture. Experiments on towing efficiencies using beam trawls were conducted by Lee and Matuda (1973) and Fujiishi and Ishizuka (1974).

Problems with by-catch, often ranging from 50–80%, prompted construction of selective trawls (Pereyra et al., 1967; High et al., 1969; Browning, 1980). A V-type vertical separator panel for eliminating by-catch via an escape chute was devised by Watson and McVea (1977) (see Fig. 5). The shrimps are selectively captured by using water-flow patterns within the trawl. Selective trawls produced 98–99% efficiencies, but catches were consistently lower than in those using standard trawls (High et al., 1969; Browning, 1980). The lower catch was the major factor in the selective trawl's lack of popularity.

The electric trawl was developed to exploit penaeid shrimps, which bur-

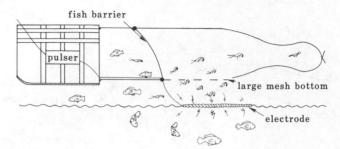

Fig. 6. Selective electric trawl diagram. (Redrawn by Yvonne Wilson, from Seidel and Watson, 1978.)

row during daylight hours (Seidel, 1969; Watson, 1976). The by-catch could be eliminated by closing the net mouth and by forcing shrimps into the trawl through a large mesh net bottom (Seidel and Watson, 1978). An electric field caused shrimps to jump into the net, while fishes were selectively frightened away (see Fig. 6). Pulses of 4 per sec, with a potential of at least 3.0 V across 100 mm parallel to the electric field, are the most helpful for forcing shrimps out of the benthos (Klima, 1968). Such an electric field caused 78–100% of burrowed shrimps to jump 75 mm within 1.90 sec. Pumped jets of water also disturb buried shrimps (Penn and Stalker, 1975).

Traps or "pots" have been used for some species, especially on rough bottoms unsuitable for trawling. Traps function by providing shelter and/or food. Various trap designs were described by Kessler (1968), Butler (1970), Tiews (1970), Inoue (1981), King (1981), and Koike et al. (1981).

Other types of shrimp collecting gear have been devised: a pushnet for collecting postlarval and juvenile shrimps in marsh grass (Allen and Inglis, 1958); a device for measuring near-bottom vertical distribution of Pandalus jordani (Beardsley, 1973), whereby segregated shrimps can be caught in a series of nets with 1 ft (0.3m)-high vertical openings positioned from the sea bed to 13 ft (4.0 m) above the bottom; an automatic pumping device for catching penaeid postlarvae (Marullo, 1973) (but small catches masked fluctuations in abundance of shrimps as caught by beam trawl). Of interest is the fact that the acoustic target strength of individual shrimps is arranged at various orientations to the direction of transmitted sound beams (Sofoulis et al., 1979).

In shallow water, shrimps traditionally have been harvested by use of cast nets, push nets, seines, and even entangling nets such as gill nets. Penaeids that use estuaries are the most vulnerable as they leave the estuary, frequently moving through narrow passes or channels on outgoing tides. Fishermen have designed a variety of fixed nets and traps to catch emigrating juvenile shrimps. The "tapos" used in Mexico are a weirlike structure designed to guide shrimps through a narrow passage where they can be trapped in nets (Edwards, 1978). In other areas butterfly nets and various floating or attached nets are used to catch shrimps, which may be swimming near the surface as they leave the bays (Kurian and Sebastian, 1976).

III. PENAEID SHRIMPS

A. Distribution of Major Genera and Location of Important Fisheries

There are at least 60 species of commercially important penaeid shrimps (see Tables II and III). Chin and Allen (1959) and Allen and Costello (1969) provided extensive annotated bibliographies on the biology of the Penaeidae. George and Rao (1967) constructed a similar bibliography of commercially important shrimps of India. Pérez Farfante (1969) reviewed the biology of the genus *Penaeus* of the western Atlantic.

The Gulf of Mexico penaeid fishery was divided into three subunits by Kutkuhn (1962a). A noncommercial fishery was composed of sport fishermen, who caught mostly immature shrimps for their personal use from shallow coastal waters. A commercial bait fishery was designated with a large number of professional fishermen, who took immature shrimps, almost exclusively from inshore waters, solely for the purpose of supplying live and dead bait to fishermen. Discussions of bait fisheries were given by Guest (1958), Inglis and Chin (1959), Chin (1960), and Joyce and Eldred (1966). The core of the shrimping industry was labeled the commercial food fishery. In that fishery, a large number of professional fisherman seek either large, mature shrimps, which inhabit offshore waters, or small immature shrimps, which live in certain inshore areas. Almost all of the commercial food harvest is destined for human consumption. The larger shrimps are mainly processed as fresh frozen, and the smaller shrimps are often dried and canned.

The most important fisheries for penaeid shrimps are in the South China Sea, near Indonesia, in the Arabian Sea, and in the Gulf of Mexico. Major new fisheries are developing in the Gulf of Carpentaria (Australia) and in the Indian Ocean (Wickins, 1976). A listing of commercially important penaeid species with fishery locations can be found in Table IV. Discussions of the catches, species composition, and other characteristics of particular fisheries were presented by Lindner (1957) for Central and South America; Jones (1969), and Kurian and Sebastian (1976) for India; Slack-Smith (1969) for Shark Bay, Western Australia; Ewald (1969) for Venezuela; Neiva (1969) for Brazil; Osborn et al. (1969) for the Gulf of Mexico; and Edwards (1978) for the Pacific coast of Mexico.

TABLE IV

Major Penaeid Fisheries by Species

Species	Location	Reference
Metapenaeopsis palmensis	Japan	Hayashi and Sakamoto (1978)
Metapenaeus affinis	India	Menon (1957), George (1961), Menon and Raman (1961), George *et al.* (1963), Subrahmanyam (1963), Kuthalingam *et al.* (1965), Mohamed (1967)
M. bennettae	Australia	Dall (1958)
M. brevicornis	India	Rajyalakshmi (1961), Ramamurthy (1967)
M. dobsoni	India	Menon (1955, 1957), Menon and Raman (1961), George (1961, 1963), George *et al.* (1963), Kuthalingam *et al.* (1965)
M. monoceros	Kenya	Brusher (1974)
	Pakistan	Zupanovic and Mohiuddin (1973)
	India	George (1959), Menon and Raman (1961)
Parapenaeopsis atlantica	Nigeria	Ajayi (1982)
P. stylifera	India	Menon (1957), George (1961), George *et al.* (1963), Kuthalingam *et al.* (1965), Mohamed (1967)
Penaeus aztecus	Gulf of Mexico	Springer and Bullis (1952), Chin (1960), Kutkuhn (1962a), St. Amant *et al.* (1966), Christmas and Gunter (1967), Gaidry and White (1973), Gunter and McGraw (1973), White and Boudreaux (1977), Juneau and Pollard (1981), Jones *et al.* (1982), Klima *et al.* (1982), Matthews (1982), Nichols (1982), Poffenberger (1982)
	Southeastern United States	Purvis and McCoy (1978)
P. brasiliensis	Venezuela	Khandker and Lares (1973)
P. duorarum	Gulf of Mexico	Springer and Bullis (1952), Iversen and Idyll (1960), Kutkuhn (1962a, 1965, 1966a), Iversen and Jones (1961), Tabb *et al.* (1962), Christmas and Gunter (1967), Klima *et al.* (1982), Matthews (1982)
P. esculentus	Australia	Hall and Penn (1979)

TABLE IV (*Continued*)

Species	Location	Reference
P. indicus	Kenya	Brusher (1974)
	Madagascar	Le Reste (1971, 1973)
	India	Menon (1957), Menon and Raman (1961), George (1961, 1962), George et al. (1963)
P. japonicus	Israel	Tom and Lewinsohn (1983)
	Kenya	Brusher (1974)
	Japan	Kitahara (1979, 1981)
P. latisulcatus	Australia	Penn (1976), Hall and Penn (1979)
P. merguiensis	Indonesia	Unar (1974)
	Australia	Lucas et al. (1979), Staples and Vance (1979)
P. monodon	Kenya	Brusher (1974)
P. notialis	Nigeria	Bayagbona et al. (1971), Ajayi (1982)
P. penicillatus	Pakistan	Zupanovic and Mohiuddin (1973)
P. plebejus	Australia	Lucas (1974)
P. semisulcatus	Kenya	Brusher (1974)
P. setiferus	Gulf of Mexico	Chin (1960), Kutkuhn (1962a,b), Christmas and Gunter (1967), Gaidry and White (1973), Gunter and McGraw (1973), Gaidry (1974), White and Boudreaux (1977), Juneau and Pollard (1981), Klima et al. (1982), Matthews (1982)
P. stylirostris	Pacific Mexico	Edwards (1978), Menz and Bowers (1980)
P. vannamei	Pacific Mexico	Edwards (1978), Menz and Bowers (1980)
Pleoticus robustus	Southeastern United States	Joyce and Eldred (1966), Thompson (1967), Roe (1969)
	Gulf of Mexico	Springer and Bullis (1952)
Sergia lucens	Japan	Omori et al. (1973)
Sicyonia brevirostris	Gulf of Mexico	Cobb et al. (1973)
Xiphopenaeus kroyeri	Gulf of Mexico	Juneau (1977)

The genus *Penaeus* occurs in tropical and subtropical waters around the world, roughly from 40°N to 40°S latitude, and it is by far the most important genus commercially (FAO, 1983a). The majority of species require nutrient-rich estuaries for their survival. However, several species live along arid coasts where freshwater runoff is sparse (Bardach et al., 1972).

Sicyonia has attracted special attention in the United States recently as a speciality market item. In the past, few *Sicyonia* reached retail markets because of their hard exoskeleton, their small size relative to *Penaeus*, and their scattered distribution, even though they have long been a favorite food of shrimp fishermen because of their fine flavor. Recent market acceptance has been related to market promotion and to demonstrations of simplified methods for cooking in the shell. *Sicyonia* is now harvested to a limited extent in Cuba and throughout the Gulf of Mexico (Cobb *et al.*, 1973; FAO, 1983a).

The genus *Metapenaeus*, which is represented by several species throughout the Indo-Pacific region, is fished heavily in Australia (Dall, 1958). Some species reproduce in brackish water, and in India this genus is taken from rice fields in substantial numbers (Bardach *et al.*, 1972).

Artemesia is an important genus caught off the Argentine coast, whereas deep-water genera of commercial importance include *Plesiopenaeus* from Spain and *Pleoticus* from Argentina and the Gulf of Mexico. Other minor genera, which are typically small and used to a limited extent for food, are *Acetes, Parapenaeopsis, Parapenaeus, Sergia*, and *Trachypenaeus* (Omori, 1974; FAO, 1983a).

Xiphopenaeus supports a sizeable fishery off the Yucatan Peninsula of Mexico and southward to Brazil. *Xiphopenaeus kroyeri* is marketed along with species of the genus *Penaeus* and is accepted interchangeably with the three species in United States markets. It may constitute up to 2% of the annual United States catch (Juneau, 1977; FAO, 1983a).

B. Life Cycle

A large number of penaeid shrimp life cycles are similar (see Fig. 7). The life cycle of *Penaeus setiferus* will be discussed as a "typical" cycle, on the basis of information from Weymouth *et al.* (1933), Pearson (1939), Anderson *et al.* (1949), Heegaard (1953), Lindner and Anderson (1954), Williams (1955a), Guest (1958), Inglis and Chin (1959), Joyce and Eldred (1966), and Etzold and Christmas (1977).

Adults spawn in offshore waters at depths of 10–15 fathoms (18.3–27.4 m). Spawning occurs from March until September, with bimodal spawning peaks in May–June, August–September in the Gulf of Mexico. At mating, the smaller male attaches an external sperm sac to the larger female. Later, the female ruptures this spermatophore to fertilize eggs as they are shed into the water. The number of eggs per spawning ranges from 500,000 to 1,000,000, with each egg being about 1/100 in. (0.03 cm) in diameter.

Fertilized eggs sink to the bottom and hatch in about 24 hours. The planktonic larvae remain offshore for about 3 weeks and develop through five naupliar, three protozoeal, and two mysis stages. Following the mysis

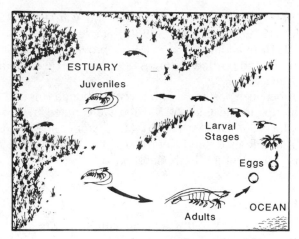

Fig. 7.   Penaeid shrimp life cycle. (From Caillouet and Baxter, 1973.)

stages are several postmysis, or postlarval, stages. Only the postlarvae that reach estuarine nursery grounds become juveniles and continue to grow. A combination of current patterns and behavioral responses make this migration to the estuary possible.

In the nursery area, ¼ in. (0.6 cm) shrimps assume a benthic habitat, and there is a direct relationship between depth and size of the shrimps. Shrimps grow at a maximum rate of about $1\frac{7}{16}$ in. (3.6 cm)/month for 4–10 weeks. Temperature may trigger a return to offshore spawning grounds at the onset of sexual maturity when the shrimps are approximately 4 in. (10.2 cm) in length. With the extended spawning season, new recruits are always entering the population, and the entire life cycle takes about 12 months.

Variations in this "typical" life cycle occur with species that complete their life cycle in deeper water offshore and with species that mature and spawn in the estuaries (Omori, 1974). In addition, a few shallow-water penaeids, for example, *Penaeus latisulcatus,* reportedly do not have to enter estuaries as postlarvae to develop (Gulland, 1970). The range of life histories of penaeid shrimps with respect to dependence on estuaries was presented by Williams (1955b), Kutkuhn (1966b), Mohamed and Rao (1971), and Staples (1980b).

Though not the particular scope of this review, a number of studies have been conducted concerning penaeid larval–postlarval distribution and ecology. A selection of these studies includes Thorson (1946, 1950), Williams (1955a,b, 1959), Renfro (1960), Idyll et al. (1962), Temple and Fischer (1965), Christmas et al. (1966), Baxter and Renfro (1967), Aldrich et al. (1968), Munro et al. (1968), Williams (1969), Jones et al. (1970), Subrahmanyam (1971), Penn (1975b), Young and Carpenter (1977), Price (1979), Staples (1979), Allen et al. (1980), Staples (1980a), and Kennedy and Barber (1981). In addition, Cook (1966) provided a generic key to the protozoeal, mysis, and postlarval stages of littoral penaeids from the northwest Gulf of Mexico.

Correlations between postlarval and juvenile abundances and subsequent catches of shrimps in the Gulf of Mexico were noted by Baxter (1962), Christmas et al. (1966), Lindner (1965), St. Amant et al. (1966), Berry and Baxter (1969), and Roessler and Rehrer (1971). Williams (1969) found no such relationship evident in North Carolina.

C. Habitat

The complex life cycle of many penaeid shrimps places them in a variety of habitats during their life history. With the "typical" penaeid life cycle, the larvae are pelagic in nearshore coastal waters. As postlarvae move into the estuarine environment, they become benthic at the extreme edges or the most shallow parts of the estuary. The postlarvae use an increasingly larger part of the estuarine habitat as they develop into juvenile shrimps.

Both in estuaries and offshore, there is some segregation of shrimp species due to differences in nutrients and subsequent discharge, depth, sediment types, vegetation cover, salinity, and temperature. Also, when several species utilize a single estuarine system, competition is partially avoided by timing of immigration and emigration (Williams, 1955a,b; Williams and Deubler, 1968; Young, 1978).

Shrimp life cycles and interactions with environmental variation and fisheries represent parts of a highly dynamic system that varies in time and space. Marked fluctuations in shrimp populations can probably be induced by yearly differences in spawning success and survival of young for recruitment. Biological and physical conditions play vital roles in these population fluctuations (Caillouet and Baxter, 1973) (see Fig. 8).

When the locations of currents carrying nutrients offshore along the Texas coast were compared with shrimp catches, a strong positive correlation was observed. Apparently dissolved nutrients and suspended inorganic and organic material from rivers and bays affect the ecology of shrimps and influence their local distribution (Lindner and Bailey, 1968). River discharge and tidal currents also affect penaeid distributions (Ruello, 1973b; Glaister, 1978b; Coles, 1979; Vance et al., 1983).

Characteristically, penaeid juveniles thrive in rich estuarine water containing high levels of dissolved nutrients and organic matter. The rapid growth of P. setiferus in ponds into which both feeds and fertilizers were introduced, testifies to the tolerance of eutrophic conditions by some spe-

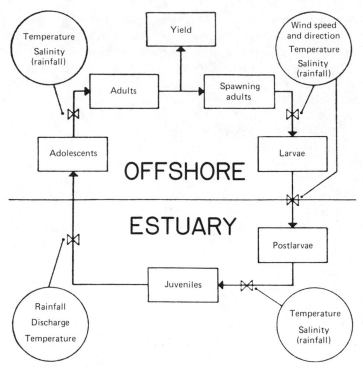

Fig. 8. Life history diagram of a penaeid shrimp showing factors influencing yearly fluctuations in catch. (From Vance *et al.,* 1983.)

cies (Bardach *et al.,* 1972). In Louisiana marshes, shrimps may thrive under conditions where high organic content of soil and water result in dissolved oxygen content of near zero during early morning hours (Gaidry and White, 1973). All species do not require rich estuarine waters, and some, such as *Penaeus duorarum,* are more abundant in the areas characterized by lower nutrient content (Williams, 1958).

With some species, there is a change in habitat preference of subadult and adult shrimps related to size or age. *Penaeus aztecus,* for example, is distributed from the shore to depths of 124 m (Springer and Bullis, 1952), and size increases generally with depth (Guest, 1958). Immediately after leaving the estuary, *P. aztecus* may be found in dense concentrations at 10 to 20 m. Within 1 month, the concentrations have dispersed to depths of 30 or 40 m. During succeeding months, as the shrimps grow and mature, they move to even greater depths. Late in the season, commercial catches may be common at depths of 60 to 70 m.

The deep-water species, for example, those in the genera *Pleoticus* and *Plesiopenaeus,* have not been studied extensively and relatively little is known of their habitat, environmental requirements, or life history. *Pleoti-cus robustus* was reported at depths of 175–300 fathoms (320–549 m) in the northern Gulf of Mexico (Anderson, 1966). Some deep-water shrimps have been found at all locations that have been fished in waters just off the continental shelf in the low and middle latitudes (Gulland, 1970).

Variation in substrate is a major factor in penaeid shrimp distributions (Branford, 1981a,b). In the Gulf of Mexico, Etzold and Christmas (1977) found that adults of *P. aztecus* and *P. setiferus* were found in mud and silt habitats, whereas *P. duorarum* inhabited sand, sand–shell, and coral–mud. Juveniles of *P. aztecus* lived in muddy sand or sandy mud and in peat substrates, especially in areas with vegetation cover and plant debris. The young of *P. setiferus* were found in softer areas of mud, whereas those of *P. duorarum* inhabited submerged vegetation. Giles and Zamora (1973) had previously studied distributional patterns of *P. aztecus* and *P. setiferus*. When given the choice, both species showed a "preference" for grass habitats as opposed to bare substrates. This behavior is possibly due to increased defense or food offered by the grass. However, when populations were mixed, *P. aztecus* inhabited the grass, while *P. setiferus* remained on the bare substrate. If such exclusion occurs in nature, there might be significant effects on the survival of *P. setiferus*.

Water volume (pore space) and compaction of the sediment may influence the degree of burrowing by certain penaeid species (Williams, 1958). Hughes (1966) found that *Penaeus semisulcatus* and *Metapenaeus mono-ceros* could burrow rapidly and were collected in shallow pools and intertidal flats. *Penaeus monodon*, which burrows partially, inhabited the edges of mangrove channels. *Penaeus indicus*, a nonburrower, was found in muddy areas where there was no need to burrow.

Annual harvests of *P. setiferus* have been shown to be positively correlated with rainfall during the previous 2-year period (Gunter and Edwards, 1969). Other correlations between rainfall and penaeid shrimp catches were noted by Hildebrand and Gunter (1952), Ruello (1973b), and Vance *et al.* (1983). Although causal relationships are not clear, the resulting low salinities in coastal bays may reduce competition from other species or may concentrate the shrimps in areas where they are more vulnerable to capture (Hoese, 1960).

Salinity can be very important in shrimp distributional patterns, and it may be a limiting factor, especially in shallow water (Barrett and Gillespie, 1973; Heaton, 1981). In the northern Gulf of Mexico, juveniles of *P. setiferus, P. aztecus,* and *P. duorarum* are most abundant at <10, 10–20, and >18‰, respectively (Gunter *et al.,* 1964). Along the Texas coast, *P. aztecus* and *P. duorarum* may enter hypersaline estuaries as postlarvae and develop through the juvenile stage at salinities well above that of seawater. These juveniles have been taken from water of 70‰ or higher, as well as from waters of a salinity less than 1‰ (Gunter, 1961).

The environmental factor of greatest importance from the harvester's viewpoint is temperature, which has a profound effect on immigration into, and emigration from, the estuaries. Temperature greatly influences movements and activity in the offshore environment, and it may affect survival of larval stages (Wiesepape et al., 1972). The water temperature at the time of capture did not influence the catch of shrimps in North Carolina (Williams and Deubler, 1968). However, penaeid catches fluctuated according to the number of hours that water temperature fell below 20°C after the first week of April in Louisiana (Barrett and Gillespie, 1973). Likewise, spring shrimp landings were found to be highly correlated with water temperatures during the previous winter, especially during the two coldest consecutive weeks (Hettler and Chester, 1982).

Combination effects of temperature and salinity can be extremely important for penaeid survival, distribution, and subsequent catch, especially when both are low in the nursery areas (Barrett and Gillespie, 1975; Barrett and Ralph, 1976; Etzold and Christmas, 1977; Hunt et al., 1980). Likewise, excessive runoff during critical periods can decrease usefulness of nursery areas, which subsequently can affect the catch (White, 1975).

D. Population Dynamics

1. REPRODUCTION AND FECUNDITY

Extremely high fecundity is characteristic of penaeids, and a female may spawn from several thousand to 1 million or more eggs depending on the species and shrimp size. Representative fecundity determinations based on female body length can be found in Table V. Martosubroto (1974) suggested that body weight was a better predictor of fecundity than total length. A linear relationship was found between number of eggs (44,000–534,000) and body weight (10.1–66.8 g) for *P. duorarum*. An exponential relationship was noted between fecundity and total length.

The eggs of penaeid shrimps are extremely small [(0.31–0.33 mm for *P. duorarum* Dobkin, 1961)], and survival is low. A tremendous biotic potential exists when high fecundity is coupled with multiple spawnings in penaeid shrimps, and recruitment to the fishery bears little, if any, relationship to the number of spawners (Etzold and Christmas, 1977). Environmental factors affecting survival and distribution are much more important than the number of eggs spawned. A relatively small population of spawners is capable of producing a large new year–class if environmental conditions are good. A positive correlation between water temperature and spawning of *P. duorarum* was found by Cummings (1961).

TABLE V

Representative Fecundities of Penaeid Shrimps[a]

Species	Number of eggs produced	Female body length (mm)
Metapenaeus stebbingi	88,000	95
	363,000	160
Penaeus latisulcatus	105,000	123
	650,000	217
P. semisulcatus	68,000	140
	731,000	200
Trachypenaeus granulosus	34,500	70
	160,000	120

[a] Data taken from Badawi (1975) and Penn (1980).

From the fisheries viewpoint, this means that high fishing mortality can be tolerated and in many situations recruitment overfishing is not a major concern, even when fishing pressure is high. However, growth overfishing can be a significant economic problem under such conditions. The relationship between size composition and ex-vessel value of shrimp catches (1959–1975) from Texas and Louisiana was investigated by Caillouet et al. (1979, 1980b) and Caillouet and Patella (1978). In Texas, shrimping regulations greatly restricted the catch of small shrimps, whereas Louisiana had few restrictions. For a given weight of catch, the ex-vessel value of P. aztecus and P. setiferus in Texas were 1.2 and 1.6 times that of Louisiana, respectively, largely due to growth overfishing in Louisiana. In contrast to value, Louisiana produced 22% more shrimps (by weight) than Texas in 1982 (Thompson, 1983).

The relationship between fecundity and number of larval stages was examined by Wickins (1976). The positive relationship supports the theory that the high fecundity of penaeids evolved in response to high mortality during the larval portion of the life cycle, related to the pelagic habitat and complexities of transport to the estuarine nursery grounds.

2. GROWTH

Growth is not a continuous process in the Crustacea but is related to the molting cycle. Important aspects of growth include magnitude of the increase in body size at molting and frequency of molting. According to Dyar's Law, the ratio of postexuvial length to initial length and that of postexuvial body weight to initial body weight are constant. In Penaeus japonicus, increases in body length represent retrogressive geometric growth, whereas increases in body weight represent moderate progressive geometric growth (Choe, 1971).

TABLE VI

Major Studies of Growth in Penaeid Shrimps

Species	Reference
Metapenaeus bennettae	Dall (1958)
M. brevicornis	Ramamurthy (1967)
M. dobsoni	Banerji and George (1967), George (1970), Mohamed and Rao (1971)
M. macleayi	Glaister (1978a)
M. monoceros	Mohamed and Rao (1971), Subrahmanyam and Ganapati (1971)
Penaeus aztecus	Williams (1955a), Zein-Eldin (1963), Loesch (1965), Zein-Eldin and Aldrich (1965), St. Amant et al. (1966), Zein-Eldin and Griffith (1966), Fontaine and Neal (1971), Ogle and Price (1976), Knudsen et al. (1977), Parrack (1979), Heaton (1981)
P. duorarum	Zein-Eldin (1963), Kutkuhn (1966a,b), Lindner (1966), Fontaine and Neal (1971)
P. japonicus	Choe (1971)
P. merguiensis	Lucas et al. (1979), Staples (1980b)
P. monodon	Delmendo and Rabanal (1956), Subrahmanyam and Ganapati (1971)
P. plebejus	Lucas (1974)
P. setiferus	Gunter (1950), Lindner and Anderson (1954), Williams (1955a), Zein-Eldin (1963), Loesch (1965), Fontaine and Neal (1971), Klima (1974), Heaton (1981)
P. vannamei	Edwards (1977), Menz and Blake (1980), Menz and Bowers (1980)
Sergestes similis	Omori (1979)

Growth rates in penaeid shrimps vary according to species, age or size, sex, population density, season, temperature, food, and other environmental factors. Thus, it would be extremely difficult to attempt comparisons of growth rates among different species. Instead, a listing of growth studies for different penaeids can be found in Table VI.

Growth in *Penaeus vannamei* is negatively correlated with size (Menz and Blake, 1980). In *Metapenaeus dobsoni,* growth has been recorded as 7.9 mm per month at 1 year of age, 1.6 mm at 2 years, and 0.3 mm at 3 years (Banerji and George, 1967). Similar results have been reported for *M. brevicornis* (Ramamurthy, 1967).

In the northern Gulf of Mexico, female penaeid shrimps grow more rapidly and reach larger sizes than do males. Within an estuary, population density of shrimps probably affects growth rates (Etzold and Christmas, 1977).

During Mexico's dry season (winter; November–May), penaeid growth is slower than it is during the wet season (summer; June–October) (Menz and Bowers, 1980). The growth rate of *P. vannamei* averaged 0.02 mm per day in March and 0.44 mm per day in September. Furthermore, growth rates for *Penaeus stylirostris* in April and August were 0.03 and 0.64 mm per day, respectively.

Growth rates of *Penaeus aztecus* increased with rising temperature up to 32.5°C, although maximum increase occurred between 17.5° and 25.0°C. Extreme, though nonlethal, temperatures strongly affected growth by decreasing greatly the molting frequency (Zein-Eldin and Aldrich, 1965; Zein-Eldin and Griffith, 1966). In aquaria, the growth rate of *P. aztecus* is usually lower than in nature (Zein-Eldin, 1963; Zein-Eldin and Aldrich, 1965). The type of food that is available may play an important role, at least during laboratory studies. When *P. aztecus* was fed *Artemia*, the growth rate was 0.60 mm per day; when fed mysids, it was 0.41 mm per day (Ogle and Price, 1976).

No methods of aging shrimps have been developed, and therefore growth rates are poorly understood. Tagging has been only a partial solution to the problem, and length–frequency analyses are of little value, except for short periods, because of extended recruitment and continuous mixing of subpopulations (Neal, 1975). Results from mark–recapture experiments have provided the most definitive information on growth. However, data typically are obtained for only short periods of time, and actual growth rates are underestimated to the extent that the mark or tag interferes with growth.

Lack of suitable marking methods (tagging or staining), rapid recruitment and mortality, the relatively short life span, continual movement and mixing, lack of an internal skeleton, and periodic molting have made marking difficult (Menzel, 1955; Neal, 1969; Bearden and McKenzie, 1972; Marullo et al., 1976; Howe and Hoyt, 1982). Kurian and Sebastian (1976) reviewed various tagging and staining methods, while additional marking studies were provided by Costello (1959), Costello and Allen (1962), Iversen (1962), Emiliani (1971), Penn (1975a), and Welker et al. (1975).

The well-known von Bertalanffy growth equation (von Bertalanffy, 1938) has been modified for fisheries management applications (Beverton, 1953; von Bertalanffy, 1964; Allen, 1966b, 1969; Campbell and Phillips, 1972; Rafail, 1973). A number of researchers have estimated growth curves for penaeid shrimps by using a von Bertalanffy model for shrimp length fitted to data for a short part of the growth curve. Examples of the fitting and use of such growth curves can be found in papers by Kutkuhn (1966a), Lindner (1966), Berry (1967), Neal (1967), McCoy (1968), Lucas (1974), Garcia (1977), Grant and Griffin (1979), Lucas et al. (1979), Parrack (1979), and Powell (1979).

3. MORTALITY

In spite of high reproductive rates, catches from penaeid fisheries characteristically fluctuate widely from year to year, but seldom show the long term declines indicative of recruitment overfishing. Apparently, environmental factors cause changes in natural mortality rates that overshadow fishing mortality (Etzold and Christmas, 1977). Presumably, the major mortalities occur during the larval stages at sea and include such factors as changes in current patterns, which may prevent entry of postlarvae into the estuaries (Vance et al., 1983). Penaeids are utilized as food by many other animals at all stages of their life, so natural mortality rates are extremely high (Tabb et al., 1962).

Mortality of larvae during the oceanic portion of their life cycle is important but is beyond man's control; it has little direct relationship to fishing or management procedures (Etzold and Christmas, 1977). Of much greater interest are the instantaneous rates of natural mortality (M) and fishing mortality (F) after shrimps reach a harvestable size. In addition, the instantaneous rate of total mortality (Z) is the slope of a regression of total mortality (fishing plus natural) estimates plotted against recapture time periods (Purvis and McCoy, 1978). These rates are very important in deciding when fishing seasons should be set to maximize economic return from the the harvest and to help prevent growth overfishing.

For this reason, considerable effort has been expended to obtain realistic estimates of these rates. Unfortunately, the lack of an aging method, difficulties in recognizing year classes or age groups in catches, and the problems of marking shrimps (Neal, 1969) have all contributed to the difficulty of estimating mortality. Purvis and McCoy (1978) listed the assumptions in the following mortality estimates: (1) there is no loss of marked shrimps by handling or marking procedures; (2) there is no movement of marked shrimp out of the fishing areas; (3) there is no loss of recaptures due to failure to return them; (4) all marked shrimps are available to the fishery. Realistically, these assumptions can only rarely be attained, and this contributes to the difficulties in obtaining mortality rates for shrimps. A detailed discussion of the theory underlying the assumptions of mortality estimates was presented by Kutkuhn (1966a).

Berry (1970) has reviewed the estimates of fishing, natural and instantaneous total mortality based on mark–recapture experiments and those made by Neal (1967) using a virtual population estimate. Berry used length-frequency distributions and effort data to refine estimates for the Florida Tortugas (P. duorarum) fishery.

In nearly every study of mortality rates of shrimps, estimates vary widely. For example, Berry (1967) obtained the following ranges for mortality rates of *P. duorarum:* $M = 0.16-0.23$, $F = 0.02-0.06$, $Z = 0.22-0.27$. Likewise, measurements often vary even more between studies. In comparison to Berry's rates, Kutkuhn (1966a) made mortality rate estimates as follows for *P. duorarum:* $M = 0.96$, $F = 0.55$, $Z = .76$, 1.51.

Mortality estimates vary according to the method of determination, time, handling during tagging, size—age of shrimps, variations in predation, fishing pressure, and environmental variations (Neal, 1967; Berry, 1970; Purvis and McCoy, 1978; Blake and Menz, 1980). Other studies of mortality rates in penaeid shrimps include Iversen (1962), Banergi and George (1967), Costello and Allen (1968), Lucas et al. (1972), Lucas (1974), and Edwards (1977).

4. STOCKING

Many fisheries agencies have considered stocking shrimps in their natural environment, because good hatchery methods have been available for producing large quantities of postlarvae at relatively low cost (Bardach et al., 1972). Logically, if shrimps can be reared in captivity through the stage at which catastrophic mortalities occur in nature, then released into the wild, large increases in fishing harvests may be realized. In support of this idea, it is known that the penaeid habitat, both estuarine and oceanic, does on occasion support population densities much larger than the "average" situation. The food organisms that are needed to support normal growth rates at exceptionally high population levels are present in many localities.

Small-scale tests of survival of stocked postlarvae under seminatural conditions in the United States have been discouraging in that survival rates were low. The only large-scale stocking of penaeids into the natural environment has been done in Japan. Oshima (1974) has presented data from 11 large-scale stocking experiments conducted in Japan. In some cases, stocking was directly into the natural environment, and in other cases, released shrimps were retained in net enclosures. Although many millions of shrimps have been released over a period of 10 years or more, conclusive evidence of an attractive cost—benefit ratio for this stocking has not been presented. In fact, in only a few instances when releases were made into the natural environment is there evidence that catches were increased by this activity.

E. Behavior and Migrations

1. BURROWING AND FEEDING

Many species of penaeids burrow into the substrate when not feeding, thus conserving energy and reducing predation (Fuss and Ogren, 1966; Kutty and Murugapoopathy, 1967). Burrowing activity is positively correlated with light, as a number of species are nocturnal (Fuss and Ogren, 1966; Moller and Jones, 1975). Emergence is likewise rhythmically synchronized with the light–dark cycle transition in *P. duorarum* (Hughes, 1967a, 1968, 1969b; Bishop and Herrnkind, 1976). In *Penaeus duorarum*, a correlation exists between burrowing and the lunar cycle (Fuss, 1964, Wickham, 1967). Burrowing activity in *P. aztecus* is inversely correlated with salinity (Lakshmi et al., 1976). Deburrowing in *P. duorarum* occurs at temperatures exceeding 33°C, irrespective of light (Fuss and Ogren, 1966). Postlarvae of *P. duorarum* burrow when the temperature is 12°–17°C and emerge when it is 18°–21.5°C, whereas postlarvae of *P. setiferus* do not burrow (Aldrich et al., 1968). Burrowing at low temperatures supports the hypothesis that most postlarvae hibernate during part of the winter. Burrowing activity also varies with size. *Penaeus vannamei* of 100–150 mm in length exhibited rhythmic burrowing, whereas shrimps of 50 mm in length did not burrow, and those of 80 mm showed intermediate behavior (Moctezuma and Blake, 1981).

The same type of burrowing activity occurs in *Metapenaeus bennettae, M. macleayi,* and *Penaeus duorarum* (Dall, 1958; Fuss, 1964; Ruello, 1973a). Initially, the pleopods are spread sideways and are vigorously fanned backwards on the substratum, thereby scouring a furrow. Simultaneously, the pereiopods are pushing the sediments away in front, laterally, and upward around the body. Several "upkicks" of the tailfan provide further thrust. In some 5–7 sec, the abdomen is below the surface of the substratum, with the head concealed except for the tips of the antennules, eyes, and distal part of the rostrum. During the next 1–2 min, the shrimps usually burrow deeper with spasmatic movements of the pereiopods and pleopods, until only the antennule tips are protruding. Some 5–10 min after burrowing begins, the antennal flagella are withdrawn into the sediment with the shrimp fully buried, often at more than a centimeter below the surface.

After burrowing, a reversed respiratory flow is initiated with the antennules and large antennal scales forming a respiratory tube. Water reaches the branchial chamber via the tube, and after passing over the gills, it escapes under the branchiostegites. Periodically, the flow is reversed to eject sediment particles and to remove deoxygenated water (Dall, 1958; Ruello, 1973a).

Shrimps when burrowed are much less vulnerable to capture by trawls. Special modifications of the otter trawl have been introduced to encourage penaeids to move up into the water column just prior to passage of the leadline of a trawl. One of these in common use is a tickler chain arranged to drag over the bottom slightly in advance of the leadline (Deshpande, 1960; Deshpande and Sivan, 1962). A second device, which has not found widespread use, is an electrical trawl designed to shock burrowed shrimps and to make them jump from their buried position into the path of the net

(Seidel and Watson, 1978). The costs of special equipment, together with the maintenance problems, have discouraged wide acceptance of the electrical trawl.

Feeding behavior is of little importance from the fishing viewpoint. However, trawls are designed to capture feeding shrimp that are standing on, or swimming actively near, the bottom. Studies of the omnivorous and detrital feeding behavior of various shrimps were provided by Ikematsu (1955), George (1959), Subrahmanyam (1963), Renfro and Pearcy (1966), Rao (1967), Ruello (1973a), Kuttyamma (1974), and Wickins (1976).

Penaeus aztecus normally feeds offshore between dusk and dawn, and most fishing occurs at night in the Gulf of Mexico (Springer and Bullis, 1952; Duronslet et al., 1972). In the deeper parts of the habitat, fishing may be done around the clock, with reasonably good catches of *P. aztecus* during both the day and night. Other species, *P. setiferus* for example, seem to be equally vulnerable during the daylight hours, and fishing is usually a daytime operation.

2. SCHOOLING AND CONCENTRATIONS

The majority of penaeid species do not exhibit schooling behavior except possibly during the emigration from the estuaries (Hughes, 1969a, 1972). Weather conditions along the northern coast of the Gulf of Mexico may favor rapid simultaneous migrations of juvenile shrimps from the estuaries into the open Gulf (Tabb et al., 1962). At these times, concentrations of shrimps occur in the passes and in the shallow offshore waters, but it is not understood whether this represents a schooling behavior.

An interesting contrast to the typical penaeid behavior is that of *M. macleayi*, which does form large schools and is particularly vulnerable to capture when schooling (Ruello, 1969, 1977). Extremely large catches of these shrimps have been made during a short period of time. The reasons for this schooling behavior are not clear, but they do not seem to be related to mating or spawning.

The subadults and adults of most species of penaeids in the ocean environment seem to move as independent individuals, and their distribution on the ocean bottom is scattered, as indicated by otter trawl catch rates. However, little solid evidence is available to support these indications; in fact, the entire area of penaeid behavior has received little scientific attention.

3. MIGRATING INTO AND OUT OF ESTUARIES

A long-standing puzzle exists regarding how postlarvae move from the ocean into the estuaries against the net water flow from the estuaries. The problems of swimming strength and orientation that are required to move from the spawning areas, where currents parallel to the shoreline are typically strong, into the estuaries seem insurmountable. Work done in southern Florida (United States) by Hughes (1967b, 1969a,c,d, 1972) has provided partial explanations for the migrations. He observed that postlarvae of *P. duorarum* were active in the water column after acclimation to a given salinity. If the salinity decreased, the postlarvae settled to the bottom until the salinity increased. This behavior would result in movement into the estuaries on incoming tides as salinity increased and a "holding" in position on the bottom on outgoing tides when salinity decreased. However, this would be effective only after the postlarvae reached areas influenced by tidal fluctuations and related changing salinities. It has been suggested that movement of postlarvae is random in areas beyond the influence of such tidal action, and perhaps the numbers of postlarvae are sufficiently large to permit loss of the high percentage that would never reach an estuary if random movement occurred.

With juveniles, the problem is less complicated because of net water flow out of the estuaries and greater swimming strength of the juveniles. Hughes found that juveniles exhibited changing rheotactic responses in a rhythm corresponding to tidal cycles. This biological rhythm was maintained even after the shrimps were removed from the natural environment. Juveniles also swam into the current or maintained their position on the bottom except when salinity was reduced, at which time they swam with the current.

Juveniles of *Penaeus duorarum* exhibit the three following kinds of movements: (1) daily back-and-forth movements with the tides, (2) short-term offshore movements to escape winter's cold, and (3) mass movements offshore in response to abnormal weather conditions associated with hurricanes (Tabb *et al.*, 1962). Migration routes of *P. duorarum* are broad, yet distribution patterns remain distinct (Costello and Allen, 1966). Juveniles of *P. duorarum* respond to a full moon by increased movement to the surface at ebb tide (91%), as compared with 75% at the surface during new and quarter lunar phases (Beardsley and Iverson, 1966; Beardsley, 1970). Lunar periodicity in catches of *P. monodon* also has been noted (Subrahmanyam, 1967).

Natural stable carbon isotope analyses are useful for tracing the movements of shrimps (Fry, 1981). Four isotopically distinct shrimp feeding grounds were isolated along the Texas coast. Best results were obtained by sampling the slower growing males soon after they began offshore migrations.

F. Ecological Importance

Penaeid shrimps are one of the predominant groups of animals in many tropical and subtropical estuarine environments. As such, they are an important link in the food chain. Their flexibility in food habits makes them opportunistic predators, grazers, and detritus feeders, roles that strengthen their importance in the ecology of estuaries (Ruello, 1973b). Offshore, penaeids gradually spread over large areas of ocean bottom. Even though typically more scattered in the ocean environment, they are still one of the predominant animals over large areas of coastal shelf (Divita et al., 1983).

From postlarval through adult stages, penaeid shrimps are an important food item for many species of fishes, crabs, and birds (Divita et al., 1983). The extremely high natural mortality rates of these shrimps are due largely to predation (Tabb et al., 1962). Although natural mortality rates apparently decline as shrimps reach subadult and adult stages and disperse themselves widely in the offshore environment, the larger shrimps still form an important food item in the diet of many fishes (Pearson, 1929; Divita et al., 1983). In estuaries, many species of fishes (fingerlings, juveniles, and adults) move with the shrimps and feed heavily on them throughout their stay in the estuaries. The life cycles of many fishes have evolved to effectively utilize the shrimps at various stages of development (Tabb et al., 1962; Divita et al., 1983). The high fecundity and rapid growth rates of shrimps are evolutionary adaptations for their survival under these circumstances (Wickins, 1976).

Considerable controversy has existed among shrimp biologists regarding the food habits of shrimps. Even though agreement exists that penaeids are opportunistic omnivores, the relative importance of animal foods in their diet has long been debated (Rao, 1967). A part of the controversy is related to the difficulty in studying the contents of the digestive tract. The very strong digestive enzymes quickly digest food items, and investigators have often found only the relatively indigestible hard parts of organisms eaten, when stomach contents were examined. Species differences are marked and may be related to habitat preferences of the species (Kuttyamma, 1974). Polyculture of penaeid shrimps has been suggested because food habits differ so greatly. Changes with age have also been documented (Ruello, 1973a).

For penaeid shrimps in the Gulf of Mexico, feeding upon detritus is important to their nutrition (Flint and Rabalais, 1981). All forms of plant and other detritus, including that found among sand grains, may be eaten. This leads one to conclude that the living coating (bacteria, protozoans, algae, and microscopic invertebrates) may be digested from the surface of these detritus particles as they move rapidly through the shrimp gut. In the offshore environment penaeids seem to become more predatory, feeding on a wide variety of mollusks, crustaceans, and polychaetes (Ruello, 1973a).

G. Fisheries Management

1. GENERAL PRINCIPLES, PROBLEMS, AND IMPLICATIONS

Any sustained harvest from a living marine resource usually requires proper fishery management. Penaeid shrimp fisheries have generally remained productive, despite intensive exploitation. The apparent biological resiliency of shrimp stocks (high reproductive capacity and turnover rate) helps sustain the continued high production. However, in spite of favorable biological and environmental conditions, continually improved technology for exploitation may require widespread regulation of harvest to insure maximum sustained production (Jones and Dragovich, 1973). A possible exception to this pattern is the Louisiana population of *P. setiferus,* for which spawning stocks have apparently been reduced sufficiently to reduce harvest over a 20-year period (Neal, 1975).

Management of penaeid shrimp populations is especially difficult, due to their unique life history, population dynamics, and character of the fisheries. Determinations of the best sizes for capture are very sensitive to mortality calculations and growth rates. The concept of how recruitment is affected by stock size and environment is poorly understood; thus it is unknown whether high levels of effort cause population instabilities. Likewise, the best biological and economic levels of effort are difficult to determine (Rothschild and Gulland, 1982).

In addition to being vulnerable to natural mortality at all stages of development, shrimps enter the commercial catch as juveniles and are heavily exploited by competing inshore and offshore fisheries. Likewise, the degree to which a given catch is distributed among all user groups is unknown. Furthermore, a varying proportion of small shrimps is wasted by being discarded (Caillouet and Baxter, 1973). Figure 9 summarizes relationships among the life cycle and components of penaeid fisheries.

In general, penaeid stocks are fully exploited with little opportunity for increasing total catches. Increases in fishing effort can cause serious biological, economic, and social problems. In such situations, the major problems of fishery management are failure to recognize problems and failure to act at the proper time. Inherent characteristics of shrimp fisheries create unique management problems: most fisheries are based on multiple species; age of individual shrimps cannot be directly calculated; stocks overlap state and international boundaries; entry into the fishery by recruitment or mesh selection is poorly defined; most fisheries are composed of competing groups using various types of gear to catch different size groups; nursery areas are separated from adult stocks and controlled by different factors; stock and recruitment are extremely difficult to determine and interactions are poorly understood; and habitats, especially nursery areas, are being depleted by pollution and the activities of man (Rothschild and Gulland, 1982).

Maximum sustainable yield is often a major goal of fisheries management. Various regulations are used in different shrimp fisheries to help maintain populations for sustained production. Growth overfishing is likely to occur, as the rapid growth of young shrimps causes an even greater increase in value. By shifting fishing mortality from the smallest sizes onto larger sizes,

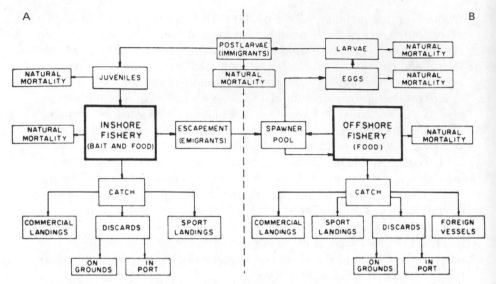

Fig. 9. Relationships among the life cycle and components of penaeid fisheries. (A) Estuarine phase of life cycle. (B) Oceanic phase of life cycle. (From Caillouet and Baxter, 1973.)

an increase in total weight and value can often be achieved. Seasonal and area closures can be valuable by protecting juveniles in nursery areas and young recruits to the fishery. Likewise, minimum size limits and regulation of mesh selection help control the catch but tend to encourage discarding. Catch quotas are not suitable for shrimps due to their short life cycles. Such quotas do not control fishing mortality and might encourage intense fishing early in the season. Fishing effort can also be regulated by limited entry, delegation of fishing rights, and taxation or license fees. Enforcement is the key to any fishery regulation but extremely difficult and costly to implement (Rothschild and Gulland, 1982).

2. METHODS OF ANALYSIS

Rothschild and Gulland (1982) compared different methods of shrimp fisheries analysis. Production models require comparatively few data and are widely used, but do not take into account effort of fishing on different size groups. Age- or size-structured models are based on yield-per-recruit calculations, but suffer from problems with estimating natural mortality. Stochastic models are useful in studying the nature of fishery "collapse" and the relationships between stock and environmental variation if recruitment is known. Predictive models are often used in connection with environmental aspects of fisheries. A biological understanding is essential before a mathematical model is possible.

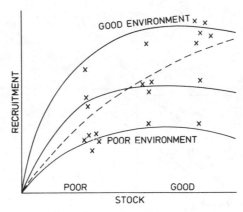

Fig. 10. Stock–recruitment relationships under different environmental conditions. (From Rothschild and Gulland, 1982.)

Recruitment variations are observed from year to year, which show no obvious connections with changes in adult stock. Actual recruitment is largely determined by fluctuations in yearly environmental conditions. Rothschild and Gulland (1982) also provided a description of the stock–recruitment relationship based on a family of curves, each corresponding to a particular set of environmental conditions (see Fig. 10). Each curve gives resulting recruitment from a given set of spawning stock under certain environmental conditions.

Often, reduced spawning stocks, which are associated with lower recruitment, cannot be explained by natural or man-made factors. If fishing is maintained at a high level, there is a risk of stock collapse by recruitment overfishing, as shown in Fig. 11. Rothschild and Gulland also illustrated that the spawning stock will be proportional to recruitment at a given fishing level (indicated by straight lines). Points at which these lines intersect the stock–recruitment curve represent positions of equilibrium. On the flat part of the curve, recruitment is independent of stock, so changes in effort have little effect on the recruitment equilibrium. However, on the steep parts of the curve, even moderate increases in effort can cause large decreases in equilibrium recruitment and possible fishery collapse.

The approaches to population dynamics used most successfully have been yield-per-recruit analyses of the Beverton and Holt type. Growth and mortality rates are estimated to predict the biomass on a "per recruit" basis rather than in actual numbers and weight. The fishing mortality estimates are then applied in the model beginning at various times to maximize harvest. The model can easily be extended to maximize income by incorporating prices for shrimps at all sizes caught. Berry (1967) has worked out a classical

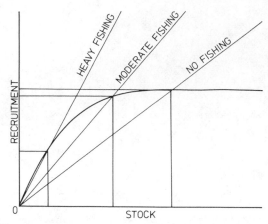

Fig. 11. Equilibrium positions under given stock–recruitment relationships for different levels of fishing effort. (From Rothschild and Gulland, 1982.)

example of this approach. Management recommendations are made in terms of the optimum time (or shrimp size) to begin fishing as a given group of shrimps reaches catchable size (see Fig. 12). Limitations of this approach are related to the difficulties in obtaining accurate estimates of growth, natural mortality, and fishing mortality rates. Also, the actual situation on the fishing grounds is that recruitment is spread over an extended period, so that sizes present vary widely at any one time.

Another possible approach is to limit the size of recruitment into a fishery by using mesh size sufficiently large to permit small or undersize shrimps to pass safely through the mesh of the trawl. Though this approach is used widely with trawl fisheries for finfishes, the shape of shrimps together with the presence of long antennae prevent effective selective fishing through trawl mesh size control. As mesh sizes are increased from a size small enough to retain all shrimps, only a small percentage of the smallest sizes escape. When mesh sizes are large enough to permit the majority of under-sized animals to escape, an unacceptable proportion of larger shrimps also escape. The relatively small size of shrimps, along with their inability to push through a small hole as a fish can, also contributes to this phenom-enon.

Neal (1975) reviewed the population dynamics studies of the Gulf of Mexico and concluded that no satisfactory approach has been devised to study penaeid shrimp population dynamics. More empirical methods must be carefully considered, for example, (1) by comparing the value of harvests in different years in which the average size harvested has been different, (2) by comparing the value of harvests from different fisheries in which manage-

SIZE WHEN HARVEST BEGINS (HEADLESS SHRIMP/ POUND)

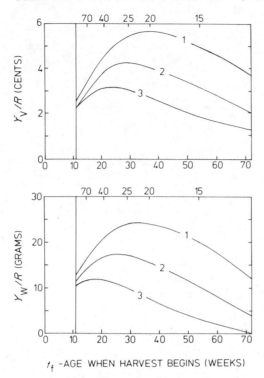

t_f -AGE WHEN HARVEST BEGINS (WEEKS)

Fig. 12. Variations of yield in value and yield in weight with age at first harvest for the following mortality rates. (Group 1) Fishing mortality 17.6%, other losses 2.4% (Group 2) Fishing mortality 20.3%, other losses 3.8% (Group 3) Fishing mortality 14.8%, other losses 5.9%. (From Berry, 1967.)

ment or fishing pressure differs, or (3) by manipulating regulations for the purposes of evaluation.

A bioeconomic simulation model for a selected portion of the brown shrimp fishery along the Texas coast has been constructed by Grant and Griffin (1979). Assumptions were made and variables selected such that the model fairly accurately simulated catches for at least 1 year (see Fig. 13–14). This approach should be a useful one for fishery management, especially as additional information on variables such as recruitment, growth rates, and natural mortality become available.

Another bioeconomic model employing 21 parameters to describe the performance of two classes of vessels exploiting several shrimp stocks was constructed by Clark and Kirkwood (1979). The model predicts the number

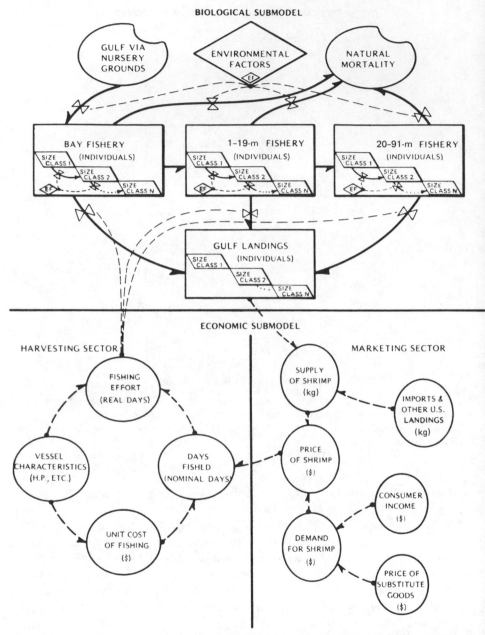

Fig. 13. Flow diagram of conceptual model of the Gulf of Mexico shrimp fishery. (From Grant and Griffin, 1979.)

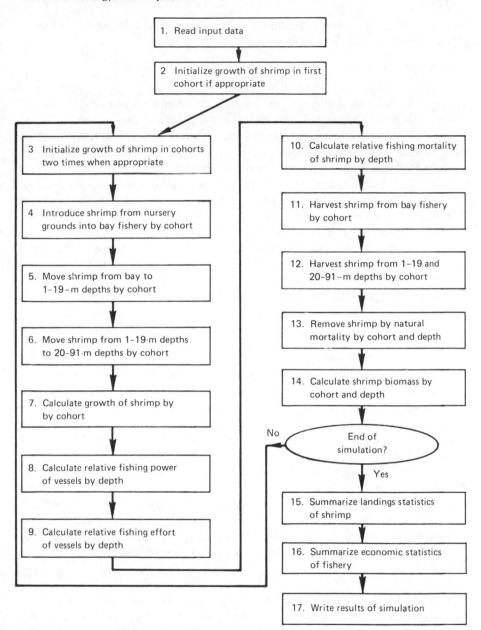

Fig. 14. Flow diagram of simulation model of the Gulf of Mexico shrimp fishery. (From Grant and Griffin, 1979.)

of vessels of each class entering the fishery under free access and compares the prediction with available data.

3. MULTIPLE SPECIES, BY-CATCH

Most analyses of shrimp fisheries are in terms of a single species or of stock, without separation of the species. Problems occur because species are biologically separated, but significant interaction also occurs.

By-catch is an important aspect of most trawl fisheries, but the proportion of by-catch discarded is probably higher in shrimp fisheries. This is largely due to the high disparity in price between shrimps and by-catch species. The fate of by-catch is highly variable, mainly due to the type of fishery and not to the ratio of fishes to shrimps. Length of the fishing trip, storage and refrigeration capacity, and regional demand for the by-catch species are all important factors. Failure to use the 1–2 M.T. of annually discarded potential protein have attracted much attention and concern. There is not only a technological problem of developing appropriate processing but the economic problem of making by-catch attractive for specialized shrimp trawlers (Rothschild and Gulland, 1982).

Small specimens of *P. aztecus* are discarded in the western Gulf of Mexico (Baxter, 1973). Discarding is encouraged by box grading by processors and the minimum size limit of 65 tails per pound (0.45 kg) in Texas (United States). Often fishermen are unable to "head" the entire catch at sea, so they naturally concentrate on the larger shrimps, those which bring the best price. Discarding accounts for 20–60% of the shrimp catch depending on the portion of the season during which the catch was made. As shrimp size increases with time, discarding is reduced.

For the shrimp by-catch, there are ecological implications (Rothschild and Gulland, 1982). An impact of shrimp fisheries on finfish fisheries exists, although to an unknown extent. Discarded fishes and shrimps might supply a food source for other shrimps, but such recycling is only a remote possibility. Possibly shrimp by-catch decreases the number of animals competing with shrimps or preying on them; yet most by-catch species do not feed heavily on shrimps. Further discussion of shrimp by-catch and its economic possibilities can be found in the section on waste utilization.

4. MANAGEMENT IN THE UNITED STATES

States bordering the Gulf of Mexico (Texas, Louisiana, Mississippi, Alabama, and Florida) collectively landed 209.9 million pounds (95.3 million kg) of shrimps in 1982, with a value of $425,748,000 (74% of the United States total in weight and 84% in value). The South Atlantic states (North Carolina, South Carolina, Georgia, and Florida) contributed 25.6 million pounds (11.6 million kg), worth $59,942,000 in 1982 (9% of United States total in weight and 12% in value) (Thompson, 1983).

Results of research on shrimp resources in the Gulf of Mexico have been summarized by Caillouet and Baxter (1973). In addition, the effects of size composition on ex-vessel value for various species of shrimps were reported as follows: *P. aztecus, P. setiferus,* Texas–Louisiana (Caillouet and Patella, 1978; Caillouet *et al.,* 1979, 1980b); *P. aztecus, P. setiferus,* Texas (Caillouet and Koi, 1983); *P. aztecus, P. setiferus, P. duorarum,* Gulf and Southeast states (Caillouet *et al.,* 1980a; Caillouet and Koi, 1980, 1981b); *P. duorarum,* Florida (Caillouet and Koi, 1981a). Estimated costs, returns, and financial analyses of shrimp vessels were provided by Griffin *et al.* (1974), Griffin and Jones (1975), Griffin and Nichols (1976), and Griffin *et al.* (1976a,b). Catch per unit effort data with economic implications were given by Nichols and Griffin (1975) and Blomo *et al.* (1978). The economic impact of Mexico's 200 mile fishing zone on the Gulf of Mexico shrimp fishery was reviewed by Griffin and Beattie (1978). Key factors in financial performance of the Gulf shrimp fishery are highly variable cost/price relationships, seasonality of production, and vessel costs (Warren and Griffin, 1980).

Use of Griffin's yield model (Griffin *et al.,* 1976b) can lead to inaccuracies if certain assumptions of stock and effort are made (Khilnani, 1981). There is available a multidisciplinary simulation model combining biological and economic aspects of Gulf shrimp fisheries in a dynamic framework (Blomo *et al.,* 1982). A multilinear regression analysis of water temperature, salinity, and the number of postlarvae of *P. aztecus* in nursery areas was able to predict the June–July commercial harvest in Mississippi (Sutter and Christmas, 1983). The model explained 80.2% of variations in harvest, or 85.4% if effort was included.

Conditional fishery status was proposed as a solution to overcapitalization in the Gulf of Mexico shrimp fishery (Blomo, 1981). The following classic symptoms of overcapitalization were noted in the open access fishery with high product prices: rapidly increasing effort, low rate of capacity utilization in engineering and economic terms, and volatility of profits for a luxury item.

Growth and mortality parameters were used to compute potential yield estimates along with length–weight relationships in the North Carolina fishery for *P. aztecus* (Purvis and McCoy, 1978). Economic yield was found to occur at a much larger shrimp size than does maximum biomass. A model of growth over time and length–weight relationships that account for stochastic variation in populations of *P. aztecus* and *P. duorarum* in North Carolina was developed by Cohen and Fishman (1980). The model has a potential use in facilitating quantification of population biomass.

Migratory timing behavior of *P. aztecus* according to a time–density model (Mundy, 1979) has been examined in a series of three theses and resulting papers at Old Dominion University, Norfolk, Virginia, as follows:

(1) quantitative description of migratory behavior and prediction of total yield through a time–density approach (Babcock, 1981, 1982; Babcock and Mundy, 1985); (2) study of environmental influences on migratory behavior that also affect growth and natural mortality (Matylewich, 1982; Matylewich and Mundy, 1985), with a determination of the relation of environmental factors to cumulative catch and catch per unit effort by use of a linearized logistic model and a multiple linear model; the time factor and water temperature proved to have an important influence on migratory timing; (3) determination of the relationship of size class distribution to migratory timing (Paula, 1983), whereby data on catch and nominal effort were stratified into count size ranges and the time density model specified for each stratum; total yield estimates from size class stratification were frequently more accurate than were yield estimates.

IV. CARIDEAN SHRIMPS

A. Location and Importance of Major Fisheries

Significant caridean shrimp fisheries occur in shallow temperate waters of both the North Atlantic and North Pacific Ocean. Important genera contributing to these fisheries are *Pandalus, Palaemon,* and *Crangon,* with several other genera making small contributions to the catch. A few smaller catches of pandalids are made in the southern temperate waters, for example, a fishery for *Heterocarpus* exists along the coast of Chile (Bahamonde and Henriquez, 1970; Struhsaker and Aasted, 1974). Extensive reviews of the biology of certain caridean shrimps have been provided by Lloyd and Yonge (1947), Forster (1951, 1959), and Allen (1959, 1960, 1963).

Several good review papers have been written that discuss caridean shrimp fisheries. Wickins (1976) reviewed the major fisheries and listed important references to each of the fisheries. Gulland (1970) presented a comprehensive discussion of existing fisheries, some biological background on important species, and an estimate of future potential for certain caridean shrimp fisheries. A bibliography for pandalid shrimp has been prepared by Scrivener and Butler (1971). Lindner (1957) had references to fisheries for carideans in Latin American waters, Kurian and Sebastian (1976) reviewed caridean fisheries of India, and Anonymous (1958) contained discussions of many other fisheries. For reference purposes, a summary of selected descriptive papers dealing with individual caridean fisheries is presented in Table VII.

Caridean shrimp catches amount to 15% of the total worldwide catch of shrimps, or about 150,000 M.T. annually (Kutkuhn, 1966b). (See Tables II

TABLE VII

Major Selected Caridean Fisheries by Species

Species	Location	References
Crangon crangon	Europe	Tiews (1970)
	North Sea	Leloup and Gilis (1967)
	England	Driver (1976)
	Federal Republic of Germany	Meyer-Waarden and Tiews (1965), Schumacher and Tiews (1979)
	Netherlands	Boddeke (1978), Boddeke and Becker (1979)
Heterocarpus gibbosus	India	Suseelan (1974)
H. reedi	Chile	Bahamonde and Henriquez (1970)
H. wood-masoni	India	Suseelan (1974)
Macrobrachium malcolmsonii	India	Ibrahim (1962)
M. rosenbergii	India	Raman (1967)
Palaemon adspersus	East German Democratic Republic	Scheer (1967)
Pandalus borealis	United States (Maine)	Scattergood (1952), Dow (1967), Wigley (1973)
	Greenland	Smidt (1965, 1967, 1969), James (1975), Carlsson (1979)
	Iceland	Sigurdsson and Hallgrimsson (1965), Skúladóttir (1979)
	Norway	Hjort and Ruud (1938), Rasmussen (1951, 1967b)
	Denmark	Jensen (1965, 1967), James (1975)
	United States (Alaska)	Harry (1964), Butler (1967), Ivanov (1967, 1972)
P. jordani	Pacific North America	Butler (1953)
	United States (Washington)	Tegelberg and Smith (1957), Magill and Erho (1963), Magoon and Tegelberg (1969)
	United States (Oregon)	Magill and Erho (1963)
	United States (California)	Dahlstrom (1965, 1973), Abramson and Tomlinson (1972)
P. kessleri	Japan	Kubo (1951)
P. montagui	United Kingdom	Simpson et al. (1970)
P. platyceros	Pacific North America	Butler (1970), Barr (1973)
Parapandalus spinipes	India	Suseelan (1974)
Plesionika ensis	India	Suseelan (1974)
P. martia	India	Suseelan (1974)

and III and Fig. 1 for information on distribution of the catch). In American waters, their value to the fisherman is much lower than for penaeid shrimps, due principally to the small size, the small proportion of edible meat, and the practice of machine peeling, which results in loss of some meat (Collins and Kelley, 1969). The value of the European catch is higher, especially where shrimps are marketed fresh for speciality markets. Some larger species of American caridean shrimps taken in traps or incidentally by fishermen searching for other species may also bring premium prices (Kessler, 1968).

In spite of the low prices paid for smaller varieties, these fisheries are important both locally to the fishing communities and internationally as a source of high-priced fishery products. The shrimps are sold canned, frozen, or fresh and constitute a significant contribution to the luxury food market for crustacean products.

B. Life Cycle

1. DESCRIPTION OF LIFE CYCLE

The life cycle of *P. borealis* was chosen as a representative of caridean shrimps, on the basis of information from Allen (1959), Smidt (1967), Meixner (1969), Barr (1970a), and Haynes (1979) (see Fig. 15). Spawning occurs from late September until mid-October in Alaska. Just prior to spawning, the female molts into a specialized "spawning shell," which is especially suited for carrying eggs. Setae on the abdominal appendages are longer and abdominal plates are deeper than normal. Soon after molting (1–2 days), the male attaches a packet of sperm on the underside of the female. The female extrudes some 2000 eggs, with fertilization occurring as the eggs pass over the sperm packet.

Eggs are carried on the pleopods for 5–6 months by the female, during which time the female does not molt or grow. Hatching occurs over a 1–2 day period. The female stands on the bottom and vigorously "fans" water with the abdominal appendages at several-minute intervals. During each fanning, larvae, which were hatched during the preceding few minutes, are washed from the egg cluster and drift away. About 2 weeks after the eggs hatch, the female molts and returns to a nonbreeding type shell.

Caridean shrimps typically pass the nauplius stage in the egg and hatch as a zoea. During the 2–3 month planktonic larval period in *P. borealis*, development proceeds through five other zoeal stages. Following the zoeae are two to three postzoeal stages. Caridean shrimps typically have fewer larval stages than penaeids, and there seems to be a direct relationship between fecundity and the number of larval stages.

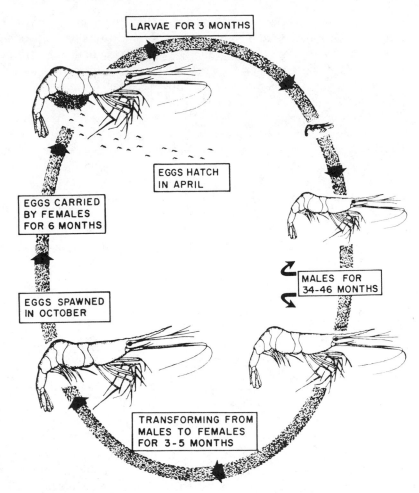

Fig. 15. Caridean (pandalid) shrimp life cycle. (From Barr, 1970a.)

As previously mentioned, female carideans carry developing embryos attached to pleopods. Brooding behavior consists of aeration by beating pleopods and by removing accumulated debris with cleaning appendages. If the cleaning appendages are removed, egg death results from one or more of the following factors: direct smothering by sediment and bacterial growth; localized pockets of anoxia due to bacterial action; pathogenic attack by microbes (Bauer, 1979).

Bauer also detailed brooding behavior in the hippolytid shrimp, *Heptacarpus pictus*. Often dead or diseased eggs are individually removed from healthy ones. Swimming, fanning, and gill currents circulate water through the egg mass. Unfortunately, the egg mass acts as a filter with its tightly crowded embryos, so cleaning is frequently necessary.

The incubation period in *Palaemon serratus* was studied by Phillips (1971). As temperature is increased to an optimum of 21°C, time to hatching decreases. Beyond the optimum temperature, eggs are disengaged from the pleopods in large clumps or are cast off completely by a premature molt. Irrigation is accomplished by extending the pereiopods, and often the telson, to arch the body and raise the abdomen. Then the pleopods are fanned back and forth. Chelate first pereiopods are prodded and poked into the egg mass in vibrating movements and cleaned by drawing them along the setose edges of the second and third maxillipeds. Cleaning periods are irregular and last 10–15 min. Egg fanning rates vary with temperature (16.3 and 60.3 strokes per min for early stages of brooding at 15° and 21°C, respectively) and with stage of embryo development (46.9 and 75.8 strokes per min at 18°C for early and late stages, respectively). The female is very sensitive to deteriorating water quality (oxygen depletion, ammonia accumulation) and egg fanning rates are increased accordingly.

Typically pandalid shrimps are protandric hermaphrodites. Depending on species and location, the functional males gradually become females (with several molts over about a 6-month period) during the second, third, or fourth year of life. Once the change is completed, the shrimp remains a female over the remainder of the 4–6-year life span (Berkeley, 1929; Yaldwyn, 1966; Barr, 1970a; McCrary, 1972; Robinson, 1972; Tegelberg, 1972; Charnov *et al.*, 1978).

Deviations from the normal hermaphroditic pattern were noted for *P. jordani*, with up to 70% of the 1-year-old shrimps maturing into females without ever functioning as males (Allen, 1959; Butler, 1972; Dahlstrom, 1973). Only partial hermaphroditism was found to occur in certain populations of *P. borealis* from British Columbia, Canada (Butler, 1965). Hermaphroditism seemed more dominant in optimum sections of ranges. In shallow water under adverse conditions, there was a tendency for early appearance of females. Sex is under hormonal control but is influenced by temperature, possibly salinity, substratum, and competition. An increase in population reproductive potential occurs when females mature early. Temperatures >8°C cause hyperacceleration of sexual development in *P. borealis* with a male phase being omitted (Ivanov, 1972). Theoretically, this could cause a scarcity of males. Thus, protandry is unfavorable for the penetration of pandalids into warm subtropical and tropical areas. Self-propagation is most effective where favorable temperatures, substrates, and circular currents are present. *Pandalus bonnieri* is dioecious (Pike, 1952), as is *Crangon vulgaris* (Lloyd and Yonge, 1947).

2. CONSEQUENCES OF REPRODUCTIVE CYCLE

Populations of fishes or shellfishes that are heavily exploited typically exhibit a year class structure markedly deficient of older year classes as compared to an unexploited population. With a species such as *P. borealis* in which females do not reproduce until 4 years old (Barr, 1970a), an inadequate supply of reproductive females could be the consequence of a high fishing mortality rate. Undoubtedly, protandric hermaphrodism results in slow recovery of overexploited populations.

The egg-carrying habit of the females also has particular significance from the standpoint of fishery management. In any fish or shellfish population, the removal of a female prior to spawning eliminates her potential contribution to the next generation. With caridean shrimps, a female's contribution can also be removed after she has spawned, while she is carrying the eggs. The consequences of this situation have not been studied carefully but may be of significance in light of the greater value of larger (females in pandalids) shrimps, some evidence of segregation of males and females permitting selective exploitation of females in certain species, and the generally low fecundity of the caridean shrimps.

C. Habitat

Little is known regarding specific caridean habitat requirements other than the water depth, bottom type, temperature ranges, and salinities, because most occur in the coastal ocean environment. Of these, only bottom type and depth preferences differ significantly among the commonly utilized species, as oceanic environmental conditions are basically constant.

The typical habitat for *Pandalus* is far enough from the coast so that temperature and salinity are essentially those of the adjacent ocean and are influenced little by rivers or estuaries (Allen, 1959). *Crangon* differs in that its distribution is typically close to estuaries (Boddeke, 1978; Boddeke and Becker, 1979). Examples of bottom-type preferences include sand or mud for *Crangon crangon* (Boddeke, 1978; Boddeke and Becker, 1979); sand, mud, gravel, or rock for *Pandalus montagui* (Allen, 1963; Warren and Sheldon, 1967); mud for *P. borealis* (Allen, 1959; Smidt, 1965; Ivanov, 1972; Browning, 1980); hard-bottom, kelp-covered areas for *Pandalus platyceros* (Browning, 1980); and green mud substrate for *P. jordani* (Dahlstrom, 1973). These indicate a wide range of substrate preferences for various species within the group.

As with penaeid shrimps, carideans live at depths from a few meters to hundreds of meters, according to species. *Crangon crangon* is not taken at depths greater than 20 m (Tiews, 1970), whereas *P. platyceros* may be taken at depths of 73–274 m (Butler, 1970).

Little is known about types of ecological habitats and faunal relationships. Larvae are pelagic, and the older shrimps are benthic, but movements and patterns of distribution have generally not been studied. Most information on habitat is obtained from trawling operations, able to collect shrimps only on smooth, unobstructed bottoms.

D. Population Dynamics

1. REPRODUCTION AND FECUNDITY

A species may evolve toward having greater numbers of smaller eggs or toward having fewer numbers of larger eggs that hatch as more advanced larvae or (as with caridean shrimps) receive an increased degree of parental care. The carideans as a group have evolved in the direction of large eggs. Wide dispersion of larval stages may not be necessary as with penaeids, and in fact, it is probably detrimental to a demersal animal living in a relatively narrow depth range along the continental shelf.

Caridean fecundity may be in the range of 1,000 to 18,000 eggs per female (see Table VIII). Harris *et al.* (1972) found a linear relationship between number of eggs and carapace length in *Pandalopsis dispar*. Eggs, attached to the pleopods, are carried roughly 3 to 5 months before hatching, depending upon the species and water temperature (Forster, 1951). The fluctuation in temperature probably influences the size of eggs in *C. crangon* (Boddeke, 1982). Large winter eggs are spawned in relatively small numbers

TABLE VIII

Representative Fecundities of Commercially Important Caridean Shrimps[a]

Species	Number of eggs produced
Crangon crangon	4,000–17,500
Heterocarpus gibbosus	10,732–17,095
H. wood-masoni	7,387–11,092
Pandalus jordani	1,000– 3,000
Plesionika ensis	1,542– 3,927
P. martia	1,152– 5,230

[a] Data taken from Tiews (1970), Dahlstrom (1973), and Suseelan (1974).

from October until March. Small summer eggs are spawned in relatively large numbers from March until September. The two types of eggs (unique among Malacostraca), year-round egg production, sex change by which all adults contribute to egg production and specific recruitment mechanisms, all of which help offset the limited nursery areas and high mortality. Spawning frequency ranges from 2 to 5 times per year in C. crangon (Dornheim, 1969; Meixner, 1969; Tiews, 1970).

This relatively low fecundity compared with other marine animals, and the late age at which reproduction begins, constitute a low reproductive potential that is characteristic of an animal that is well-adapted to its environment and not subject to frequent, catastrophic mortalities at any stage of the life cycle. Thus, caridean shrimp populations are probably not capable of rebounding quickly following heavy fishing mortalities.

2. GROWTH, AGE, AND SEX STRUCTURE OF POPULATIONS

Descriptions of the age structure of populations and the variations in year class strength for P. borealis are presented by Jensen (1965), Sigurdsson and Hallgrimsson (1965), Smidt (1965), and Ivanov (1967). These papers provide examples of sex ratios, as well as estimates of the rates of growth of this species in several different locations. Pandalus borealis usually does not reach a useful commercial size until its third or fourth year, although some more southern species, such as P. jordani, may grow faster and reach harvestable size by their second year (Dahlstrom, 1965, 1973). For P. borealis, individuals in their fourth and fifth year of life are most important in commercial catches, and relatively few are taken after they reach 6 years of age (Barr, 1970a; McCrary, 1972; Charnov et al., 1978). When fishing is heavy, males are much more abundant in a population than females (Dahlstrom, 1965). Temperature may regulate growth and maturity in P. borealis; as temperature decreases, maturation time increases, and correspondingly the growth rate decreases (Smidt, 1965). Probably the only factor affecting sex change is the growth rate (attainment of a certain size) (Rasmussen, 1967a).

Crangon crangon is a faster growing species for which fairly good growth data are available (Dornheim, 1969; Meixner, 1969; Tiews, 1970; Boddeke and Becker, 1979). This species reaches a generally acceptable commercial size at 1 year of age, although some may be taken at the age of 6 months (Tiews, 1970). Growth rates in C. crangon are influenced by food, temperature, and size (Meixner, 1969, 1970). Meixner also noted that growth rates were higher in females than in males and listed three distinct phases of growth during the average 17-day molting period. The first 3 days are characterized by no changes in size. For the next 13 days, growth is slight but continuous. Finally on the last day, rapid growth occurs with exuviation. A sex ratio in C. crangon (>18 mm) of about 5 : 7 (males : females), was determined by Dornheim (1969).

3. MORTALITY

Relatively little is known regarding mortality of caridean shrimps other than what can be deduced from length–frequency distributions or from tracing dominant year classes in the catch. After pandalid shrimps reach harvestable size, fishing mortality is probably greater than natural mortality in heavily utilized stocks. On the contrary, mortality due to predation is several times greater than fishing mortality for C. crangon (Tiews, 1970), and fishing mortality is not thought to be a major source of mortality for P. montagui (Simpson et al., 1970).

Estimates of mortality rates for P. jordani off the California coast during periods of fishing and of no fishing were presented by Dahlstrom (1973). During a 6-year period, the total mortality rates ranged from 48 to 70% per year with a mean of 61%. Natural mortality rates during the first, second, and third winter of life were 64, 24, and 57%, respectively.

Individuals of 1½-year-old P. borealis experienced very high mortality rates of about 50% (Allen, 1959). Mortality occurs more at the egg-laying than the egg-hatching period. Evidence was also presented for an increase in mortality at an age of 31 months. Such mortality rates are significant considerations for fisheries, especially where overfishing occurs.

E. Behavior and Migrations

1. FEEDING

Caridean shrimps generally feed on or near the bottom. An exception is P. jordani, which may move to midwater depths during the night, presumably to feed (Pearcy, 1972). Shallow-water species, for example C. crangon (Dornheim, 1969) and P. montagui (Warren and Sheldon, 1967), feed more actively at night, whereas species living at greater depths are active both day and night, on the basis of catches with traps. Movements along the coasts or to deeper or shallower water may be associated with searching for food. These movements may cover distances of up to 10 or 20 km over periods of several months (Boddeke, 1976).

Omnivorous feeding habits of Palaemon adspersus var. fabricii were discussed by Inyang (1977a,b). A higher percentage of animal than plant material was consumed, with crustaceans and detritus contributing the highest amounts. Food conversion efficiencies of 10.4–12.4% were calculated (dry weight increase per dry weight Artemia consumed × 100).

2. ABUNDANCE AND CONCENTRATIONS

The only data available on abundance concern the catch per unit time with trawling gear, and these data provide little specific information on density. Generally caridean shrimps seem to be scattered in distribution and do not exhibit schooling behavior characteristics of some penaeids. *Pandalus jordani* is considered a schooling species, in that relatively dense concentrations of animals of all sizes and both sexes occur in limited areas (Dahlstrom, 1965, 1973). Photography was used to show the rather evenly spaced distributions of *P. jordani* during the day with orientation toward the north, the usual direction of bottom currents off Oregon (United States) (Pearcy, 1970, 1972). The average daily tidal migrations have been noted for *Crangon crangon* (Hartsuyker, 1966; Al-Adhub and Naylor, 1975; Van Der Baan, 1975; Janssen and Kuipers, 1980); and for *P. borealis* (Barr, 1970b).

Movements to shallow areas in springtime or by younger animals have been observed, as have movments to deeper waters in the fall by egg-bearing females or adults of breeding ages (Boddeke *et al.*, 1977; Kanneworff, 1979). These migrations are usually of a few kilometers. The autumn–winter and spring migrations of *C. crangon* (Boddeke, 1976; Boddeke *et al.*, 1977) are highly flexible in time, duration, distance, and participation, with relative differences in water temperature triggering the movements. Sensitivity varies greatly; berried females are generally more sensitive than nonberried ones, and fertile males are more sensitive than spent individuals. Seasonal migrations probably ensure maximum reproductive potential by efficient distribution of stock. Indirect evidence of rapid, coordinated movements exists for several species, which may be taken in large numbers at a site on one day, then seem to be absent from the site a few days later (Hagerman, 1970; Beardsley, 1973). Feeding migrations may be unrelated to seasonal movements, and the observations of various authors on migrations are not in complete agreement (Allen, 1966a; Rodriguez and Naylor, 1972).

F. Ecology, Predator–Prey Relationships

Although environmental conditions are relatively constant in the offshore habitats of caridean shrimps, evidence is mounting that temperatures may have an important impact on these shrimp populations. Exceptionally low temperatures may reduce fecundity or affect survival of larval shrimps, resulting in year class failure in extreme situations. Exceptionally low temperatures ($-1.6°–0°C$) for extended periods may even be lethal to adult pandalids (Ivanov, 1972; P. Holmes, personal communication). The effects of water column stability or instability and fluctuations in current patterns are being examined carefully because of their apparent impact on pandalid populations. Very high (>2.5°C, March; >13.5°C, June) or extremely variable temperature caused decreased landings of *P. borealis* in Maine (Dow,

1967). The possibility of estimating landings of C. *crangon* based on positive correlations with total rainfall and inverse correlations with air temperature during the previous year was determined by Driver (1976).

Many species of animals feed on larval, juvenile, and adult stages of these shrimps. Lists of predators are presented by Smidt (1965, 1967), Tiews (1970), and Dahlstrom (1973). Approximately 60% of the juvenile production of C. *crangon* are consumed within the immediate area (Kuipers and Dapper, 1981). Tiews (1978a,b) noted an average loss of C. *crangon* from 1954 to 1973 was 110,000 million shrimps eaten by predatory fishes per year in the German Bight, with age-group 0 especially preyed upon (Tiews, 1978a,b). Estimated 0.9–4.5 times as many shrimps are destroyed annually by predatory fishes as by the fishery. A highly significant negative correlation was found between losses to predators and shrimp catches the following year. The role of caridean shrimps in the food chain has apparently not been studied thoroughly, but carideans are assumed to be a major food item for numerous demersal species.

Food items utilized by caridean shrimps range from detritus to crustaceans, mollusks, polychaetes, and other benthic invertebrates. A thorough discussion of the foods of C. *crangon* is presented by Tiews (1970). The average food consumption of C. *crangon* makes them the most important carnivore population of the Wadden Sea, the Netherlands (Kuipers and Dapper, 1981).

The impact of shrimp harvest on other species was discussed by Boddeke (1978). When fishing for C. *crangon* was heavy, adverse effects were seen in shrimp predator populations with decreased catches and a decrease in average size. Shrimp trawling also destroys benthic nesting areas of certain fishes.

G. Management of Populations

Fox (1972) reviewed population dynamics of pandalid shrimps and described a computer simulation model. Application of yield models to a California population of P. *jordani* was discussed by Abramson (1968) and Abramson and Tomlinson (1972). Two types of yield models were utilized to analyze the fishery data. The Schaefer form of the Stock Production Model has the advantage of requiring only catch and effort data, which are relatively inexpensive to obtain. Also density-dependent effects are included, even though they are treated grossly, and the population response to density is assumed to be instantaneous. The Dynamic Pool Model or Yield Per Recruit Model utilizes catch data by age categories, both in weight and numbers, to estimate mortality, growth, and recruitment parameters.

Cohort analysis was used for estimates of mortality rates in C. *crangon*

(Schumacher and Tiews, 1979), while Skúladóttir (1979) employed a method using least squares, based on catch per unit effort for stocks of *P. borealis*, along with cohort length analysis for assessing maximum sustainable yield.

A characteristic of caridean shrimp stocks is their susceptability to overexploitation and slow recovery from overfishing. Their slow growth, sensitivity to adverse environmental conditions, age of reproduction, and low fecundity all contribute to this feature (Smidt, 1965; Rasmussen, 1967a; Boddeke, 1982). The importance of this fact from a fishery viewpoint is that a given resource may support an active fishery involving many boats for a short period of time, or it may support a few boats over extended periods. Typically, these fisheries demonstrate a "boom–bust" cycle, during which good catches are made by many boats in a localized area and following which catches decline drastically, so that fishing in that area is not profitable for several years or longer.

Few population models have been attempted and the resource manager has little information on which to base a management strategy, because of the difficulties in estimating population size, catch per unit effort, spawner–recruit relationships, or even growth or mortality rates. In addition, the practical management options open to him are few. Closed seasons and closed areas (determined perhaps on the basis of catch quotas) are virtually the only effective and efficient management tools available. Limitation of fishing vessels can sometimes be used, depending on the situation (Boddeke, 1978).

Regulation of mesh size in nets is useful in harvesting only larger animals in many populations of fishes, but it has limited applicability when used with shrimps (Dahlstrom, 1965). The vulnerability of a wide range of sizes of shrimps to large mesh nets as a result of their body shape and long antennae is of special importance for caridean shrimps. The total range of sizes within a population is small, and year-classes are normally mixed on a given fishing ground. It is practically impossible to catch selected age groups of a population for this reason (Smidt, 1965; Ivanov, 1967). Therefore, males, females, young, and old are all taken together.

A long historical record of catch and effort information may be the most useful data available for use in regulation of fishing, because the manager most likely does not have adequate biological data on the species being harvested. A Pandalid Shrimp Workshop held in February, 1979, by the State of Alaska (P. Holmes, personal communication) provided some insight into management practices now in use. Quotas based on biomass estimates and catch–effort data are being used by some countries, even where knowledge of stock sizes and recruitment are poorly understood. Closed seasons are widely used to protect egg-bearing females, to reduce fishing during bad weather, or to comply with traditions or politics. Mesh restrictions are widely applied to reduce catches of undersized shrimps. Limited entry of new boats into the fishery is used in a number of countries.

According to Holmes, the value and impact of these regulatory practices are widely questioned, and many researchers feel efforts to manage for maximum sustained yield have been unsuccessful. It does seem to be generally accepted that fast maturing species with shorter life spans require less stringent management than do the slower-growing longer-lived species.

V. STOMATOPOD AND ZOOPLANKTON FISHERIES

A. Stomatopod Fisheries

Stomatopods, or mantis shrimps, are caught incidentally with penaeid shrimps throughout the tropics. They are fished commercially in Japan (Kubo et al., 1959) and in the Mediterranean, especially Italy, where 4300 M.T. were taken in 1971 (Forest, 1973; Do Chi, 1978) (see Table II for 1981 production). In the Mediterranean, fishing is done with trawls, trammel nets, and pots. Although mantis shrimps are considered a delicacy in Japan and Italy, the portion of edible meat in the abdomen is small. *Squilla mantis,* the commercial species in the Mediterranean, is commonly 18–20 cm in length, with a maximum size of 25 cm (Forest, 1973), whereas some tropical species may be slightly larger. Large concentrations of mantis shrimps do not occur. Catches, therefore, are small, and marketing of the fresh catch is done strictly on a local basis.

B. Zooplankton Fisheries

1. GENERAL PRINCIPLES

Marine planktonic crustaceans constitute about 11% of the crustacean fishery (Omori, 1978). Complexities of zooplankton biology and socioeconomic factors create unique problems for the fisheries, which were summarized by Omori (1978). The life span of most zooplanktonic organisms is less than 2 years; thus the turnover rate is rapid, with high production rates, but standing stocks are not always great. The location and acoustic detection of zooplankton swarms is difficult but essential for the fisheries. Swarms must be utilized due to low standing stocks, but swarming is often seasonal, and the complexities are not fully understood. There is a general lack of satisfactory methods for large-scale zooplankton fishing and processing. Often, labor costs are high, the catch price is low, or local demand is limited. Large-scale marketing is not feasible due to high fishing costs. Zooplankton is a good source of direct protein but is not widely accepted as food, so, at present, zooplankton must be used mainly as highly nutritious additives and not directly for human consumption.

Clarke (1939) estimated that about 750 g of plankton would be required to sustain a person (3000 cal) per day. The following early studies that were concerned with the utilization of plankton for food: Hardy (1941), Juday (1943), and Shropshire (1944). Jackson (1954) concluded that plankton harvesting was not economically feasible until areas of increased density could be located and more economical methods introduced. At present, certain technology is feasible, but demand is not sufficient to meet the high cost of harvesting zooplankton (Omori, 1978).

The major groups of crustacean zooplankton from a fishery standpoint are mysids (order Mysidacea); sergestid shrimps [order Decapoda, suborder Dendrobranchiata, superfamily Sergestoidea, family Sergestidae (Bowman and Abele, 1982)]; copepods (order Copepoda); and euphausiids, or krill (order Euphausiacea, family Euphausiidae). Table IX contains species of economically important crustacean zooplankton.

2. MYSIDS

Marine species of the Mysidacea are occasionally harvested and used for food. The group has worldwide distribution and includes a large number of species adapted to various habitats. Little is known about the taxonomy of mysids in the areas fished, thus no catch statistics are available. The species that typically school in shallow marine waters are most readily taken by fishermen. Fixed and scoop nets are often used, especially in estuaries that utilize tidal currents (Parsons, 1972).

Several mysid species of mixed groups are fished locally in India, Thailand, China, Korea, Japan, and throughout Southeast Asia. Most of the catch is used to make paste and sauces that are used with rice cakes. In estuarine and brackish lakes of Japan, catches of *Neomysis intermedia* are boiled, dried in the sun, and often made into the preserved, cooked food *tsukudani* (Omori, 1978; Mauchline, 1980). *Praunus flexuosus* is collected in the British Isles and made into a paste called *cherve*, which is used as bait for mullets. In India, *Mesopodopsis orientalis* and *Gangemysis assimilis* are harvested as mixed species and sold as *kada chingri* in Calcutta. Elsewhere in India, *M. orientalis* is eaten as *sridhar* (Mauchline, 1980).

3. SERGESTID SHRIMPS

Various species of *Acetes* are fished locally from North and South America, East Africa, India, Japan, Korea, China, and throughout Southeast Asia. *Acetes* lives in estuaries and coastal waters, and at certain times, it forms aggregates near the shore, where fishing is accomplished with push nets and

TABLE IX

Species of Zooplankton of Economic Value[a]

Species	Body length (mm)	Locality	Commercial exploitation[b]
Copepoda			
Calanus finmarchicus	3–5	Norway	+
Calanus plumchrus[c]	4–7	Canada, Japan	+
Mysidacea			
Neomysis intermedia	9–11	Japan ⎫	
Neomysis japonica	10–12	Japan ⎬	+++
Acanthomysis mitsukurii	8–9	Japan ⎭	
Various species		China, Korea, Southeast Asia	++(?)
Euphausiacea			
Euphausia pacifica[c]	15–20	Canada, Japan, Korea	+++
Euphausia superba	50–55	Antarctica	++++
Meganyctiphanes norvegica[c]	25–48	France, Monaco, Norway	+
Decapoda			
Acetes americanus americanus	7–20	Brazil, Surinam	+
Acetes chinensis	30–40	China, Korea, Taiwan	++++
Acetes erythraeus	16–30	China, Southeast Asia, East Africa, India	+++
Acetes indicus	16–31	Southeast Asia, India	++++
Acetes intermedius	20–24	Philippines, Taiwan	+
Acetes japonicus	12–29	China, Korea, Japan	++
Acetes serrulatus	14–20	China	+
Acetes sibogae sibogae	18–32	Southeast Asia	++
Acetes vulgaris	20–34	Southeast Asia	++
Sergia lucens	35–45	Japan	+++

[a] Modified from Omori (1978).

[b] +, < 100 tons/year; ++, 100 to 1,000 tons/year; +++, 1,000 to 10,000 tons/year; ++++, > 10,000 tons/year.

[c] Used primarily for feed and bait.

fixed bag nets usually during daylight hours. *Acetes* accounts for some 15% of total shrimp catch. Only a small amount is marketed fresh, and most is dried, salted, or fermented with salt to make a highly esteemed shrimp paste and sauce (Omori, 1978).

Sergia lucens is found only in Suruga Bay, Japan. A fishery has been operating since 1894 using a two-boat purse seine at night at depths of 20–50 m. The season lasts from March until early June and October to December. The summer season is closed to protect spawning. The fishery is controlled with yearly forecasts of stock size before commencement of the fishery. Most is marketed as dried shrimps for human consumption. *Suboshi-ebi* is made from sun- or machine-dried shrimps; *kamaage-ebi* is processed from shrimps boiled in saltwater; *niboshi-ebi* is dried after boiling and sometimes artificially colored red; while *muki-ebi* is processed similar to *niboshi-ebi*, but each individual carapace is removed (Omori *et al.*, 1973; Omori, 1978).

The enormous stocks of *S. lucens* are not affected by fishing pressure (Omori *et al.*, 1973). The fishable life span does not exceed 1 year, so that catch depends on the success of individual year classes entering the fishery. There are positive correlations between catch for different year classes and water temperature. *Sergia lucens* accounts for about 1% of annual world shrimp catch.

4. COPEPODS

Calanus finmarchicus and *Calanus plumchrus* are the most abundant copepod species in the North Atlantic and North Pacific, respectively. *Calanus finmarchicus* has been harvested since about 1960 in a small-scale commercial fishery in certain fjords of western Norway. The fishery operates from April until July using either anchored or towed fine-meshed nets. The catch is deep-frozen and used as supplementary food in salmonid fish culture, as a source of the red pigment asthaxanthin, or canned for pet fish food. Some experimentation, with "red feed" products for human consumption, has been undertaken in Norway. *Calanus plumchrus* has been fished from the Fraser River Estuary in British Columbia, Canada, and off Kinkazen, Japan, for many years. Most is marketed as pet food or ground bait (Parsons, 1972; Wiborg, 1976; Omori, 1978).

Herdman (1891) initially attested to the high quality flavor of copepods, likening them to lobsters. The dry weight content of 15–35% (average 20%) is very similar to krill in composition (10–40% protein, 12–47% fat, 3% chitin, 3.6% ash). Wickstead (1967) noted that copepods could provide a balance of protein for deficiencies, but not an entire diet. Wickstead further stated that copepods have all the essential amino acids, in about the same proportions as hen's eggs, and on a weight of protein per unit weight basis copepods were similar to beef steak.

5. EUPHAUSIIDS (KRILL)

a. Major Economic Species and Fishery Potential. Of the several groups of shrimplike animals, the euphausiids have generated the greatest interest and hold the most promise for becoming an important fishery resource. Krill are distributed worldwide in ocean waters, and a total of 86 species has been described (El-Sayed and McWhinnie, 1979). Major reviews of euphausiid biology were presented by Marr (1962), Mauchline and Fisher (1969), and Mauchline (1980). McWhinnie *et al.* (1981) prepared an annotated bibliography of about 2600 euphausiid papers from 1830 to 1979. Eddie (1977) reviewed harvesting techniques, and Grantham (1977) reviewed krill utilization, while Suzuki (1981) reviewed krill processing, utilization, and chemical composition.

The biochemical composition of krill determines technological and nutritional properties of the various products. Likewise, the specific composition directly influences selection of processing and product options (Grantham, 1977). One of the most important characteristics of krill is the high protein concentration (43–52% by dry body weight, Raymont *et al.*, 1971; Clarke, 1980). Mauchline and Fisher (1969), Grantham (1977), and Mauchline (1980) reviewed biochemical composition of krill. Krill composition varies with season, age, location, sex, diet, and various physiological conditions (Grantham, 1977).

Japan has been harvesting *Euphausia pacifica* commercially for years along the west coast off Kyushu and Honshu and in the coastal Sea of Japan. The only east coast fishery is off Honshu, near the island of Kinkayan, with the principal market located at Onagawa. Small vessels are generally employed with side trawls using nets of 8 m in length and 2-m wide at the mouth, with 5-mm mesh. Some of the catch is boiled, dried in the sun, and made into a type of rice cake called *tsukudani.* Most of the catch is frozen and used in the culture of sea breams and rainbow trouts (Parsons, 1972). This same species is also fished along the east coast of Korea and off British Columbia, Canada (Omori, 1978). In Canada, *E. pacifica* is used for fish food in culture operations (Parsons, 1972). In Norway, France, and Monaco, *Meganyctiphanes norvegica* is harvested and used mainly for bait.

The large krill resource of the Antarctic Ocean is estimated to be between 183–1350 million M.T. (Bakus *et al.,* 1978), with other estimates as high as 3.5–6 billion M.T. (El-Sayed and McWhinnie, 1979). Acoustic densities up to 60 g per square meter (200 M.T. per square nautical mile) were found by O'Sullivan (1983b). Technical developments in harvesting methods in recent years have put this resource within economic reach of the fishing industry. During the same period, major advances in krill processing technology and utilization have been made. Although present harvests are low, it is possible that improved markets and increased demand for protein will result in increasing utilization of the resource (Eddie, 1977).

The species of principal importance to the developing Antarctic fishery, in the order of their importance, are *Euphausia superba, Euphausia crystallorophias, Thysanoessa macura,* and *Euphausia vallentini* (Bakus *et al.,* 1978). Other locally important species include *Euphausia frigida* and *Euphausia triacantha* (Hempel and Marschoff, 1980). The largest and most important species, *E. superba,* averages 20–30 mm in length and may reach a max-

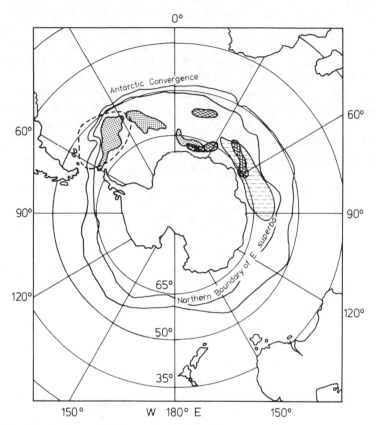

Fig. 16. Distribution and fishing grounds of *Euphausia superba*. (---) Fishing grounds of Soviet Union expeditions. Shaded areas indicate fishing grounds of Japanese expedition. (▨) 1972–1973. (▧) 1973–1974. (▤) 1974–1975. (Redrawn by Yvonne Wilson, from Bakus *et al.*, 1978.)

imum of 60–70 mm (Bakus *et al.*, 1978; El-Sayed and McWhinnie, 1979), with an average weight of about 1 g (O'Sullivan, 1983a). Distribution and fishing grounds of *E. superba* can be found in Fig. 16.

Although a wide variety of fishing gear has been tested experimentally and on a pilot commercial scale, the evolving Antarctic industry has settled upon a midwater trawl as the most efficient type of gear. The most successful net consisted of a 5 × 5-m mouth with about a 5-mm mesh, towed at 1.5–3 knots (Parsons, 1972). Little effort is made by the krill to avoid nets, so no herding devices are necessary (Eddie, 1977). Sonar is used specifically to locate swarms for maximizing the catch rate and estimating biomass (Nast, 1982).

The fishing vessels of the Soviet Union process krill on board by washing, pressing, and treating in a thermal coagulator for 10 min. The coagulated protein is separated, frozen, and marketed as *okean*. This krill paste contains 24% protein and 71% moisture. The yield of paste is 20–30% of the weight of krill harvested. Most of the Japanese catch is boiled immediately and marketed as frozen blocks or as dried food (Parsons, 1972).

Based upon experience of those countries now fishing this resource, catches of 200–300 M.T. per vessel per day are possible, and higher catch rates may occur as fishermen become experienced in harvesting this resource. Some fishing companies are projecting average harvests of 500 M.T. per day (Eddie, 1977).

The small and variable size of krill is a complication with respect to harvesting and processing. Unusually fine nets must be used to capture the animals, which puts extreme stresses on the gear and requires powerful boats to pull large nets. Processing must be completely mechanized to be economical, and spoilage problems are of major concern. The general advantages of shrimplike animals with respect to processing and handling also apply to krill (Grantham, 1977; O'Sullivan, 1983a). It might seem that spoilage would be slow with Antarctic krill that are harvested from water near 0°C. However, any potential advantages are overshadowed by the powerful enzymes in the krill head region, which are very active at existing temperatures (Eddie, 1977; Bakus et al., 1978). Further information on krill handling, processing, and utilization can be found in a following section on food technology.

Together with the problems and difficulties of harvesting and processing of krill, commercial prospects must be viewed in light of the remoteness of the krill resource from markets, port facilities, and industrial services required by the fishing vessels. Extreme weather conditions and ice are common problems even during the 5-month season when fishing is possible. No navigational aids are available to fisherman and oceanographic charts are incomplete (Eddie, 1977; Bakus et al., 1978).

Nevertheless, several countries have begun or intend to begin commercial fishing on a regular basis (see Table II for 1981 production). Japanese scientists have estimated that 100 million M.T. could be harvested on a sustained basis annually, a quantity greater than the entire present annual world harvest of fishery products. Present annual production of krill as estimated from predation by other animals is between 200–330 million M.T. per year (Bakus et al., 1978).

b. Development and Distribution. Two developmental strategies employed by euphausiids were described by Mauchline (1980). Some 67% of the species lay eggs freely in the sea. A nauplius larva hatches followed by a second naupliar stage, a single metanauplius, three calyptopis larval stages,

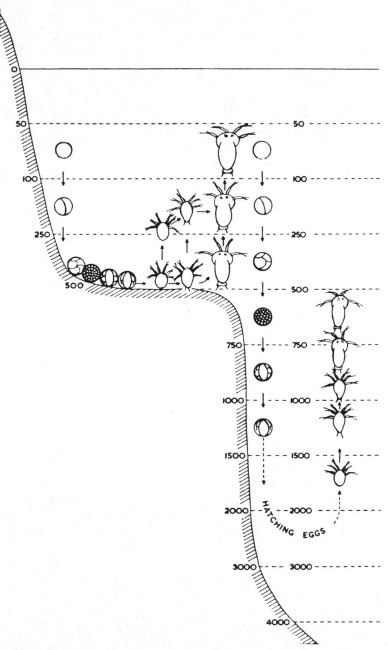

Fig. 17. Development of sinking euphausiid eggs in shelf and oceanic waters. (From Marr, 1962, with permission of the Institute of Oceanographic Sciences, Wormley, Godalming, Surrey, England.)

three to six furcilia stages, a single cyrtopia, and postlarva. The remaining 33% of the species carry eggs attached to thoracic appendages. Hatching occurs as second nauplii, which then develop in a similar sequence as that previously discussed for egg-shedding species.

Large numbers of krill eggs were found in February around the tip of the Antarctic Peninsula (Hempel et al., 1979). The eggs hatch in 5–6 days (about 150 hours). After egg laying, the eggs sink at a rate of about 10 m/hour, so hatching occurs at about 1500 m in the open ocean (Marschall, 1983). Over the continental shelf, eggs sink to the bottom and hatch; however, the highest survival occurs in eggs that remain afloat, due to benthic predation (Marr, 1962; Mauchline, 1980) (see Fig. 17).

As development occurs in eggs and early nauplii, lower depths are attained (Hempel, 1978; Hempel et al., 1979). As development continues, there is an increased tendency to move closer to the surface. Calyptopis larvae occur from intermediate depths to the neuston. Later stages concentrate in the neuston (Hempel and Hempel, 1978; Hempel et al., 1979).

Larvae of E. superba were traced from spawning in shelf waters around the tip of the Antarctic Peninsula in a northeast direction up to South Georgia (Hempel, 1981). Horizontal larval distribution of various species off the Antarctic continent was also studied by Hempel and Marschoff (1980), Witek et al. (1981), Hempel and Hempel (1982), and Siegel (1982a). Voronina (1974) hypothesized that krill larvae were concentrated in the Weddell Sea region largely because the high level of the upper boundary of dense bottom water prevents the eggs from sinking to great depths. Outside of coastal areas, larvae seem unable to cover the great distances separating them from upper water layers. The usual tendency of eggs to sink to a great depth prevents a larger occupation of the Antarctic area. The distribution of E. superba was found to be closely connected with the distribution of phytoplanktonic food, and concentrations occurred where the West Wind Drift mixed with the Weddell Current (Makarov et al., 1970). Figure 18 shows horizontal concentrations of the Antarctic krill.

Diurnal vertical migration was traced in larvae and adults of E. suberba by Nast (1978). Reviews of vertical distribution of euphausiids were presented by Mauchline and Fisher (1969), Vinogradov (1970), and Mauchline (1980).

c. *Schooling and Swarming Behavior.*   The schooling behavior of krill is extremely important from the fishery viewpoint. These schools or swarms form narrow bands averaging 40–60 m, with a maximum of about 500 m in length (El-Sayed and McWhinnie, 1979), but usually only 3–4 m in width (Hamner et al., 1983). Normally, swarms occur in the upper 100 m and frequently in the upper 10 m, sometimes at the surface (Bakus et al., 1978; Witek et al., 1981). The density of krill in these swarms may be from 5–33

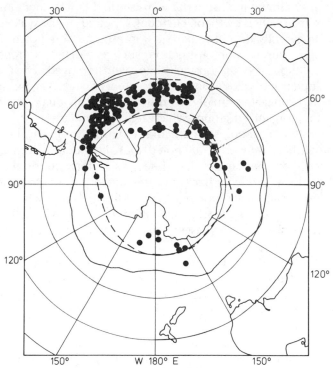

Fig. 18. Principal concentrations of Antarctic krill. (Redrawn by Yvonne Wilson, from Bakus *et al.*, 1978.)

kg/m^3 (Bakus *et al.*, 1978) or 20,000–30,000 individuals/m^3 (Hamner *et al.*, 1983). The swarms are typically monospecific and usually contain about equal numbers of each sex, but sometimes are composed almost entirely of gravid females (Marr, 1962). Schools are segregated into size groups possibly by lifetime integrity of the group or segregation of swimming speeds (Hamner *et al.*, 1983).

The early observations on swarming of krill were reviewed by Komaki (1967) and Mauchline (1980). Possible reasons for such behavior include mixing of water masses, convergence and downwelling, predator–prey relationships, aggregation due to currents, and internal demands of maturation or reproduction (Komaki, 1967; Lyubimova *et al.*, 1973; Mezykowski and Rakusa-Suszewski, 1979). Oxygen was found to be the limiting factor for size and density of swarms (Kils, 1978), and krill maintained a high rate of swimming at 8 times body length per second, with backward (tail-flipping) escape responses of 11 times body length per sec over at least a 60-cm distance (Kils, 1979).

Details of swarming in *E. superba* were studied by Hamner *et al.* (1983). Schools are maintained by rheotactic cues supplied by the wake preceding them and apparently do not require vision. The authors noted that directional changes caused by an obstruction, such as a diver's arm, persisted at the same location after the obstacle had been removed. Each individual precisely copied the detour made by its predecessors. Large-scale disturbances were propagated rapidly, as disoriented individuals randomly tail-flipped among organized groups which reacted similarly.

Molting within a school is a particularly difficult problem, because krill are cannibalistic. Hamner *et al.* noted that individuals swim out of the school, shed their shell in less than 1 sec, swim forward for about 1 min, then sink and slip suddenly back into the school. The authors found that external stimuli can trigger molting. If the individuals of a school are frightened, random tail-flipping results, but a number of krill will instantaneously and synchronously molt. The cast exuviae act as visual decoys for predators as the school escapes.

During tight schooling and rapid swimming, no feeding was observed. Feeding would disrupt oriented swimming, and such feeding would be random. Krill typically are active foragers that feed on concentrated patchy food when the school is dispersed (Hamner *et al.*, 1983).

d. Feeding Behavior. The feeding behavior is important because krill that have recently fed on green phytoplankton are undesirable as material for some krill products (Eddie, 1977). Although the feeding–schooling behavior is poorly understood, Eddie suggested that krill may disperse to feed, then swarm to rest and to digest their food. Whether nor not feeding is synchronous over large areas is not clear, and feeding may occur either once or twice a day. Some investigators have identified feeding activity peaks in the upper 25 m of the water column before noon and midnight (Makarov *et al.*, 1970; Bakus *et al.*, 1978).

The history of research into food and feeding methods of euphausiids was reviewed by Nemoto (1972). Other reviews of feeding methods and food of krill include Mauchline and Fisher (1969) and Mauchline (1980). Various conflicting views have been presented on functions of euphausiid feeding appendages, and Berkes (1975) re-examined, though inconclusively, many of these opinions. Berkes concluded that encounter feeding was probably important for krill. Previously, no evidence of hunting or stalking behavior was found in krill (Mauchline and Fisher, 1969). More than 80% of ingested carbon was assimilated in laboratory studies using *E. pacifica* (Lasker, 1966). Crustacean nauplii were "preferred" over unicellular algae, with particular size being extremely important. Lasker found that 1-year-old krill must capture 100–200 nauplii per day to achieve carbon requirements, whereas 2-year-old individuals need 200–300 nauplii per day.

The feeding mechanism of *E. superba* was studied by Hamner et al. (1983). When not feeding, each euphausiid swims rapidly, pressing the six pairs of thoracic endopodites together to form a keel. During feeding, the thoracic appendages are extended to form a basket. When fully expanded, setae of adjacent endopodites overlap, forming narrow suture zones. When feeding begins, the thoracic exopodites stop beating and flatten laterally against the suture zones forming a water-tight basket. Phytoplankton is collected on the inner setae, and water is expelled by compressing the endopodites. The entire process is repeated 1 to 5 times per sec.

Occasionally, when the euphausiid is not feeding, the thoracic basket is briefly opened, presumably testing for food while the animal is swimming rapidly in tight circles. Also area-intensive foraging occurs at the surface, as the euphausiid rolls upside down against the surface with rapid spinning turns and quickly sweeps the surface with the feeding appendages.

Another reported mode of feeding takes place when the krill swims horizontally just beneath the surface and holds one branch of the antenna out of the water. Floating particles are flicked out of the surface film and are sometimes eaten.

In the laboratory, krill were observed raking algae, which live attached to sea ice, into the food basket; this behavior has implications for year-round feeding (Hamner et al., 1983). Further overwintering strategies, when phytoplankton is scarce, include absorbing body fats, cannibalism, and adopting a carnivorous diet (O'Sullivan, 1983b).

e. *Ecological Relationships.* *Euphausia superba* is an extremely important link in complex Antarctic trophic interactions. Krill were found as the main food in the stomachs of 31 species of fishes constituting 12 families (Permitin, 1970). Krill predators include whales, seals, squids, decapod crustaceans, various other invertebrates, and birds, in addition to fishes, with sei whales eating krill exclusively (Mauchline, 1980). The most heavily preyed upon size-class of krill was found to be the 27- to 50-mm group (Mackintosh, 1974). Considine (1982) summarized krill consumption in the Antarctic (in million M.T. annually): baleen whales, 33; crabeater seals, 100; leopard, Ross, and fur seals, 4; penguins (7 species and 19 other bird species), 39; marine fishes and squids, 100–200; total 276–376. Trophic relationships in the Antarctic Ocean were discussed by El-Sayed and Mc-Whinnie (1979). Only one herbivore, *E. superba,* supports five diverse groups with many species of predators. The system manages because each species exploits a different segment of the krill population and thus reduces competition. The inherent weakness of such a system is the great dependency on a single organism, so any major changes in krill populations by harvesting could affect the entire ecosystem.

f. *Population Dynamics.* There is much controversy concerning population dynamics of krill. Much conjecture exists about brood sizes, with literature values ranging from 310–800 to 2,110–14,086 eggs (Mauchline, 1980). New evidence is beginning to show that previous fecundity measurements have likely been underestimated in numbers of eggs produced and number of spawning periods. Ross and Quetin (1983) found that 15.6% of *E. superba* spawned each day during the summer, with an average spawning frequency of 6.7 days. Females appear to be continually producing eggs, with multiple spawnings over the entire season (Ross *et al.*, 1982).

Age studies of krill were performed by measuring accumulation of the catabolic pigment lipofuscin (Ettershank, 1983). Adults were found to survive at least 3 years, with some living 4 years. Toward the end of autumn and when starved, certain adults reverted to more juvenile states, which added 1 or more years to life span. Thus, krill appear to be longer-lived and have a more extended breeding season than previously suspected. The huge Antarctic stocks may represent an accumulation of year-classes of long-lived species of relatively low fecundity. Uncontrolled harvest under such conditions could be disastrous; much more information is needed concerning reliable estimates of replacement rates, longevity, and reproductive potential.

The average growth rate of *E. superba* in laboratory studies varied inversely with temperature: 0.043 mm/day at 5°C, 0.020 mm/day at 0°C (Poleck and Denys, 1982). A 3–8 day molting frequency depending on temperature and not on food was noted by Lasker (1966). Controversies also exist as to the age at maturity of krill. *Euphausia superba* reaches about one-half the size at maturity after 1 year, with maturity achieved after 2 years (Mackintosh, 1972). The maturation period of *E. superba* was listed as 3 years by Ivanov (1970) and Fevolden (1979). The carapace growth rate decreases with maturity in males, so the best measurements for growth are total length and body length (Siegel, 1982b).

g. *Management and Perspectives.* McElroy (1982) noted that the growth rate of the krill fishery was increasing rapidly, but more remarkable was the size of recent catches as compared to virtually no market for krill products. Reasons for the rise of the krill fishery include the following: the rapid growth rate in the world fishery catch recently came to a sudden and unforeseen halt; most traditional fisheries were approaching maximum potential; and prospects of krill catch were enormous. Other factors that helped promote the krill fishery were an ever increasing world population and need for protein sources, restrictions on conventional fishing grounds, threats of decreases in agricultural productivity due to excessive land use, soil erosion, climatic changes, and increased cost of producing crops (McWhinnie et al., 1981).

The danger of overexploitation in the near future seems virtually nonexistent when all factors are considered. The size and extent of the resource is tremendous. Fishing is presently marginally economical even on the highest concentrations of krill because of the location of the resource. Investment requirements to develop the fishery are substantial, and therefore, only a relatively slow development can be anticipated. Even rapid improvements in processing technology and related increases in demand for krill products would not seem to pose a threat of overexploitation.

VI. FOOD TECHNOLOGY

A. Handling at Sea

Shrimps are typically caught in an otter trawl pulled along the bottom for up to several hours depending on shrimp densities. When pulled in, the net is raised and the bottom is opened, which releases the catch onto the deck. Shrimps are often separated from fishes and other organisms, with use of small wooden rakes to spread the catch on the deck. The typical method of handling penaeid shrimps at sea is to "head" them (separate the abdomen from the cephalothorax, which is discarded) and to hold the "tails" on the boat in ice (Green, 1949a). Headless shrimps usually have longer storage life than whole shrimps, because strong digestive enzymes in the cephalothorax will begin to damage the edible abdominal muscle within 24 hours in whole shrimps even if they are packed in ice (Lightner, 1974). When catches are very good, shrimps may be stored "heads-on" and then "headed" on shore. The quality of the rock shrimp, Sicyonia brevirostris, was found to be influenced by handling (Bieler et al., 1973). Atypical of penaeids, storing rock shrimps with "heads-on" aids in quality retention by maintaining lower bacterial counts and higher solids content. Postmortem biochemical changes in shrimp muscle were studied by Flick and Lovell (1972) and Lightner (1974).

The spoilage rate of shrimps depends on the speed with which chemical reactions and bacterial growth that cause spoilage proceed and usually depends on the temperature of the system (Pedraja, 1970). Three basic temperature controlling mechanisms to reduce the rate of spoilage were discussed by Ronsivalli and Baker (1981): chilling, the bacterial spread is slowed down without freezing; superchilling, some freezing occurs at temperatures ranging from $-3.0°$ to $-1.0°C$; freezing, bacteria immobilized with product frozen solid as $\leq -17.8°C$.

One of the most widely used methods of storing the shrimp catch is to surround it with crushed or flaked ice. Crushed ice may cause wounds in the shrimps, so softer flaked ice is often used. Typically equal amounts of ice and shrimps by weight are mixed. Ice has the advantage of removing large amounts of heat as it melts without changing the temperature at $0°C$ (high latent heat of fusion). Generally, headless and whole shrimps are stored in separate bins, with alternating layers of ice (each about 6-inches thick) and shrimps. Each day, ice is added at the edges of the bin, but the shrimps are not otherwise disturbed unless excessive melting occurs. The optimum temperature of $0°C$ is usually reached after about 1-day's storage in ice. A well-insulated hold will keep shrimps fresh for about 9 days. Ice can become contaminated by bacteria if prepared from low quality water or if unclean surfaces are contacted. Surface contamination can be removed by washing with chlorinated water (5–10 ppm) (Green, 1949a; Kurian and Sebastian, 1976).

An extension of the use of ice is the use of chilled seawater, which maximizes the surface area. The shrimp catch is surrounded by a mixture of ice and water, which brings down the temperature more quickly and maintains it more uniformly than does ice alone (Ronsivalli and Baker, 1981).

Refrigerated seawater, cooled by mechanical refrigeration, is not as limited as is chilled seawater in heat removal or temperature control (Ronsivalli and Baker, 1981). The addition of glucose syrup or corn syrup can bring down the temperature even more. Refrigerated seawater helps shrimps retain their shape, avoids the expense of ice, and wards off blackening, or melanosis (Kurian and Sebastian, 1976). Lee and Kolbe (1982) noted that refrigerated seawater spray is widely used in the Pacific Northwest and, in addition to above advantages, reduces fuel costs for hauling ice and reduces labor costs of mixing shrimps and ice. The major disadvantage is exposing the entire hold to a common environment of recirculating seawater. Any temperature flucutations or contaminations are quickly spread. Advantages of using carbon dioxide-modified refrigerated seawater spray, which extended shrimp keeping quality for several additional days, were discussed by Barnett et al. (1978) and Bullard and Collins (1978).

Fresh seafood is more acceptable, has a higher value, and is cheaper to produce than are frozen products. However, frozen foods have a much longer shelf life than fresh products. Freezing ($<17.8°C$) is most rapid when the shrimps are brought into contact with refrigerants, including refrigerated air and liquid refrigerants [brine, ammonia, mono- and dichlorodifluoromethane (Freon), liquid nitrogen, and carbon dioxide] (Ronsivalli and

Baker, 1981). Advantages of using liquid nitrogen for freezing rather than refrigerated air (lower costs, faster processing, improved quality) were discussed by Brown (1966), and freezing with freon caused no weight loss in shrimps (Morgan, 1974).

The bacteriology of shrimps was studied by Holmes and McClesky (1949) and Green (1949a,b,c,d). The majority of bacteria of freshly caught shrimps was found in the surface slime and head. An average of 42,000 bacteria/g was noted for fresh shrimps. Heading and washing reduced the bacterial count by 75% and 40%, respectively. Spoiled headless shrimps averaged 288 million bacteria/g. Harrison and Lee (1969) performed a microbiological evaluation of P. jordani. The following bacterial composition was found: Acinetobacter–Moraxella, 46.8%; Flavobacterium, 21.4%; Pseudomonas, 9.9%; gram-positive cocci, 7.1%; and Bacillus, 4.4%

Bailey et al. (1956) reviewed the following chemical, physical, and microbiological tests for determining the quality of ice-stored shrimps: trimethylamine nitrogen, volatile acids, bacterial count, pH, amino nitrogen, hydration of water-insoluble proteins, and B vitamin content. Results were best when tests were used in combinations and not individually. Estimation of trimethylamine and total volatile nitrogen were found to be unreliable; total bacterial plate count and pH inside the shrimps were better measures of spoilage (Iyengar, 1960).

A major problem with ice-stored shrimps is loss of solid material or "driploss." Jacob et al. (1962) found that Metapenaeus affinis and M. dobsoni initially contained 24.21% solids. After 8 days of ice storage, washed-whole, headless, and peeled-deveined shrimps had 16.67, 17.10, and 13.86% solids, respectively. All picked up moisture from the ice, in amounts related to solids lost. A decrease in amino nitrogen due to a loss of free amino acids through leaching may be a contributing factor in flavor loss of shrimps after prolonged ice storage (Velankar and Govindan, 1959; Velankar et al., 1961; Pedraja, 1970).

Another problem, associated with shrimp storage, is blackening or "melanosis" due to bacterial contamination. Blackening occurred after 2 days in whole shrimps, with 50, 75, and 100% after 5, 6, and 8 days, respectively (Jacob et al., 1962). Headless shrimps showed no blackening for 6 days, with only 3% after 9 days, because beheading helps reduce the bacterial load that causes blackening. Likewise, shelled shrimps showed a much higher bacterial count than unshelled ones, because the carapace helps resist contamination.

A sodium bisulfite dip has been found to reduce bacterial spoilage of iced shrimps and to prevent blackening (Pedraja, 1970). Some fishermen use this dip, while others mix the sodium bisulfite with the shrimps and ice as they are packed in the hold. Sulfur dioxide residue (10–30 ppm) and formaldehyde (4–6 ppm) were found in shrimps treated with sodium bisulfite (0.5%) (Yamanaka et al., 1977). The acceptability level for sulfur dioxide is 100 ppm, but formaldehyde is not supposed to be present. A nonenzymatic reaction generates formaldehyde from trimethylamine oxide in shrimp muscle. Levels of sulfur dioxide are greatly reduced by prolonged freezing and cold storage. Blackening can also result from enzymatic reactions, but covering iced shrimps with water helps prevent blackening, which is an oxidative change (Kurian and Sebastian, 1976).

Vessels with on-board freezers are used when longer trips are desirable or when fishing grounds are remote from freezing or processing facilities. General use of freezer boats has not been shown to be economical, although at one time many were being outfitted for the Gulf of Mexico fishery. Maintenance costs of on-board freezers are high, refrigeration techniques are not always available, and crews do not like the longer trips. Nevertheless, boats with freezers are used for some fisheries in remote areas.

Pandalid shrimps are caught in much cooler waters than are penaeids, where spoilage is less of a problem. The typical larger catches of P. borealis, or similar species, are not headed at sea and are handled en masse without ice when short trips are possible (Stern, 1958). Separation of small fishes and invertebrates from the shrimps is a problem that has not been solved and requires considerable labor. Storage for 1 or 2 days in the vessel's hold or on deck is considered desirable to "loosen" the shell for machine peeling, although it does result in some loss of product quality (Stern, 1958). Smaller catches or catches of the select larger varieties may be treated with more care and iced or refrigerated on the vessel.

At present, the processing of krill is normally done at sea because of the remoteness of the fishing area from onshore processing facilities. A number of problems with processing krill are related to their small size, and to the fact that they must be handled very efficiently and in very large quantities if the fishing activity is to be profitable.

One of these problems is the separation of krill from other species caught at the same time. If major size differences exist between the krill and the unwanted species, physical separation methods can be used, but animals similar in size to the krill, such as salps, frequently cannot be separated economically. In such cases, the catch may be discarded or used for meal rather than for more valuable products (Eddie, 1977).

A second problem is caused by powerful enzymes in the stomach and hepatopancreas that result in deterioration of the product soon after harvest, even at the low temperatures on board vessels working in the Antarctic. If krill are to be used for human consumption, processing must be initiated within 4 hours at 0–7°C, within 1 hour at 10°C after dumping the catch on deck, and within 12 hours for krill meal production (Eddie, 1977; Grantham, 1977; O'Sullivan, 1983a).

Fevolden and Eidså (1981) studied the bacteriology of E. superba and found low bacterial concentrations on fresh krill (500–200,000/g). Exponential bacterial growth was noted after low temperature storage, with Moraxella, Pseudomonas, and Alteromonas as the dominant bacteria. Due to the low initial bacterial concentrations and high autolytic activity, Fevolden and Eidså suggested that rapid krill degradation was produced by enzymatic activity.

The rapid breakdown of krill is attributed to the weak physical structure of the cephalothorax (Konagaya, 1980). Once viscera are destroyed by handling, visceral enzymes bring about additional digestive degradation of organs and surrounding tissues. Studies have confirmed that the enzymatic autolysis in krill is due to proteases (Seki et al., 1977; Konagaya, 1980; Shibata, 1983a,b).

Norwegian and German studies reported that Antarctic krill contain large enough quantities of fluoride to be dangerous for human consumption. Sclerosis and bone disease could result from eating $\frac{1}{2}$ pound (0.2 kg) daily (Anonymous, 1980). Fluoride values (in ppm) for E. superba were found as follows: whole animal, 1950; carapace, 2840; pleon muscle, 325; boiled meat, 780 (Schneppenheim, 1980). Likewise, the highest fluoride concentrations were located in the krill carapace in studies by Szewielow (1981). Thus the fluoride content of krill products largely depends on the degree of separation of meat from the carapace. Calyptopis larvae of E. superba contain fluoride in >50% greater concentrations than do adults (Hempel and Manthey, 1981). It is highly probable that the larvae are extracting fluoride from the water for some as-yet-unknown biological function. Krill fluoride was compared to sodium fluoride supplements in chicken feed, and no significant nutritional differences were found (Soevik and Braekkan, 1979).

Concentrations of dimethyl sulfide were found to be much higher in krill than in shrimps (Tokunaga et al., 1977). Heat treatment greatly reduced levels of dimethyl sulfide, which suggested an enzymatic formation mechanism. High levels of prostaglandins (PGF_2) and PGE_2) were found in E. superba (Mezykowski and Ignatowska-Switalska, 1981). Such prostaglandins are cyclic unsaturated carboxylic acids with high hormonal activity. Animals fed a long-term krill diet experienced decreases in breeding performance and body weight. Catechol oxidase possibly causes the darkening, which results from krill handling and processing (Ohshima and Nagayama, 1980).

Many technological developments have emerged in attempts to separate the more valuable abdominal muscle of krill from the cephalothorax, which contains high levels of autolytic enzymes and fluoride. A Japanese patent has been granted for a centrifugation process, whereby the hepatopancreas and other internal organs containing the damaging chemicals can be removed (Grantham, 1977). The application of a juice extractor, which separates krill into a juice fraction containing much of the proteins and a pulp fraction with most of the shells, was proposed by Toyama and Yano (1979). A de-carapace treatment for *E. superba,* using a high pressure water jet applied to a krill–water suspension, with recovery yield of 53.4% was developed by Yanagimoto et al. (1979) and Kobayashi and Yanagimoto (1981). The turbulence-agitation method as a practical extension of the water jet de-carapace treatment was introduced by Yanagimoto et al. (1982). Using the new process, >80% of the krill were processed without damage to the meat.

Processing methods of krill for removal of the cephalothorax by beheading or removal of internal organs by mild centrifugation were reviewed by Grantham (1977). Krill roller-peelers are used mainly in Japan and Chile to separate the tail meats from the cephalothorax. Rapidly alternating contrarotating inclined parallel rollers, assisted by copious amounts of water, remove the head, carapace, thoracic appendages, and antennae. Yields of 15–29% have been noted.

Another developing technique was discussed by Grantham for intact flesh removal, yielding the highest value shell-free intact tail meats. Frozen attrition involves centrifugal degutting, washing, cooking in 85°C seawater, cooling, further centrifugation to remove liquid and individually quick freezing the krill to < −35°C. Following freezing, the krill are then "sandblasted" with high velocity jets of hard inert plastic pellets for about 3 min in a perforated rotating drum. Shells, eyes, gills, antennae, thoracic appendages, and plastic pellets are removed through the drum perforations. About a 15% yield of high quality whole flesh can be achieved.

Methods for minced flesh removal and for the production of krill concentrates, isolates, and meal were likewise noted by Grantham (1977). Mechanical sieving or straining krill material through perforations, meshes, or slots yields high-protein, minced flesh products, but they are of inferior quality to intact flesh. Minced flesh also contains undesirable blood, organ remains, and shell fragments. Typical yields of minced flesh range from 29 to 85%.

Krill protein concentrate (KPC) is produced by three different methods. The krill are cooked, pressed, and dried, forming a hygienic meal (KPC type B). Krill protein concentrate type A is made by shell removal, by solvent extraction of fat and water, or by proteolysis, separation, and drying to a hydrolysate. Functional krill protein (FKP) is produced by solvent extraction, by "functionalizing" KPC, or by solubilization, separation, and concentrating under nondenaturing conditions. Krill meal is processed by cooking, pressing, and drying (Grantham, 1977).

The processing of krill meal is especially difficult for several reasons. Due to their small size, the system tends to clog, cooking times are longer for krill than fishes, and steam is also needed. Fat is difficult to remove, with meals containing up to 20% fat being produced. Storage of such unstable products is a problem and antioxidants are needed. Antioxidants in common use include hydroxyanisole, butylated hydroxytoluene, citric acid, ethoxyquin, and tocopherol. Frozen products can likewise be protected by glazing. In addition, only low yields (13–18%) of krill meal can be produced using conventional methods (Grantham, 1977).

The detailed processing methods used vary depending upon the desired product. New products being made experimentally from krill are discussed under the new products section below.

B. Handling and Preservation Ashore

When the shrimping vessels arrive back at the dock, the storage bins are opened, and the shrimp/ice mixture is shoveled into baskets, which are emptied into a washing tank. The shrimps are then placed on a conveyor belt and carried to a work table where whole shrimps are beheaded. If the shrimps are to be marketed fresh, they are packed with ice in wooden crates and refrigerated until they are trucked to dealers. The other shrimps are frozen, canned, peeled, or breaded, depending on the desired product (Green, 1949a).

The largest portion of penaeid landings is washed, sorted, and frozen, and the frozen tails product is the standard item of international trade. The penaeid tail, unpeeled, comprises 60% of the whole weight of the animal (Bayagbona et al., 1971; Cobb et al., 1973). A listing of various shrimp products can be found in Table X.

A more complicated process is required for the "butterfly," peeled, and deveined product or breaded shrimps. Peeling and deveining (removal of the gut and ovary) along with heading are processes frequently done by hand but may also be done with machines. The breading process typically uses headless frozen shrimps, which are thawed, peeled, deveined, and washed prior to breading. The shrimps are spread on a conveyor belt by hand and passed through liquid batter and a heavy sprinkling of dry breading material. The breading step is repeated until approximately equal weights of shrimps and breading material are obtained (Surkiewicz et al., 1967). Both "butterfly" and breaded shrimps are usually frozen for marketing. Shrimp flesh is affected less by freezing and thawing than is most fish flesh.

TABLE X

Shrimp Products[a]

Product	1981 Production (metric tons)	Producers (% of total catch)
Frozen shrimps (whole)	205,264	India (26)
		Mexico (18)
		Malaysia (8)
		Australia (6)
		Hong Kong (6)
		Brazil (5)
		Ecuador (5)
		Norway (5)
		Faeroe Is. (3)
		Panama (3)
		Cuba (2)
		El Salvador (2)
		Surinam (2)
		14 others (9)
Frozen shrimp tails (shell on)	48,051	United States (92)
		Canada (8)
Frozen shrimps (peeled)	33,607	United States (95)
		Canada (5)
Frozen shrimps (peeled, deveined, breaded)	38,636	United States (100)
Canned shrimps (airtight containers)	10,647	United States (57)
		Norway (14)
		Mexico (10)
		Malaysia (9)
		Denmark (4)
		Iceland (3)
		Brazil (1)
		India (1)
		Japan (1)
Canned shrimps (nonairtight containers)	2,867	Denmark (91)
		Thailand (9)
Shrimp meal	486	United States (100)

[a] Compiled from various sections of FAO (1983b), with permission of the Food and Agriculture Organization of the United Nations, Rome, Italy.

The causes and prevention of weight loss in frozen shrimp products were discussed by Rao and Novak (1977). Desiccation, which causes weight loss, results from surface evaporation before freezing or from sublimation of ice after freezing. The amount of evaporation is directly proportional to product thickness, to time required for freezing, and to temperature. Evaporation and sublimation can be reduced by decreasing voids and spaces in the packages. The product should be stored under constant temperature, in relatively impermeable containers and under humid conditions, so product moisture is not lost to the air.

Shrimps must be completely defrosted for peeling and breading processes. Water defrosting is most common, in which 5 lb (2.3 kg) boxes are immersed in 80°F (26.7°C) overflowing water tanks for 1–3 hours. The surface thaws first, resulting in a decrease of surface thermal conductivity and an increased time to thaw the rest of the product. With microwave heating, the frozen mass is placed in an alternating electrical energy field. Individual molecules oscillate, causing molecular friction and uniform heat generation. The use of microwave defrosting is especially suited for raw, headless shrimps, because defrosting time and cost are reduced, while product wholesomeness and efficiency are increased (Bezanson et al., 1973). In addition, no significant differences were found in vitamin values of microwave cooked, boiled, and uncooked shrimps (Rao and Novak, 1975).

Radiation pasteurization methods have been employed to reduce bacterial spoilage in shrimps (Scholz et al., 1962; Awad et al., 1965; Kumta et al., 1970; Savagaon et al., 1972b). The application of 350 krad of radiation to steamed shrimps (100°C, 15 min) increased the shelf life from 1–2 days in untreated shrimps to 6–8 weeks in treated ones (Savagaon et al., 1972a). Previously, microbial analyses of frozen raw and cooked shrimps were compared by Silverman et al. (1961).

Concentrations of the red pigment astacene decrease during freeze drying and this pigment change could be used as an objective index of product quality (Lusk et al., 1964). The effects of heat treatment on color and machine peelability of P. borealis were studied by Collins and Kelley (1969). Increased color retention, better texture, and improved flavor were found with heat treatment. Glucose oxidase was successfully used to decrease oxidation and loss of color in P. borealis (Kelley, 1971), while an extraction method was developed for estimating a carotenoid index as a measure of quality in P. borealis (Kelley and Harmon, 1972).

Pandalid shrimps are often machine peeled and canned or frozen. Machine peeling reduces labor costs but also results in a reduced yield of marketable product. Yields of peeled flesh of about 20% of the whole weight can be expected.

All types of shrimps are sold fresh for markets near the landing sites. The exoskeleton forms a useful barrier to spoilage bacteria, so that the shelf life of fresh headed shrimps is 1 week or more when they are held at low temperatures (Kurian and Sebastian, 1976).

In Europe and Australia, it is common to boil small shrimps before marketing. They then may be retailed whole or peeled for canning or for freezing. These procedures are laborious, and result in yields of about 25% for caridean shrimps.

Blackening sometimes occurs in the canning process (Kurian and Sebastian, 1976). Hydrogen sulfide, formed from sulfur-bearing amino acids in the shrimps, reacts with metals, such as iron and copper, in the can contents. Iron blackening can be eliminated by keeping the pH acidic. Copper blackening is not affected by pH, and copper normally averages from 0.3 to 1.0 mg/100 g shrimps. Mild, moderate, and excessive blackening occur at 1.58, 2.79, and 6.39 mg copper/100 g shrimps, respectively.

C. Waste Utilization

Until recently, shrimp heads and processing wastes have been discarded. Some groups of people eat the head contents or use them to add flavor to shrimp dishes; however, for the most part, heads and exoskeletons are waste. Meyers and Rutledge (1973) categorized shrimp processing wastes into crude meal, proteinaceous material, specific organic compounds, and chitin–chitosan. A summary of uses for the shrimp processing waste can be found in Fig. 19.

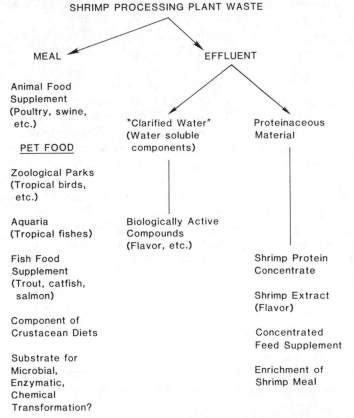

Fig. 19. Uses of shrimp processing waste. (Redrawn by Yvonne Wilson, from Meyers and Rutledge, 1973, with permission of the Marine Technology Society, Washington, D.C.)

Typically, the dried waste material (heads and shells) is made into a crude meal of varying chemical composition depending on processing methods. Problems exist with incorporation into diet formulations until the meal can be standardized. At present, such crude meal preparations are important in trout and salmon nutrition, along with shrimp and other crustacean culture (Meyers and Rutledge, 1973).

According to Meyers and Rutledge, a major concern in crustacean waste utilization is reclamation of protein. Procedures using fish flesh separators followed by solvent treatment result in a shrimp protein concentrate of 81% protein. Similar processes have yielded 42% protein krill mash. Shrimp shells have been used to produce nutrient rich additives for pet and snack foods along with flavorings. Head waste has potential use for shrimp meal, and shrimp heads contain natural pigments useful in maintaining bright colors in certain captive birds.

Meyer and Rutledge also noted that specific organic compounds are important waste products. Nucleotides and amino acids from effluents are useful as flavorings.

Research on use of crustacean wastes have led to the production of chitosan, a deacylated derivative of chitin. Chitosan has some unusual qualities that add luster to plastics and make them useful as filtration media such that greater industrial or commercial uses may materialize.

The multiple uses of chitin were summarized (Anonymous, 1973; Grantham, 1977; McWhinnie et al., 1981) as follows: papermaking additive to improve wet-strength properties of newsprint; additive to baby food formulations; coagulants and flocculants in water, sewage treatment; additive to stomach antacids; controlled, long-term release of herbicides and insecticides; textile finishes; water-base paint emulsions; new synthetic fiber; film manufacture; manufacture of specialty adhesives; ion exchange resins and membranes for chromatography and electrodialysis/desalination; inert tobacco extenders; microbiological media; nonthrombogenic wound healing accelerators and prostheses; stabilizers, thickeners, encapsulators and emulsifiers in food, pharmaceuticals, and cosmetics.

Previously, the recovery and use of dissolved shrimp protein was neglected, with prime attention given to macroscopic solids recovery (Perkins and Meyers, 1977). Using shrimp landings from the Gulf of Mexico for 1976, Perkins and Meyers estimated a total potential waste of 136.3 million lb (61.8 million kg) or 75.9% of the heads-on weight. Based on about 3.3 gal (12.5 liters) of water per pound (0.45 kg) of shrimps processed, as much as 600 million gal. (2.27 billion liters) of water were used in Gulf shrimp processing. The waste waters also contained a total potential dissolved and suspended microscopic waste load of 4.7 million lb (2.1 million kg).

A different type of waste produced by the shrimp fishing industry is millions of M.T. of small fishes and invertebrates captured with the shrimps and killed during capture with the otter trawls. This "trash fish" catch has great potential value for human food or other fish meal, but the value per ton is very small in comparison with the shrimps being taken. The vessel typically has limited storage space and ice capacity, and deck hands are usually busy handling and heading the shrimp catch. As a consequence of these circumstances the bycatch, often including smaller sizes or less desirable species of shrimps (Berry and Baxter, 1969), is discarded at sea in nearly every shrimp fishery around the world. Typically only a few of the larger, select species of fishes (flounders, snappers) taken are held for sale on shore (Juhl and Drummond, 1977). The challenge of finding practical and economical methods for utilization of the valuable shrimp by-catch has perplexed many fishery workers, but we seem no closer to a solution today than we were 20 years ago. The loss of this portion of the fishery harvest is especially tragic where the need for animal protein in human diets is great.

The first study of the problem of by-catch was conducted in the Gulf of Mexico during 1972 (Juhl and Drummond, 1977). The croaker, *Micropogon undulatus,* was found to dominate the by-catch in both weight and numbers. Depending on size, croakers can be utilized as bait, petfood, meal, minced fishes, and food fishes. In addition to loss of protein, finfish discards are lost to other fisheries, and estuarine discards could accumulate, causing environmental deterioration.

The portion of by-catch returned alive was very small, consisting of crabs

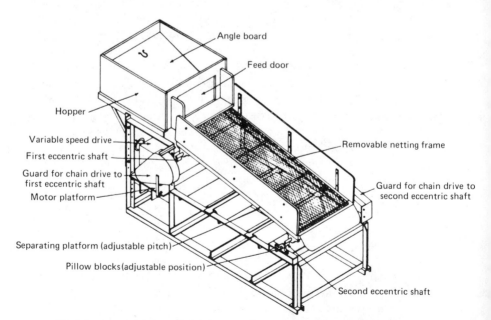

Fig. 20. Isometric view of shrimp separator operation. (From Corbett, 1970.)

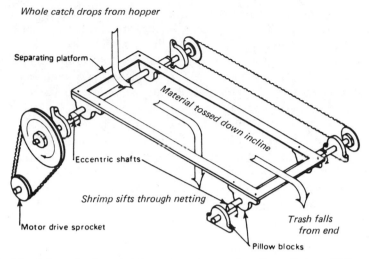

Fig. 21. Schematic view of shrimp separator operation. (From Corbett, 1970.)

and mollusks (Juhl and Drummond, 1977). During the long trawling period of several hours, most organisms die from suffocation. Changes in hydrostatic pressure caused by hauling nets to the surface may cause severe damage. In addition, sorting takes 15–30 min, and few fishes can live that long out of the water. Those that do survive are often attacked by predators as they are shoveled overboard. Cannibalism in shrimp traps could result in losses up to 15%, with larger individuals affected most (Struhsaker and Aasted, 1974).

A method for separating *P. borealis* from by-catch with 95% recovery of shrimps and elimination of 90% of the trash was proposed by Corbett (1970) (see Figs. 20 and 21). The catch is shoveled into the hopper. The angle board forces the catch against the feed door, which, when lifted, allows the catch to tumble onto the separating platform. Two eccentric shafts, supporting each other, provide up and down movement to separate the shrimps and a throwing action, which is needed to move the shrimps and trash along the netting. Wooden boxes are placed underneath to catch the shrimps as they fall through the netting.

VII. PERSPECTIVES

A. Future Demand and Supply

The worldwide demand for shrimps is very strong and is increasing. Shrimps are considered a luxury food in most countries of the world and have universal taste appeal. Over the last 20 years, prices have increased steadily at a faster rate than inflation. The maximum total supply from wild stocks will probably not be much greater than present harvests because few underexploited stocks exist, whereas numerous stocks have been overexploited. Nevertheless Gulland (1970) estimated potential harvests of 1,492,000 M.T., or approximately double his estimated harvest for the mid-1960s. He noted the apparent presence of unexploited stocks below the edge of the continental shelf nearly everywhere in the lower and middle latitudes. The economics of harvesting these deep-water stocks is only marginal at present but may improve if demand continues to outstrip supply. Gulland's estimates do not include krill.

Demand projections are closely related to the affluence of people around the world. World aggregate consumption may be 3,200,000 M.T. by the year 2000 based on consumption, population, and demand statistics (Anonymous, 1970). Even large increases in the supply through aquaculture will not satisfy this growing demand. The demand–supply picture will very likely result in continuing high prices and probably in some additional increases in prices. Again krill have been omitted from these calculations. Krill is a substantially different product from the penaeid and pandalid shrimps. Although demand may develop for krill if supplies can be delivered to consumers at reasonable prices, krill probably will not compete directly with other shrimp products.

Income and shrimp price are very influential determinants of shrimp consumption in the United States (Cleary, 1970). Together, these two factors account for 90% of the change in per capita consumption since 1950. Each 1% gain in per-capita real income tends to be accompanied by a 1.77% increase in per capita shrimp consumption. Each 1% rise (relative to a general price level) in the retail price of shrimps tends to be accompanied by a 0.46% decrease in per capita shrimp consumption. The status and problems of the United States shrimping industry can be viewed in terms of the two major factors influencing the industry: resource availability and market demand (Hutchinson, 1978).

B. Fishery Trends

1. HARVESTING COSTS

The harvesting sector of the United States shrimping industry has the most problems (Hutchinson, 1978). Reasons include a direct dependency on domestic shrimp resources and domestic market demand, and a sensitivity to price changes.

Fishing vessels are extremely expensive and consume fuel at high rates while fishing or running to or from the fishing grounds. Recent increases in fuel costs have had a drastic impact on the costs of building, operating, and maintaining fishing vessels. The extreme example is the cost of harvesting krill where large vessels are required because of weather conditions, and where transport for long distances is required to reach markets. Shrimp consumers seem to be willing, so far, to pay additional costs for this luxury food. However, energy costs can be expected to increase even more in the near future and the long-term impact on shrimp fisheries cannot be predicted. The marginally profitable fishing operations such as those for deep-water penaeids, for krill, and for some remote or sparse populations of pandalid shrimps may not be profitable in the future. It may be found profitable to harvest some pandalid populations, like forests, less frequently, with heavy cropping of the resource.

The new 200-mile fishery jurisdiction, implemented by most coastal countries, recently has placed nearly every shrimp resource (except krill) within the jurisdiction of some country. This offers a new opportunity for wise management of resources, some of which were formerly common property exploited on a first-come, first-served basis. Hopefully, better management strategies will gradually come into practice under this new regime. These improved practices can potentially limit economic overfishing, the utilization of too many boats and men to harvest a resource. Shrimp fisheries are characterized by overcapitalization and misuse of the resources, boats, fuel, and manpower used to conduct the harvest. This wastage is often encouraged by management agencies in the name of conservation or in order to spread the value of the harvest among more fishermen. Such waste can be avoided more easily under the new extended fishery jurisdictions.

2. EFFECTS OF ENVIRONMENTAL CHANGES

Caridean shrimps are largely continental shelf dwelling animals that live at moderate depths and are therefore generally not influenced greatly by environmental changes. An exception may be the *Crangon* populations in Europe, which live in more shallow waters than do the other caridean shrimps. Generally however, it can be said that pollution and other environmental changes are probably having little impact on caridean shrimp populations.

The penaeid populations present quite a different story, however. The penaeids that are dependent upon an estuarine phase are particularly subject to three types of human activity. These are pollution of the estuaries with wastes of a biological or chemical nature; the chemical control of arthropod pests, such as mosquitoes, fire ants, or agricultural insect pests in or near the estuaries; and the reduction of freshwater flow into estuaries by diversion of runoff for irrigation to domestic or industrial uses.

The effects on shrimps of the usual pollution of estuaries through waste disposal have not been well-documented. Indications are that penaeids are not often adversely affected by wastes of biological origin and may be fairly resistant to the chemical and industrial wastes commonly dumped into estuaries in the United States. Shrimps do accumulate contaminants such as heavy metals and some hydrocarbons, and their food value is undoubtedly affected by such pollution (Butler, 1962; Broad, 1965).

Arthropod-specific pesticides developed to control insect pests are also highly toxic to their cousins, the shrimps. Control of salt marsh mosquitoes with pesticides can cause mortalities among shrimp populations. These mortalities are not easily detected because, unlike fishes, which have air bladders, dead shrimps do not float to the surface of the water.

In many coastal areas, the natural flow of freshwater into the estuaries is being reduced. Construction of dams to provide freshwater for irrigation or other uses is commonplace. These dams reduce the flow of freshwater as well as the movement of nutrients into the estuaries and their impact is greatest during dry seasons, when conditions in many coastal estuaries and lagoons may have been marginal for penaeids even prior to diversion of the freshwater flow. The impact of these changes are not known and scientific analysis of the consequences of resulting environmental changes is just beginning. One of the most striking examples of freshwater diversion is the Aswan Dam on the Nile River, which is not only reducing the abundance of shrimp and fish resources, but is resulting in the erosion of the Nile Delta at an alarming rate (Halim, 1960; Inman et al., 1976).

The problem of shrimping as a source of suspended sediment in Corpus Christi Bay, Texas, (United States) was discussed by Schubel et al. (1979). They found that sediment disturbed in the bay each year by shrimp trawling was 10–100 times greater than sediment dredged per year for maintenance of shipping channels.

3. MANAGEMENT PROBLEMS

A few special management problems are faced by shrimp resource managers that were not mentioned in the population dynamics section. The effects of pollution, pesticides, and reduced freshwater flow are some of these problems that were described in the section on environmental changes. Little can be done to alleviate existing problems, so resource managers can only attempt to document their effects on shrimps and to use this documentation in decisions regarding the environmental costs of the damaging practices.

Interest in shrimp farming is resulting in widespread distribution of more desirable penaeid species. So far, no known transplanted populations have become established in the wild. A greater risk of species movement is that of introducing diseases to which native populations have not developed resistance. No major research is being conducted on this topic, and interest in such research will probably not develop unless catastrophic mortalities occur.

The destruction of estuarine penaeid shrimp nursery grounds for aquaculture, agriculture, or the restriction of access to the sea by roads or railroads are making continuing incursions into the total area available for growth of juvenile shrimps. Filling of land along the edges of estuaries for housing, industrial development, or storm protection may also destroy the shallow "edge" of the estuarine environment that is utilized by small penaeid shrimps. Although only a fraction of a percentage of the total nursery area in a locality may be lost or damaged in any 1 year, the long-term effects of this "progress" will be significant changes in the abundance of penaeid shrimps.

C. New Products

Some new shrimp products have been recently introduced. Frozen shrimp blocks of peeled shrimps, usually smaller varieties, are used by the processing industry to produce larger shrimp-shaped pieces, often breaded, for the market. Minced shrimp flesh may be mixed with fish products to create shrimp-flavored products. Individually quick-frozen shrimps are also being sold as a speciality product.

Shrimp chips are a popular product in Southeast Asia. The ingredients, shrimp flesh, tapioca, eggs, coloring, and possibly other secret ingredients, are blended into a loaf, steamed, sliced into fine chips and air dried. This product is fried by the consumer and puffs up to form a light cracker.

Shrimp paste is another special product made in Southeast Asia. Pastes of varying consistency are made of small shrimps, such as *Acetes* or mysid shrimps, salted, and fermented by means of special procedures (Omori, 1978).

The development of new products from krill and similar small, planktonic crustaceans is probably the most active area of investigation with respect to the shrimplike animals. The vast unutilized resources in the Antarctic are the impetus for these investigations.

Whole krill has been successfully marketed in Japan and throughout the Indo-Pacific region, but is generally viewed as inferior to established shrimp products. Raw krill has been satisfactorily accepted as *sashimi*, while some is used in fish dishes such as *tsukudani*. In addition, much of the krill catch is boiled at sea before freezing, and some is sun-dried (Grantham, 1977; McWhinnie et al., 1981). Cryogenic ground krill has been used in the production of certain types of soy sauce (Nakamura et al., 1979).

The meat of whole krill tails is the most likely product for acceptance in the western world. Japan and Poland have used frozen attrition processes, in which good pigment retention and flavor are achieved but the meat is fragmented. Roller-peeling processes used in Chile yield firm, compact meats but lack flavor and pigmentation, due to large amounts of water used in processing. Individual krill tails are sold raw, frozen cooked, canned, or dried. In Chile, intact tail meats are moulded into plate-frozen blocks, cut into portions and breaded as "krill sticks," with about a 20% yield (Grantham, 1977). Krill tails have also been combined with minced fish flesh in fish cakes (McWhinnie et al., 1981).

Coagulated krill paste has been produced in the Soviet Union, Japan, the Federal Republic of Germany, Poland, and Chile. By use of the Soviet Union's paste process, krill mince is successfully heat coagulated under constant agitation. Large, pink, curdlike flakes of 32% solids are produced, and the uncoagulated water contains 13% solids, which can be reduced to protein concentrate or meal. A vibrating screen conveyor is used to separate the flakes from the liquid, and then the flakes are ground into paste with a 20–30% yield. Often the paste is formed into blocks and frozen. Okean is the frozen paste product sold in the Soviet Union as a flavoring in butter, cheese, vegetable, and mayonnaise products. The paste is also used in salads, pâté, stuffed eggs, tomato products, pelmeni (Siberian dumplings), meat pies, and zrazy (fish balls). Krill paste is likewise used in various manufactured products in the Soviet Union. "Shrimp butter" is composed of 50–80% krill paste, 15–25% butter, and 1–25% flavoring; korall is made from 10–20% krill paste and 80–90% cheese (Grantham, 1977). Medicinal values for krill paste have been claimed, ranging from treating stomach hyperacidity to treating hardening of the arteries (El-Sayed and McWhinnie, 1979).

Various sausages are being produced, with use of krill in an intermediate form between pastes and moulded blocks (Grantham, 1977). Krill mince has the necessary emulsification–gelification properties required for sausages and is produced from raw and cooked krill. The cooked mince is firmer, less fibrous, lighter in color, and more stable than are the uncooked varieties. Krill mince is useful as an extender and thickener for fish cakes and Scandinavian fish balls, in addition to sausages.

Krill protein concentrate is one of the most important food aids for low cost use in developing countries. Such protein concentrates contain about 80% crude protein, but yields are only about 8% (Kuwano and Mitamura, 1977; Kuwano et al., 1979). Krill protein concentrate might be more acceptable (orange/pink color, shrimplike taste) than fish concentrates (dark color, strongly flavored) (Grantham, 1977). Hydrolysis of krill has yielded stock-seasoning products used in soups, gravies, and sauces in Japan, Norway, the Federal Republic of Germany, Chile, and New Zealand (Grantham, 1977; McWhinnie et al., 1981).

Krill protein, in fresh, undenatured form, has functional properties that are necessary for many retexturing and reforming techniques being developed for soy protein and other fibrillar proteins. A kneaded sea urchin analog has been produced in Japan along with okiebi, a product based on krill surimi (frozen minced meat) and onion (Grantham, 1977; Suzuki, 1981).

One of the prime considerations for krill is the production of meals for animal feeds. Growth rates comparable to those obtained with the use of fish meal has been found for pigs, poultry, fur animals, and fishes in Japan, the Soviet Union, Poland, Chile, and the Federal Republic of Germany (Grantham, 1977). Yields of krill meal range from 6 to 10% compared with 20% for menhaden in the fish meal industry. Krill protein has 55% protein and a high oil content. The oil is undesirable in the meal but may have uses as a separate product (Eddie, 1977; Bakus et al., 1978).

Krill fat has an unusual composition, high in polyunsaturated fatty acids. The absence of waxes and the particular composition of nonglyceride fractions suggests other applications, in addition to food and industrial uses for fish oils (Grantham, 1977).

D. Research Priorities

On the basis of information presented in this chapter, several critical research needs can be identified. These are topics that have received little attention by biologists and economists but are nevertheless directly related to maintenance and to improved use of shrimp fishery resources. They represent major gaps in our knowledge of shrimp population management and utilization.

In spite of the importance of the shrimp fisheries, we have a poor understanding of how much can be harvested from particular populations on a sustained basis. It is clear that the classical methods for determining maximum sustained yield are not appropriate for most shrimps, particularly the penaeids. New approaches are needed and a comparative approach is suggested. Comparisons can be made between yields and size composition of harvests for similar populations managed in different ways, for single populations managed differently through time, and for various populations in ecologically different environments. Long-term series of landing, effort, and size composition data are needed for these comparisons, and data collection and reporting methods will have to be standardized. Unfortunately, few governments or scientists have been willing to invest needed money and time in the collection of data required for this purpose.

The shrimp fisheries around the world are striking examples of overcapitalized fisheries. Excess harvesting capacity in terms of boats and gear is typical, because of the relatively high price of shrimps. This condition is viewed as acceptable by many shrimp fishery managers, because it spreads the value of the harvest among many people and provides additional employment. The unnecessary expenditures for vessels, equipment, and fuel, in addition to the wasted manpower, are a high price for a country to pay for distribution of income. As fuels, building materials, equipment, and skilled manpower are becoming more expensive, the need for economists, fishery biologists, and sociologists to plan research that addresses this problem specifically becomes more pressing.

Several of the critical research needs of the shrimp fisheries relate to the estuarine portion of the life cycles of penaeid shrimps. The habitat is being damaged or reduced through pollution, destruction of mangroves, construction of coastal ponds for aquaculture, diversion of freshwater for irrigation, and other uses, including construction of roads and railroads, landfill, dredging for navigation, storm protection, poldering for agriculture, insect control, and other activities. The relationships between the numbers of juvenile shrimps in the estuaries and the number of shrimps harvested are virtually unknown. Baxter (1962) and Berry and Baxter (1969) did some pioneering work on this topic, but little more has been done since that time. Resource managers needing data to help evaluate the contribution of estuaries to shrimp fisheries do not have the required information, and because users competing for estuaries often can point to well-defined benefits in terms of employment and products, the decision is often made in favor of competing users, with some unknown loss to the fishery. This is a worldwide problem requiring research attention.

A set of related problems researchers should address is the impact of specific activities such as insect control, industrial pollution of various types, freshwater diversion, and related modifications of the estuaries on penaeid growth and survival. Data on effects of individual activities are urgently needed to evaluate potential impact of development and to modify adverse conditions.

Shrimp fishery managers seem oblivious to the dangers of introduction of disease through transfers of species and stocks from region to region. One needs only to recollect the fate of the European crayfish to realize that crustacean populations, as well as other animals, are subject to catastrophic epizootics. Prevention of the spread of disease with other cultured fishes and shellfishes has been largely accomplished through the introduction of disease-free eggs rather than by the transfer of adults. Research on the treatment of eggs and nauplii of shrimps to prevent the spread of disease is needed urgently. On the basis of this work, it would then be possible to enforce improved methods of transportation as a precaution against introduction of new diseases in wild shrimp populations.

Research needed to permit economical utilization of krill is underway. However, the potential impact of developing economical means for processing krill onboard vessels to a low-weight product that can be transported practically is tremendous. In spite of the obstacles to utilization, this Antarctic resource should not be forgotten. Work on processing methods is the most pressing research need for krill. In October, 1982, the "First International Symposium on the Biology of the Antarctic Krill *Euphausia superba*" was held at Wilmington, North Carolina (United States). The proceedings from the symposium have appeared in a special supplement to the *Journal of Crustacean Biology*, vol. **4**, spec. no. 1, 1984.

ACKNOWLEDGMENTS

We are grateful for the critical reviews and helpful comments given by Drs. Anthony J. Provenzano, Jr. and Phillip R. Mundy and for necessary reference material provided by Drs. Provenzano and Mundy and Mr. Mario Paula, Department of Oceanography, Old Dominion University, Norfolk, Virginia. Dr. Charles W. Caillouet, Jr., National Marine Fisheries Service, Southeast Fisheries Center, Galveston, Texas; Dr. Samuel P. Meyers, Department of Food Science, Louisiana State University, Baton Rouge, Louisiana; Mr. Wayne Burton, Ms. Jan Mitchell and Ms. Diana Ballin of the Interlibrary Loan Office, Old Dominion University, Norfolk, Virginia, all supplied useful reference material. We thank Ms. Yvonne Wilson of the Center for Instructional Development, Old Dominion University, Norfolk, Virginia, for expert assistance in figure preparation. We also appreciate the time and effort given by Mrs. Yvonne Maris in typing and proofreading this chapter.

REFERENCES

Abramson, N. J. (1968). A probability sea survey plan for estimating relative abundance of ocean shrimp. *Calif. Fish. Game* **54**, 257–269.

Abramson, N. J., and Tomlinson, P. K. (1972). An application of yield models to a California ocean shrimp population. *Fish. Bull.* **70**, 1021–1041.

Ajayi, T. O. (1982). The maximum sustainable yields of the inshore fish and shrimp resources of the Nigerian continental shelf. *J. Fish. Biol.* **20**, 571–577.

Al-Adhub, A. I I. Y., and Naylor, E. (1975). Emergence rhythms and tidal migrations in the brown shrimp *Crangon crangon* (L.). *J. Mar. Biol. Assoc. U. K.* **55**, 801–810.

Aldrich, D. V., Wood, C. E., and Baxter, K. N. (1968). An ecological interpretation of low temperature responses in *Penaeus aztecus* and *P. setiferus* postlarvae. *Bull. Mar. Sci.* **18**, 61–71.

Allen, D. M., and Costello, T. J. (1969). Additional references on the biology of shrimp, family Penaeidae. *Fish. Bull.* **68**, 101–134.

Allen, D. M., and Inglis, A. (1958). A pushnet for quantitative sampling of shrimp in shallow estuaries. *Limnol. Oceanogr.* **3**, 239–241.

Allen, D. M., Hudson, J. H., and Costello, T. J. (1980). Postlarval shrimp (*Penaeus*) in the Florida Keys: Species, size, and seasonal abundance. *Bull. Mar. Sci.* **30**, 21–33.

Allen, J. A. (1959). On the biology of *Pandalus borealis* Kröyer, with reference to a population off the Northumberland coast. *J. Mar. Biol. Assoc. U. K.* **38**, 189–220.

Allen, J. A. (1960). On the biology of *Crangon allmani* Kinahan in Northumberland waters. *J. Mar. Biol. Assoc. U. K.* **39**, 481–508.

Allen, J. A. (1963). Observations on the biology of *Pandalus montagui* [Crustacea: Decapoda]. *J. Mar. Biol. Assoc. U. K.* **43,** 665–682.

Allen, J. A. (1966a). The rhythms and population dynamics of decapod Crustacea. *Oceanogr. Mar. Biol. Annu. Rev.* **4,** 247–265.

Allen, J. A. (1966b). A method of fitting growth curves of the von Bertalanffy type to observed data. *J. Fish. Res. Board Can.* **23,** 163–179.

Allen, K. R. (1969). Application of the Bertalanffy growth equation to problems of fisheries management: A review. *J. Fish. Res. Board Can.* **26,** 2267–2281.

Anderson, W. W. (1966). The shrimp fishery of the southern United States. *U S. Fish Wildl. Serv., Fish. Leafl. No. 589,* 8 pp.

Anderson, W. W., King, J. E., and Lindner, M. J. (1949). Early stages in the life history of the common marine shrimp, *Penaeus setiferus* (Linnaeus). *Biol. Bull. (Woods Holes, Mass.)* **96,** 168–172.

Anonymous (1958). Foreign shrimp fisheries other than Central and South America. *U. S. Fish Wildl. Serv., Spec. Sci. Rep. Fish. No. 254,* 71 pp.

Anonymous (1970). Basic economic indicators, shrimp, Atlantic and Gulf. *U. S. Bur. Commer. Fish., Div. Econ. Res., Work. Pap. No. 57,* 70 pp.

Anonymous (1973). Shellfish shells salvaged for commercial use. *Mar. Fish. Rev.* **35**(3-4), 5.

Anonymous (1980). German (FRG) scientist casts doubt on krill. *Fisheries* **5**(3), 51.

Awad, A. A., Sinnhuber, R. O., and Anderson, A. W. (1965). Radiation pasteurization of rain and chlortetracycline treated shrimp. *Food Technol.* **19,** 182–184.

Babcock, A. M. (1981). A quantitative description of migratory behavior of the brown shrimp (*Penaeus aztecus*) with applications in fisheries management. M.S. Thesis, Old Dominion Univ., Norfolk, Virginia.

Babcock, A. M. (1982). A quantitative measure of migratory timing illustrated by applications to the commercial brown shrimp fishery. *Dep. Oceanogr., Old Dominion Univ., Norfolk, Va, Tech. Rep. No. 82-4,* 184 pp.

Babcock, A. M., and Mundy, P. R. (1985). A quantitative measure of migratory timing applied to the evaluation of a commercial brown shrimp fishery in North Carolina. *North Am. J. Fish. Manage.* **5,** no. 2A, 181–196.

Badawi, H. K. (1975). On maturation and spawning in some penaeid prawns of the Arabian Gulf. *Mar. Biol. (Berlin)* **32,** 1–6.

Bahamonde, N., and Henriquez, G. (1970). Sinopsis de datos biologicos sobre el camaron nailon *Heterocarpus reedi* Bahamonde, 1955. *FAO Fish. Rep.* **54,** 1607–1627.

Bailey, M. E., Fieger, E. A., and Novak, A. F. (1956). Objective tests applicable to quality studies of ice stored shrimp. *Food Res.* **21,** 611–620.

Bakus, G. J., Garling, W., and Buchanan, J. E. (1978). The Antarctic krill resource: Prospects for commercial exploitation. *U. S. Dep. State, Natl. Tech. Inf. Serv., Springfield, Va, Tetra Tech. Rep. TC-903,* p. 149.

Banerji, S. K., and George, M. J. (1967). Size distribution and growth of *Metapenaeus dobsoni* (Miers) and their effect on the trawler catches off Kerala. *Proc. Symp. Crustacea, Ser. 2, Mar. Biol. Assoc. India, Pt. II,* pp. 634–648.

Bardach, J. E., Ryther, J. H., and McLarney, W. O. (1972). "Aquaculture: The Farming and Husbandry of Freshwater and Marine Organisms." Wiley (Interscience), New York.

Barnett, H. J., Nelson, R. W., Hunter, P. J., and Groninger, H. (1978). Use of carbon dioxide dissolved in refrigerated brine for the preservation of pink shrimp (*Pandalus* spp.). *Mar. Fish. Rev.* **40**(9), 24–28.

Barr, L. (1970a). Alaska's fishery resources—The shrimps. *U.S. Fish Wildl. Serv., Fish. Leafl. No. 631,* 10 pp.

Barr, L. (1970b). Diel vertical migration of *Pandalus borealis* in Kachemak Bay, Alaska. *J. Fish. Res. Board Can.* **27,** 669–676.

Barr, L. (1973). Studies of spot shrimp, *Pandalus platyceros* at Little Port Walter, Alaska. *Mar. Fish. Rev.* **35**(3-4), 65–66.

Barrett, B. B., and Gillespie, M. C. (1973). Primary factors which influence commercial shrimp production in coastal Louisiana. *La. Wildl. Fish. Comm., New Orleans, Tech. Bull. No. 9,* 28 pp.

Barrett, B. B., and Gillespie, M. C. (1975). 1975 environmental conditions relative to shrimp production in coastal Louisiana. *La. Wildl. Fish. Comm., New Orleans, Tech. Bull. No. 15,* 22 pp.

Barrett, B. B., and Ralph, E. J. (1976). 1976 environmental conditions relative to shrimp production in coastal Louisiana. *La. Wildl. Fish. Comm., New Orleans, Tech. Bull. No. 21,* 20 pp.

Bauer, R. T. (1979). Antifouling adaptations of marine shrimp (Decapoda : Caridea): Gill cleaning mechanisms and grooming of brooded embryos. *J. Linn. Soc. London, Zool.* **65,** 281–303.

Baxter, K. N. (1962). Abundance of postlarval shrimp—one index of future shrimping success. *Proc. Gulf Caribb. Fish. Inst.* **15,** 79–87.

Baxter, K. N. (1973). Shrimp discarding by the commercial fishery in the western Gulf of Mexico. *Mar. Fish. Rev.* **35**(9), 26.

Baxter, K. N., and Renfro, W. C. (1967). Seasonal occurrence and size distribution of postlarval brown and white shrimp near Galveston, Texas, with notes on species identification. *Fish. Bull.* **66,** 149–158.

Bayagbona, E. O., Sagua, V. O., and Afinowi, M. A. (1971). A survey of the brown shrimp resources of Nigeria. *Mar. Biol. (Berlin)* **11,** 178–189.

Bearden, C. M., and McKenzie, M. D. (1972). Results of a pilot shrimp tagging project using internal anchor tags. *Trans. Am. Fish. Soc.* **101,** 358–362.

Beardsley, A. J. (1973). Design and evaluation of a sampler for measuring the near-bottom vertical distribution of pink shrimp, *Pandalus jordani. Fish. Bull.* **71,** 243–253.

Beardsley, G. L. (1970). Distribution of migrating juvenile pink shrimp, *Penaeus duorarum duorarum* Burkenroad, in Buttonwood Canal, Everglades, National Park, Florida. *Trans. Am. Fish. Soc.* **99,** 401–408.

Beardsley, G. L., and Iversen, E. S. (1966). Studies on the distribution of migrating juvenile pink shrimp in Buttonwood Canal, Everglades National Park. *Proc. Gulf Caribb. Fish. Inst.* **18,** 17 (Abstr.).

Berkeley, A. A. (1929). Sex reversal in *Pandalus danae.* Am. Nat. **63,** 571–573.

Berkes, F. (1975). Some aspects of feeding mechanisms of euphausiid crustaceans. *Crustaceana* **29,** 266–270.

Berry, R. J. (1967). Dynamics of the Tortugas pink shrimp population. Ph.D. Thesis, Univ. of Rhode Island, Kingston.

Berry, R. J. (1970). Shrimp mortality rates derived from fishery statistics. *Proc. Gulf Caribb. Fish. Inst.* **22,** 66–78.

Berry, R. J., and Baxter, K. N. (1969). Predicting brown shrimp abundance in the northwestern Gulf of Mexico. *FAO Fish. Rep.* **57,** 775–798.

Beverton, R. J. H. (1953). Some observations on the principles of fishery regulation. *J. Cons. Int. Explor. Mer* **19,** 56–68.

Bezanson, A., Learson, R., and Teich, W. (1973). Defrosting shrimp with microwaves. *Proc. Gulf Caribb. Fish. Inst.* **25,** 44–55.

Bieler, A. C., Matthews, R. F., and Koburger, J. A. (1973). Rock shrimp quality as influenced by handling procedures. *Proc. Gulf Caribb. Fish. Inst.* **25,** 56–61.

Bishop, J. M., and Herrnkind, W. F. (1976). Burying and molting of pink shrimp, *Penaeus duorarum* (Crustacea: Penaeidae), under selected photoperiods of white light and UV-light. *Biol. Bull. (Woods Hole, Mass.)* **150,** 163–182.

Blake, B. F., and Menz, A. (1980). Mortality estimates for *Penaeus vannamei* Boone in a Mexican coastal lagoon. *J. Exp. Mar. Biol. Ecol.* **45,** 15–24.

Blomo, V. (1981). Conditional fishery status as a solution to overcapitalization in the Gulf of Mexico shrimp fishery. *Mar. Fish. Rev.* **43**(7), 20–24.

Blomo, V., Griffin, W. L., and Nichols, J. P. (1978). Catch–effort and price–cost trends in the Gulf of Mexico shrimp fishery: Implications on Mexico's extended jurisdiction. *Mar. Fish. Rev.* **40**(8), 24–28.

Blomo, V. J., Nichols, J. P., and Grant, W. E. (1982). Dynamic modeling of the eastern Gulf of Mexico shrimp fishery. *Am. J. Agric. Econ.* **64,** 475–482.

Boddeke, R. (1976). The seasonal migration of the brown shrimp *Carngon crangon*. *Neth. J. Sea Res.* **10,** 103–130.

Boddeke, R. (1978). Changes in the stock of brown shrimp (*Crangon crangon* L.) in the coastal area of the Netherlands. *Rapp. P.-v. Reun. Cons. Int. Explor. Mer* **172,** 239–249.

Boddeke, R. (1982). The occurrence of winter and summer eggs in the brown shrimp (*Crangon crangon*) and the pattern of recruitment. *Neth. J. Sea Res.* **16,** 151–162.

Boddeke, R., and Becker, H. B. (1979). A quantitative study of the fluctuations of the stock of brown shrimp (*Crangon crangon*) along the coast of the Netherlands. *Rapp. P.-v Reun. Cons. Int. Explor. Mer* **175,** 253–258.

Boddeke, R., Dijkema, R., and Siemelink, M. E. (1977). The patterned migration of shrimp populations: A complete study of *Crangon crangon* and *Penaeus brasiliensis*. *FAO Fish. Rep.* **200,** 31–50.

Bowman, T. E., and Abele, L. G. (1982). Classification of the Recent Crustacea. *In* "Biology of Crustacea" (L. G. Abele, ed.), Vol. 1, pp. 1–27. Academic Press, New York.

Branford, J. R. (1981a). Sediment and the distribution of penaeid shrimp in the Sudanese Red Sea. *Estuarine Coastal Shelf. Sci.* **13,** 349–354.

Branford, J. R. (1981b). Sediment preferences and morphometric equations for *Penaeus monodon* and *Penaeus indicus* from creeks of the Red Sea. *Estuarine Coastal Shelf. Sci.* **13,** 473–476.

Broad, A. C. (1965). Environmental requirements of shrimp. *In* "Biological Problems in Water Pollution, Third Seminar, 1962" (C. M. Tarzwell, ed.), United States Department of Health, Education and Welfare, Washington.

Brown, D. C. (1966). The use of liquid nitrogen in the shrimp industry. *Proc. Gulf Caribb. Fish. Inst.* **18,** 26–40.

Brown, F. A. (1961). Physiological rhythms. *In* "The Physiology of Crustacea" (T. H. Waterman, ed.), Vol. II, pp. 401–430. Academic Press, New York.

Browning, R. J. (1980). "Fisheries of the North Pacific: History, Species, Gear and Processes," rev. ed. Alaska Northwest Publ. Co., Anchorage.

Brusher, H. A. (1974). The magnitude, distribution and availability of prawn (Penaeidae) resources in coastal and estuarine waters of Kenya, 1970. *J. Mar. Biol. Assoc. India* **16,** 335–348.

Bullard, F. A., and Collins, J. (1978). Physical and chemical changes of pink shrimp, *Pandalus borealis,* held in carbon dioxide modified refrigerated seawater compared with pink shrimp held on ice. *Fish. Bull.* **76,** 73–78.

Bullis, H. R., and Floyd, H. (1972). Double-rig twin shrimp trawling gear used in Gulf of Mexico. *Mar. Fish. Rev.* **34**(11–12), 26–31.

Butler, P. A. (1962). Effects on commercial fisheries. *In* "Effects of pesticides on fish and wildlife: A review of investigations during 1960." *U. S. Fish Wildl. Serv. Bur. Sport Fish. Wildl., Circ. No.* **143,** 20–24.

Butler, T. H. (1953). The appearance of a new commercial shrimp in a newly developed fishery. *Fish. Res. Board Can. Prog. Rep. Pac. Coast. Stn.* **94,** 30–31.

Butler, T. H. (1965). Growth, reproduction, and distribution of pandalid shrimps in British Columbia. *J. Fish. Res. Board Can.* **21,** 1403–1452.

Butler, T. H. (1967). Shrimp exploration and fishing in the Gulf of Alaska and the Bering Sea. *Fish. Res. Board Can. Tech. Rep. No. 180,* 49 pp.

Butler, T. H. (1970). Synopsis of biological data on the prawn *Pandalus platyceros* Brandt, 1851. *FAO Fish. Rep.* **57,** 1289–1316.

Butler, T. H. (1972). Sex change in *Pandalus jordani* and *Pandalus hypsinotus* in British Columbia. *Proc. Natl. Shellfish. Assoc.* **62,** 2 (Abstr).

Caillouet, C. W., and Baxter, K. N. (1973). Gulf of Mexico shrimp resource research. *Mar. Fish. Rev.* **35**(3-4), 21–24.

Caillouet, C. W., and Koi, D. B. (1980). Trends in ex-vessel value and size composition of annual landings of brown, pink, and white shrimp from the Gulf and South Atlantic coasts of the United States. *Mar. Fish. Rev.* **42**(12), 18–27.

Caillouet, C. W., and Koi, D. B. (1981a). Trends in ex-vessel value and size composition of reported annual catches of pink shrimp from the Tortugas fishery, 1960–1978. *Gulf Res. Rep.* **7,** 71–78.

Caillouet, C. W., and Koi, D. B. (1981b). Trends in ex-vessel value and size composition of reported May–August catches of brown shrimp and white shrimp from the Texas, Louisiana, Mississippi, and Alabama coasts, 1960–1978. *Gulf Res. Rep.* **7,** 59–70.

Caillouet, C. W., and Koi, D. B. (1983). Ex-vessel value and size composition of reported May–August catches of brown shrimp and white shrimp from 1960 to 1981 as related to the Texas closure. *Gulf Res. Rep.* **7,** 187–203.

Caillouet, C. W., and Patella, F. J. (1978). Relationship between size composition and ex-vessel value of reported shrimp catches from two Gulf coast states with different harvesting strategies. *Mar. Fish. Rev.* **40**(2), 14–18.

Caillouet, C. W., Patella, F. J., and Jackson, W. B. (1979). Relationship between marketing category (count) composition and ex-vessel value of reported annual catches of shrimp in the eastern Gulf of Mexico. *Mar. Fish. Rev.* **41**(5-6), 1–7.

Caillouet, C. W., Koi, D. B., and Jackson, W. B. (1980a). Relationship between ex-vessel value and size composition of annual landings of shrimp from the Gulf and South Atlantic coasts. *Mar. Fish. Rev.* **42**(12), 28–33.

Caillouet, C. W., Patella, F. J., and Jackson, W. B. (1980b). Trends toward decreasing size of brown shrimp, *Penaeus aztecus,* and white shrimp, *Penaeus setiferus,* in reported annual catches from Texas and Louisiana. *Fish. Bull.* **77,** 985–989.

Cambell, N. A., and Phillips, B. F. (1972). The von Bertalanffy growth curve and its application to capture–recapture data in fisheries biology. *J. Cons. Int. Explor. Mer* **34,** 295–299.

Carlsson, D. M. (1979). Research and management of the shrimp, *Pandalus borealis,* in Greenland waters. *Rapp. R.-v. Reun. Cons. Int. Explor. Mer* **175,** 236–239.

Charnov, E. L., Gotshall, D. W., and Robinson, J. G. (1978). Sex ratio: Adaptive response to population fluctuations in pandalid shrimp. *Science* **200,** 204–206.

Chin, E. (1960). The bait shrimp fishery of Galveston Bay, Texas. *Trans. Am. Fish. Soc.* **89,** 135–141.

Chin, E., and Allen, D. M. (1959). A list of references on the biology of shrimp (family Penaeidae). *U. S. Fish Wildl. Serv., Spec. Sci. Rep. Fish. No. 276,* 143 pp.

Choe, S. (1971). Body increases during molt and molting cycle of the oriental brown shrimp *Penaeus japonicus. Mar. Biol. (Berlin)* **9,** 31–37.

Christmas, J. Y., and Gunter, G. (1967). A summary of knowledge of shrimps of the genus *Penaeus* and the shrimp fishery in Mississippi waters. *Proc. Symp. Crustacea, Ser. 2, Mar. Biol. Assoc. India. Pt. IV,* pp. 1442–1447.

Christmas, J. Y., Gunter, G., and Musgrave, P. (1966). Studies of annual abundance of postlarval penaeid shrimp in the estuarine waters of Mississippi, as related to subsequent commercial catches. *Gulf Res. Rep.* **2,** 177–212.

Clark, C. W., and Kirkwood, G. P. (1979). Bioeconomic model of the Gulf of Carpentaria prawn fishery. *J. Fish. Res. Board Can.* **36,** 1304–1312.

Clarke, A. (1980). The biochemical composition of krill, *Euphausia superba* Dana, from South Georgia. *J. Exp. Mar. Biol. Ecol.* **43,** 221–236.

Clarke, G. L. (1939). Plankton as a food source for man. *Science* **89,** 602–603.

Cleary, D. P. (1970). World demand for shrimp and prawns may outstrip supply during next decade. *Comm. Fish. Rev.* **32**(3) 19–22.

Cobb, S. P., Futch, C. R., and Camp, D. K. (1973). The rock shrimp, *Sicyonia brevirostris* Stimpson, 1871 (Decapoda, Penaeidae). *Mem. Hourglass Cruises* **3,** 1–38.

Cohen, M., and Fishman, G. S. (1980). Modeling growth-time and weight–length relationships in a single year-class fishery with examples for North Carolina pink and brown shrimp. *Can. J. Fish. Aquat. Sci.* **37,** 1000–1011.

Coles, R. G. (1979). Catch size and behaviour of pre-adults of three species of penaeid prawns as influenced by tidal current direction, trawl alignment, and day and night periods *J. Exp. Mar. Biol. Ecol.* **38,** 247–260.

Collins, J., and Kelley, C. (1969). Alaska pink shrimp, *Pandalus borealis:* Effects of heat treatment on color and machine peelability. *Fish. Ind. Res.* **5,** 181–189.

Considine, D. M., ed. (1982). "Foods and Food Production Encyclopedia." Van Nostrand-Reinhold Princeton, New Jersey.

Cook, H. L. (1966). A generic key to the protozoean, mysis and postlarval stages of the littoral Penaeidae of the northwestern Gulf of Mexico. *Fish. Bull.* **65,** 437–447.

Corbett, M. G. (1970). Machine for separating northern shrimp, *Pandalus borealis,* from fish and trash in the catch. *Fish. Ind. Res.* **6,** 53–62.

Costello, T. J. (1959). Marking shrimp with biological stains. *Proc. Gulf Caribb. Fish. Inst.* **11,** 1–6.

Costello, T. J., and Allen, D. M. (1962). Survival of stained, tagged, and unmarked shrimp in the presence of predators. *Proc. Gulf Caribb. Fish. Inst.* **14,** 16–19.

Costello, T. J., and Allen, D. M. (1966). Migrations and geographic distribution of pink shrimp, *Penaeus duorarum,* of the Tortugas and Sanibel Grounds, Florida. *Fish. Bull.* **65,** 449–459.

Costello, T. J., and Allen, D. M. (1968). Mortality rates in populations of pink shrimp, *Penaeus duorarum,* on the Sanibel and Tortugas Grounds, Florida. *Fish. Bull.* **66,** 491–502.

Cummings, W. C. (1961). Maturation and spawning of the pink shrimp, *Penaeus duorarum* Burkenroad. *Trans. Am. Fish. Soc.* **90,** 462–468.

Dahlstrom, W. A. (1965). The California ocean shrimp fishery. *Rapp. P.-v. Reun. Cons. Int. Explor. Mer* **156,** 112–115.

Dahlstrom, W. A. (1973). Status of the California ocean shrimp resource and its management. *Mar. Fish. Rev.* **35**(3–4), 55–59.

Dall, W. (1958). Observations on the biology of the greentail prawn, *Metapenaeus masterii* (Haswell) (Crustacea Decapoda: Penaeidae). *Aust. J. Mar. Freshwater Res.* **9,** 111–134.

Delmendo, M. N., and Rabanal, H. R. (1956). Cultivation of "sugpo" (jumbo tiger shrimp), *Penaeus monodon* Fabricius, in the Philippines, *Proc. Indo-Pac. Fish. Comm.* **6,** 424–431.

Deshpande, S. D. (1960). On the fishing experiments conducted with a 10 ft. beam-trawl net. *Indian J. Fish.* **7,** 174–186.

Deshpande, S. D., and Sivan, T. M. (1962). On the effect of tickler chain on the catch of a 10 ft. beam-trawl net. *Indian J. Fish.* **9B,** 91–96.

Divita, R., Creel: M., and Sheridan, P. F. (1983). Foods of coastal fishes during brown shrimp,

Penaeus aztecus, migration from Texas estuaries (June–July 1981). *Fish. Bull* **81**, 396–404.

Dobkin, S. (1961). Early developmental stages of pink shrimp, *Penaeus duorarum* from Florida waters. *Fish. Bull.* **61**, 321–349.

Do Chi, T. (1978). Modeles cinetiques et structuraux en dynamique des populations exploitées. Application aux squilles, *Squilla mantis* (L.) (Crustacée Stomatopode) du Golfe du Lion. Thesis, Laboratoire d'Hydrobiologie Marine, Universite des Sciences et Techniques du Lauguedoc, Monpellier, France.

Dorheim, H. (1969). Beitrage zur Biologie der Garnele *Crangon crangon* (L.) in der Kieler Bucht. *Meeresforschung* **20**, 179–215.

Dow, R. L. (1967). Temperature limitations on the supply of northern shrimp (*Pandalus borealis*) in Maine (U.S.A.) waters. *Proc. Symp. Crustacea, Ser. 2, Mar. Biol. Assoc. India, Pt. IV*, pp. 1301–1304.

Driver, P. A. (1976). Prediction of fluctuations in the landings of brown shrimp (*Crangon crangon*) in the Lancashire and Western Sea Fisheries District. *Estuarine Coastal Mar. Sci.* **4**, 567–573.

Duronslet, M. J., Lyon, J. M., and Marullo, F. (1972). Vertical distribution of postlarval brown, *Penaeus aztecus*, and white, *P. setiferus*, shrimp during immigration through a tidal pass. *Trans. Am. Fish. Soc.* **101**, 748–752.

Eddie, G. O. (1977). "The Harvesting of Krill." Southern Ocean Fisheries Survey Programme, FAO, Rome.

Edwards, R. R. C. (1977). Field experiments on growth and mortality of *Penaeus vannamei* in a Mexican coastal lagoon complex. *Estuarine Coastal Mar. Sci.* **5**, 107–121.

Edwards, R. R. C. (1978). The fishery and the fisheries biology of penaeid shrimp on the Pacific coast of Mexico. *Oceanogr. Mar. Biol. Annu. Rev.* **16**, 145–180.

El-Sayed, S.-Z., and McWhinnie, M. A. (1979). Antarctic krill—protein of the last frontier. *Oceanus* **22**, 13–20.

Emiliani, D. A. (1971). Equipment for holding and releasing penaeid shrimp during marking experiments. *Fish. Bull.* **69**, 247–251.

Ettershank, G. (1983). How old is that krill? *Aust. Fish.* **42**(7), 22–23.

Etzold, D. J., and Christmas, J. Y. (1977). A comprehensive summary of the shrimp fishery of the Gulf of Mexico United States: A regional management plan. *Gulf Coast Res. Lab., Ocean Springs, Miss., Tech. Rep. Ser. No. 2, Part 2*, 20 pp.

Ewald, J. J. (1969). The Venezuelan shrimp industry. *FAO Fish. Rep.* **57**, 765–774.

FAO (1983a). "1981 Yearbook of Fishery Statistics—Catches and Landings," Vol. 52, Part 1. Food and Agriculture Organization of the United Nations, Rome.

FAO (1983b). "1981 Yearbook of Fishery Statistics—Fishery Commodities," Vol. 53, Part 2. Food and Agriculture Organization of the United Nations, Rome.

Fevolden, S. E. (1979). Investigations on krill (Euphausiacea) sampled during the Norwegian Antarctic Research Expedition 1976–77. *Sarsia* **64**, 189–198.

Fevolden, S. E., and Eidså, G. (1981). Bacteriological characteristics of Antarctic krill (Crustacea, Euphausiacea). *Sarsia* **66**, 77–82.

Fitzgibbon, T. (1976). "The Food of the Western World." Quadrangle/The New York Times Book Company, New York.

Flick, G. J., and Lovell, R. T. (1972). Post-mortem biochemical changes in the muscle of Gulf shrimp, *Penaeus aztecus. J. Food Sci.* **37**, 609–611.

Flint, R. W., and Rabalais, N. N. (1981). Gulf of Mexico shrimp production: A food web hypothesis. *Fish. Bull.* **79**, 737–748.

Fontaine, C. T., and Neal, R. A. (1971). Length–weight relations for three commercially important penaeid shrimp of the Gulf of Mexico. *Trans. Am. Fish. Soc.* **100**, 584–586.

Forest, J. (1973). Crustaceans. *In* "FAO Species Identification Sheets for Fishery Purposes. Mediterranean and Black Sea (Fishing Area 37) No. 2" (W. Fisher, ed.), Food and Agriculture Organization of the United Nations, Rome.

Forster, G. R. (1951). The biology of the common prawn, *Leander serratus* Pennant. *J. Mar. Biol. Assoc. U. K.* **30,** 333–360.

Forster, G. R. (1959). The biology of the prawn, *Palaemon (=Leander) serratus* (Pennant). *J. Mar. Biol. Assoc. U. K.* **38,** 621–627.

Fox, W. W. (1972). Dynamics of pandalid shrimp: A review and management considerations. *Proc. Natl. Shellfish. Assoc.* **62,** 3–4 (Abstr.).

Fry, B. (1981). Natural stable carbon isotope traces Texas shrimp migrations. *Fish. Bull.* **79,** 337–345.

Fujiishi, A., and Ishizuka, T. (1974). Basic studies on the shrimp drag. II. Field experiments using a 15-meter shrimp drag with a coner-net. *Bull. Jpn. Soc. Sci. Fish.* **40,** 993–997.

Fuss, C. M. (1964). Observations on burrowing behavior of the pink shrimp, *Penaeus duorarum* Burkenroad. *Bull. Mar. Sci. Gulf Caribb.* **14,** 62–73.

Fuss, C. M., and Ogren, L. H. (1966). Factors affecting activity and burrowing habits of the pink shrimp, *Penaeus duorarum* Burkenroad. *Biol. Bull. (Woods Hole, Mass.)* **130,** 170–191.

Gaidry, W. J. (1974). Correlations between inshore spring white shrimp population densities and offshore overwintering populations. *La. Wildl. Fish. Comm., New Orleans, Tech. Bull. No. 12,* 18 pp.

Gaidry, W. J., and White, C. J. (1973). Investigations of commercially important penaeid shrimp in Louisiana estuaries. *La. Wildl. Fish. Comm., New Orleans, Tech. Bull. No. 8,* 154 pp.

Garcia, S. (1977). Biologie et dynamique des populations de crevettes roses (*Penaeus duorarum notialis* Pérez-Farfante, 1967) En Cote d'Ivoire. *Trav. Doc. O.R.S., T.O.M. No. 79,* 271 pp.

George, M. J. (1959). Notes on the bionomics of the prawn *Metapenaeus monoceros,* Fabricius. *Indian J. Fish.* **6,** 268–279.

George, M. J. (1961). Studies on the prawn fishery of Cochin and Alleppey coast. *Indian J. Fish.* **8,** 75–95.

George, M. J. (1962). Observations on the size groups of *Penaeus indicus* (Milne-Edwards) in the commercial catches of different nets from the backwaters of Cochin. *Indian J. Fish.* **9,** 468–475.

George, M. J. (1963). Postlarval abundance as a possible index of fishing success in the prawn *Metapenaeus dobsoni* (Miers). *Indian J. Fish.* **10,** 135–139.

George, M. J. (1970). Synopsis of biological data on the penaeid prawn *Metapenaeus dobsoni* (Miers, 1878). *FAO Fish. Rep.* **57,** 1335–1357.

George, P. C. and Rao, P. V. (1967). An annotated bibliography of the biology and fishery of the commercially important prawns of India. *Proc. Symp. Crustacea, Ser. 2, Mar. Biol. Assoc. India, Pt. V.,* pp. 1521–1547.

George, S. J., Raman, K., and Nair, P. K. (1963). Observations on the offshore prawn fishery of Cochin. *Indian J. Fish.* **10,** 460–499.

Giles, J. H., and Zamora, G. (1973). Cover as a factor in habitat selection by juvenile brown (*Penaeus aztecus*) and white (*P. setiferus*) shrimp. *Trans. Am. Fish. Soc.* **102,** 144–145.

Glaister, J. P. (1978a). Movement and growth of tagged school prawns, *Metapenaeus macleayi* (Haswell) (Crustacea: Penaeidae) in the Clarence River region of northern New South Wales. *Aust. J. Mar. Freshwater Res.* **29,** 645–657.

Glaister, J. P. (1978b). The impact of river discharge on distribution and production of the school prawn *Metapenaeus macleayi* (Haswell) (Crustacea: Penaeidae) in the Clarence River region, northern New South Wales. *Aust. J. Mar. Freshwater Res.* **29,** 311–323.

Grant, W. E., and Griffin, W. L. (1979). A bioeconomic model of the Gulf of Mexico shrimp fishery. *Trans. Am. Fish. Soc.* **108**, 1–13.

Grantham, G. J. (1977). "The Utilization of Krill." Southern Ocean Fisheries Survey Programme, Food and Agriculture Organization of the United Nations, Rome.

Green, M. (1949a). Bacteriology of shrimp. I. Introduction and development of experimental procedures. *Food Res.* **14**, 365–371.

Green, M. (1949b). Bacteriology of shrimp. II. Quantitative studies on freshly caught and iced shrimp. *Food Res.* **14**, 372–394.

Green, M. (1949c). Bacteriology of shrimp. III. Quantitative studies on frozen shrimp. *Food Res.* **14**, 384–394.

Green, M. (1949d). Bacteriology of shrimp. IV. Coliform bacteria in shrimp. *Food Res.* **14**, 395–400.

Griffin, W. L., and Beattie, B. R. (1978). Economic impact of Mexico's 200-mile offshore fishing zone on the United States Gulf of Mexico shrimp fishery. *Land Econ.* **54**, 27–38.

Griffin, W. L., and Jones, L. L. (1975). Economic impact of commercial shrimp landings on the economy of Texas. *Mar. Fish. Rev.* **37**(7), 12–14.

Griffin, W. L., and Nichols, J. P. (1976). An analysis of increasing costs to Gulf of Mexico shrimp veseel owners: 1971–75. *Mar. Fish. Rev.* **38**(3), 8–12.

Griffin, W. L., Lacewell, R. D., and Hayenga, W. A. (1974). Estimated costs, returns, and financial analysis: Gulf of Mexico shrimp vessels. *Mar. Fish. Rev.* **36**(12), 1–4.

Griffin, W. L., Lacewell, R. D., and Nichols, J. P. (1976a). Optimum effort and rent distribution in the Gulf of Mexico shrimp fishery. *Am. J. Agric. Econ.* **58**, 644–652.

Griffin, W. L., Wardlaw, N. J., and Nichols, J. P. (1976b). Economic and financial analysis of increasing costs in the Gulf shrimp fleet. *Fish. Bull.* **74**, 301–309.

Guest, W. C. (1958). The Texas shrimp fishery. *Tex. Game Fish Comm., Austin, Bull. No. 36*, 23 pp.

Gulland, J. A. (1970). The fish resources of the oceans. *FAO Fish. Tech. Pap. No. 97 (FIRS/T97)*, 425 pp.

Gunter, G. (1950). Seasonal population changes and distributions as related to salinity, of certain invertebrates of the Texas coast, including the commercial shrimp. *Publ. Inst. Mar. Sci. Univ. Tex.* **1**(2), 7–51.

Gunter, G. (1961). Habitat of juvenile shrimp (family Penaeidae). *Ecology* 42, 598–600.

Gunter, G., and Edwards, J. C. (1969). The relation of rainfall and fresh-water drainage to the production of the penaeid shrimps (*Penaeus fluviatilis* Say and *Penaeus aztecus* Ives) in Texas and Louisiana waters. *FAO Fish. Rep.* **57**, 875–892.

Gunter, G., and McGraw, K. (1973). Some analyses of twentieth century landing statistics of marine shrimp of the South Atlantic and Gulf states of the United States. *Gulf Res. Rep.* **4**, 191–204.

Gunter, G., Christmas, J. Y., and Killebrew, R. (1964). Some relations of salinity to population distributions of motile estuarine organisms, with special reference to penaeid shrimp. *Ecology* **45**, 181–185.

Hagerman, L. (1970). Locomotory activity patterns of *Crangon vulgaris* (Fabricius) (Crustacea, Natantia). *Ophelia* **8**, 255–266.

Halim, Y. (1960). Observation on the Nile, bloom of phytoplankton in the Mediterranean. *J. Cons. Int. Explor. Mer* **26**, 57–67.

Hall, N. G., and Penn, J. W. (1979). Preliminary assessment of effective effort in a two species trawl fishery for penaeid prawns in Shark Bay, Western Australia. *Rapp. P.-v. Reun. Cons. Int. Explor. Mer* **175**, 147–154.

Hamner, W. M., Hamner, P. P., Strand, S. W., and Gilmer, R. W. (1983). Behavior of Antarctic krill, *Euphausia superba:* Chemoreception, feeding, schooling, and molting. *Science* **220**, 433–435.

Hardy, A. C. (1941). Plankton as a source of food. *Nature (London)* **147**, 695–696.

Harris, C., Chew, K. K., and Price, V. (1972). Relation of egg number to carapace length of sidestripe shrimp (*Pandalopsis dispar*) from Dabob Bay, Washington. *J. Fish. Res. Board Can.* **29**, 464–465.

Harrison, J. M., and Lee, J. S. (1969). Microbiological evaluation of Pacific shrimp processing. *Appl. Microbiol.* **18**, 188–192.

Harry, G. Y. (1964). The shrimp fishery of Alaska. *Proc. Gulf Caribb. Fish. Inst.* **16**, 64–71.

Hartsuyker, L. (1966). Daily total migrations of the shrimp, *Crangon crangon* L. *Neth. J. Sea Res.* **3**, 52–67.

Hayashi, K.-I., and Sakamoto, T. (1978). Taxonomy and biology of *Metapenaeopsis palmensis* (Haswell) (Crustacea, Decapoda, Penaeidae) collected from the Kii Strait, Central Japan. *Bull. Jpn. Soc. Sci. Fish.* **44**, 709–714.

Haynes, E. (1979). Description of larvae of the northern shrimp, *Pandalus borealis,* reared *in situ* in Kachemak Bay, Alaska. *Fish. Bull.* **77**, 157–173.

Heaton, T. C. (1981). Growth and recruitment of two penaeids in the Bay of St. Louis, Mississippi during 1979. M.S. Thesis, Old Dominion Univ., Norfolk, Virginia.

Heegaard, P. E. (1953). Observations on spawning and larval history of the shrimp, *Penaeus setiferus* (L.), *Publ. Inst. Mar. Sci. Univ. Tex.* **3**, 73–105.

Hempel, G., and Manthey, M. (1981). On the fluoride content of larval krill (*Euphausia superba*). *Meeresforschung* **29**, 60–63.

Hempel, I. (1978). Vertical distribution of eggs and nauplii of krill (*Euphausia superba*) south of Elephant Island. *Meeresforschung* **27**, 119–123.

Hempel, I. (1981). Euphausiid larvae in the Scotia Sea and adjacent waters in summer 1977/78. *Meeresforschung* **29**, 53–60.

Hempel, I., and Hempel, G. (1978). Larval krill (*Euphausia superba*) in the plankton and neuston samples of the German Antarctic Expedition 1975/76. *Meeresforschung* **26**, 206–216.

Hempel, I., and Hempel, G. (1982). Distribution of euphausiid larvae in the southern Weddell Sea. *Meeresforschung* **29**, 253–266.

Hempel, I., and Marschoff, E. (1980). Euphausiid larvae in the Atlantic sector of the Southern Ocean. *Meeresforschung* **28**, 32–47.

Hempel, I., Hempel, G., and Baker, A. de C. (1979). Early life history stages of krill (*Euphausia superba*) in Bransfield Strait and Weddell Sea. *Meeresforschung* **27**, 267–281.

Herdman, W. A. (1891). Copepods as an article of food. *Nature (London)* **44**, 273–274.

Hettler, W. F., and Chester, A. J. (1982). The relationship of winter temperature and spring landings of pink shrimp, *Penaeus duorarum,* in North Carolina. *Fish. Bull.* **80**, 761–768.

High, W. L., Ellis, I. E., and Lusz, L. D. (1969). A progress report on the development of a shrimp trawl to separate shrimp from fish and bottom-dwelling animals. *Comm. Fish. Rev.* **31**(3), 20–33.

Hildebrand, H. H., and Gunter, G. (1952). Correlation of rainfall with the Texas catch of white shrimp, *Penaeus setiferus* (Linnaeus). *Trans. Am. Fish. Soc.* **82**, 151–155.

Hjort, J., and Ruud, J. T. (1938). Deep-sea prawn fisheries and their problems. *Hvalrad. Skr. Norsk. Vidensk.-Akad. Oslo, It.,* 144 pp.

Hoese, H. D. (1960). Juvenile penaeid shrimp in the shallow Gulf of Mexico. *Ecology* **41**, 592–593.

Holmes, D., and McCleskey, C. S. (1949). Bacteriology of shrimp. V. Effect of peeling, glazing and storage temperature on bacteria in frozen shrimp. *Food Res.* **14**, 401–404.

Howe, N. R., and Hoyt, P. R. (1982). Mortality of juvenile brown shrimp *Penaeus aztecus* associated with streamer tags. *Trans. Am. Fish. Soc.* **111**, 317–325.

Hughes, D. A. (1966). Investigations of the "nursery areas" and habitat preferences of juvenile penaeid prawns in Mozambique. *J. Appl. Ecol.* **3**, 349–354.

Hughes, D. A. (1967a). Factors controlling the time of emergence of pink shrimp *Penaeus duorarum* Burkenroad. *FAO World Sci. Conf. Biol. Cult. Shrimps Prawns, Mexico City, FAO Exp. Pap. No. 56*, 1–12.

Hughes, D. A. (1967b). On the mechanisms underlying tide-associated movements of *Penaeus duorarum* Burkenroad. *FAO World Sci. Conf. Biol. Cult. Shrimps Prawns, Mexico City, FAO Exp. Pap. No. 48*, 1–8.

Hughes, D. A. (1968). Factors controlling emergence of pink shrimp (*Penaeus duorarum*) from the substrate. *Biol. Bull. (Woods Hole, Mass.)* **134**, 48–59.

Hughes, D. A. (1969a). Evidence for the endogenous control of swimming in pink shrimp, *Penaeus duorarum*. *Biol. Bull. (Woods Hole, Mass.)* **136**, 398–404.

Hughes, D. A. (1969b). Factors controlling the time of emergence of pink shrimp, *Penaeus duorarum*. *FAO Fish. Rep. No. 67*, 971–981.

Hughes, D. A. (1969c). On the mechanisms underlying tide-associated movements of *Penaeus duorarum* Burkenroad. *FAO Fish. Rep.* **57**, 867–874.

Hughes, D. A. (1969d). Responses to salinity change as a tidal transport mechanism of pink shrimp, *Penaeus duorarum*. *Biol. Bull. (Woods Hole, Mass.)* **136**, 43–53.

Hughes, D. A. (1972). On the endogenous control of tide-associated displacements of pink shrimp, *Penaeus duorarum* Burkenroad. *Biol. Bull. (Woods Hole, Mass.)* **142**, 271–280.

Hunt, J. H., Carroll, R. J., Chinchilli, V., and Frankenberg, D. (1980). Relationship between environmental factors and brown shrimp production in Pamlico Sound, North Carolina. *N. C. Dep. Nat. Resour. Commun. Dev., Div. Mar. Fish., Spec. Sci. Rep. No. 33*, 29 pp.

Hutchinson, R. W. (1978). Status and problems of the American shrimp industry. *Mar. Fish. Rev.* **40**(9), 29–31.

Ibrahim, K. H. (1962). Observations on fishery and biology of the freshwater prawn *Macrobrachium malcolmsonii* Milne Edwards of River Godavari. *Indian J. Fish.* **9**, 433–467.

Idyll, C. P., Jones, A. C., and Dimitrou, D. (1962). Production and distribution of pink shrimp larvae, and postlarvae. *U.S., Fish Wildl. Serv., Circ. No. 161*, pp. 93–94.

Ikematsu, W. (1955). On the life-history of *Metapenaeus joyneri* (Miers) in Ariake Sea. *Bull. Jpn. Soc. Sci. Fish.* **20**, 969–978.

Inglis, A., and Chin, E. (1959). The bait shrimp industry of the Gulf of Mexico. *U. S. Fish Wildl. Serv., Fish Leafl. No. 480*, 14 pp.

Inman, D. L., Aubrey, D. G., and Pawka, S. S. (1976). Application of nearshore processes to the Nile Delta. *In* "Proceedings of Seminar on Nile Delta Sedimentology, Alexandria, 25-29 October, 1975." United Nations Educational Scientific and Cultural Organization–United Nations Development Programme, Rome, Italy.

Inoue, M. (1981). Relation between the catch and periods of immersion for the pots of pink shrimp. *Bull. Jpn. Soc. Sci. Fish.* **47**, 577–583.

Inyang, N. M. (1977a). Effects of some enviromental factors on growth and food consumption of the Baltic palaemonid shrimp, *Palaemon adspersus* var. *fabricii* (Rathke). *Meeresforschung* **26**, 30–41.

Inyang, N. M. (1977b). Notes on food of the Baltic palaemonid shrimp, *Palaemon adspersus* var. *fabricii* (Rathke). *Meeresforschung* **26**, 42–46.

Ivanov, B. G. (1967). The biology and distribution of the northern shrimp (*Pandalus borealis* Kr.) in the Bering Sea and the Gulf of Alaska. *FAO Fish. Rep.* **57**, 799–810.

Ivanov, B. G. (1970). On the biology of the Antarctic krill *Euphausia superba*. *Mar. Biol. (Berlin)* **7**, 340–351.

Ivanov, B. G. (1972). Geographic distribution of the northern shrimp *Pandalus borealis* Kr. (Crustacea, Decapoda). *Proc. Natl. Shellfish. Assoc.* **62**, 9–14.

Iversen, E. S. (1962). Estimating a population of shrimp by the use of catch per effort and tagging data. *Bull. Mar. Sci. Gulf Caribb.* **12**, 350–398.

Iversen, E. S., and Idyll, C. P. (1960). Aspects of the biology of the Tortugas pink shrimp, *Penaeus duorarum*. *Trans. Am. Fish. Soc.* **89,** 1–8.

Iversen, E. S., and Jones, A. C. (1961). Growth and migration of the Tortugas pink shrimp, *Penaeus duorarum*, and changes in catch per unit of effort of the fishery. *Fla. Board Cons. Tech. Ser.* **34,** 5–28.

Iyengar, J. R., Visweswariah, K., Moorjani, M. N., and Bhatia, D. S. (1960). Assessment of the progressive spoilage of ice-stored shrimp. *J. Fish. Res. Board Can.* **17,** 475–485.

Jackson, P. (1954). Engineering and economic aspects of marine plankton harvesting. *J. Cons. Int. Explor. Mer* **20,** 167–174.

Jacob, S. S., Iyer, K. M., Nair, M. R., and Pillai, V. K. (1962). Quality studies on round, headless and peeled and deveined prawns held in ice storage. *Indian J. Fish.* **9,** 97–107.

James, D. G. (1975). "Some Aspects of the Shrimp Industries of Denmark and Greenland," pp. 174–179. First Australian National Prawn Seminar, Australian Fisheries Council.

Janssen, G. M., and Kuipers, B. R. (1980). On tidal migration in the shrimp *Crangon crangon*. *Neth. J. Sea Res.* **14,** 339–348.

Jensen, A. J. C. (1965). *Pandalus borealis* in the Skagerak (length, growth, and changes in the stock and fishery yield). *Rapp. P.-v. Reun. Cons. Int. Explor. Mer* **156,** 109–111.

Jensen, A. J. C. (1967). The *Pandalus borealis* in the North Sea and Skagerak. *Proc. Symp. Crustacea, Ser. 2, Mar. Biol. Assoc. India, Pt. IV,* pp. 1317–1319.

Jones, A. C., and Dragovich, A. (1973). Investigations and management of the Guianas shrimp fishery under the U.S.–Brazil agreement. *Proc. Gulf Caribb. Fish. Inst.* **25,** 26–33.

Jones, A. C., Dimitriou, D. E., Ewald, J. J., and Tweedy, J. H. (1970). Distribution of early development stages of pink shrimp, *Penaeus duorarum*, in Florida waters. *Bull. Mar. Sci.* **20,** 634–661.

Jones, A. C., Klima, E. F., and Poffenberger, J. R. (1982). Effects of the 1981 closure on the Texas shrimp fishery. *Mar. Fish. Rev.* **44**(9-10), 1–4.

Jones, S. (1969). The prawn fishery resources of India. *FAO Fish. Rep.* **57,** 735–747.

Joyce, E. A., and Eldred, B. (1966). The Florida shrimping industry. *Fla. Board Conserv. Educ. Ser.* **15,** 7–47.

Juday, C. (1943). The utilization of aquatic food resources. *Science* **97,** 456–458.

Juhl, R., and Drummond, S. B. (1977). Shrimp bycatch investigation in the United States of America: A status report. *FAO Fish. Rep.* **200,** 213–226.

Juneau, C. L. (1977). A study of the seabob, *Xiphopeneus kroyeri* (Heller) in Louisiana. *La. Dep. Wildl. Fish., New Orleans, Tech. Bull. 24,* 24 pp.

Juneau, C. L., and Pollard, J. F. (1981). A survey of the recreational shrimp and finfish harvests of the Vermilion Bay area and their impact on commercial fishery resources. *La. Dep. Wildl. Fish., New Orleans, Tech. Bull. 33,* 40 pp.

Kanneworff, P. (1979). Density of shrimp (*Pandalus borealis*) in Greenland waters observed by means of photography. *Rapp. P.-v. Reun. Cons. Int. Explor. Mer* **175,** 134–138.

Kelley, C. (1971). Glucose oxidase reduces oxidation in frozen shrimp. *Comm. Fish. Rev.* **33**(2), 51–52.

Kelley, C. E., and Harmon, A. W. (1972). Method of determining carotenoid contents of Alaska pink shrimp and representative values for several shrimp products. *Fish. Bull.* **70,** 111–113.

Kennedy, F. S., and Barber, D. G. (1981). Spawning and recruitment of pink shrimp, *Penaeus duorarum*, off eastern Florida. *J. Crustacean Biol.* **1,** 474–485.

Kessler, D. W. (1968). Experimental trawling and pot fishing for giant Alaskan prawn. *Comm. Fish. Rev.* **30**(1), 41–44.

Khandker, N. A., and Lares, L. B. (1973). Observations on the fishery and biology of pink spotted shrimp, *Penaeus brasiliensis* Latreille, of Margarita Island, Venezuela. *Proc. Gulf Caribb. Fish. Inst.* **25,** 156–162.

Khilnani, A. (1981). Use of Griffin's yield model for the Gulf of Mexico shrimp fishery. *Fish. Bull.* **78**, 973–977.

Kils, U. (1978). Performance of Antarctic krill, *Euphausia superba*, at different levels of oxygen saturation. *Meeresforschung* **27**, 35–47.

Kils, U. (1979). Swimming speed and escape capacity of Antarctic krill, *Euphausia superba*. *Meeresforschung* **27**, 264–266.

King, M. G. (1981). Deepwater shrimp resources in Vanuata: a preliminary survey off Port Vila. *Mar. Fish. Rev.* **43**(12), 10–17.

Kitahara, T. (1979). Daily variation in catch of the Kuruma prawn fisheries along the Buzen District. *Bull. Jpn. Soc. Sci. Fish.* **45**, 1267–1273.

Kitahara, T. (1981). Relations of fluctuations in standing crop of Kuruma prawn among the fisheries along the Buzen District. *Bull Jpn. Soc. Sci. Fish.* **47**, 585–591.

Klima, E. (1968). Shrimp-behavior studies underlying the development of the electric shrimp-trawl system. *Fish. Ind. Res.* **4**, 165–181.

Klima, E. F. (1974). A white shrimp mark–recapture study. *Trans. Am. Fish. Soc.* **103**, 107–113.

Klima, E. F., Baxter, K. N., and Patella, F. J. (1982). A review of the offshore shrimp fishery and the 1981 Texas closure. *Mar. Fish. Rev.* **44**(9-10), 16–30.

Knake, B. O. (1958). Operation of North Atlantic type otter trawling gear. U.S. Fish Wildl. Serv., Fish. Leafl. No. 445, 15 pp.

Knudsen, E. E., Herke, W. H., and Mackler, J. M. (1977). The growth rate of marked juvenile brown shrimp, *Penaeus aztecus*, in a semi-impounded Louisiana coastal marsh. *Proc. Gulf Caribb. Fish. Inst.* **29**, 144–159.

Ko, K. S., Suzuki, M., and Kondo, Y. (1970). An elementary study on behavior of common shrimp to moving net. *Bull. Jpn. Soc. Sci. Fish.* **36**, 556–562.

Kobayashi, T., and Yanagimoto, M. (1981). The de-carapace treatment of Antarctic krill *Euphausia superba* by water jet. *Bull. Jpn. Soc. Sci. Fish.* **47**, 1069–1074.

Koike, A., Okawara, M., and Takeuchi, S. (1981). Catching efficiency of the double framed pots for pink shrimp. *Bull. Jpn. Soc. Sci. Fish.* **47**, 457–461.

Komaki, Y. (1967). On the surface swarming of euphausiid crustaceans. *Pac. Sci.* **21**, 433–448.

Konagaya, S. (1980). Protease activity and autolysis of Antarctic krill. *Bull. Jpn. Soc. Sci. Fish.* **46**, 175–183.

Kubo, I. (1951). Bionomics of the prawn, *Pandalus kessleri* Czerniarski. *J. Tokyo Univ. Fish.* **38**, 1–26.

Kubo, I., Hori, S., Kumemura, M., Naganawa, M., and Soedjono, J. (1959). A biological study on a Japanese edible mantis-shrimp, *Squilla oratoria* De Haan. *J. Tokyo Univ. Fish.* **45**, 1–25.

Kuipers, B. R., and Dapper, R. (1981). Production of *Crangon crangon* in the tidal zone of the Dutch Wadden Sea. *Neth. J. Sea Res.* **15**, 33–53.

Kumta, U. S., Mavinkurve, S. S., Gore, M. S., Sawant, P. L., Gangal, S. V., and Sreenivasan, A. (1970). Radiation pasteurization of fresh and blanched tropical shrimps. *J. Food Sci.* **35**, 360–363.

Kurian, C. V., and Sebastian, V. O. (1976). "Prawns and Prawn Fisheries of India." Hindustan Publ. Corp., Delhi, India.

Kuthalingam, M. D. K., Ramamurthy, S., Menon, K. K. P., Annigeri, G. G., and Surendranatha Kurup, N. (1965). Prawn fishery of the Manganese Zone with special reference to the fishing grounds. *Indian J. Fish.* **12**, 546–554.

Kutkuhn, J. H. (1962a). Gulf of Mexico commercial shrimp populations—trends and characteristics, 1956–59. *Fish. Bull.* **62**, 343–402.

Kutkuhn, J. H. (1962b). Recent trends in white shrimp stocks of the northern Gulf. *Proc. Gulf Caribb. Fish. Inst.* **14**, 3–16.

Kutkuhn, J. H. (1965). Dynamics of a penaeid shrimp population and management implications. *Rapp. P.-v. Reun. Cons. Inst. Mer* **156**, 120–123.

Kutkuhn, J. H. (1966a). Dynamics of a penaeid shrimp population and management implications. *Fish. Bull.* **65**, 313–338.

Kutkuhn, J. H. (1966b). The role of estuaries in the development and perpetuation of commercial shrimp resources. *Am. Fish. Soc. Spec. Publ. No. 3*, pp. 16–36.

Kutty, M. N., and Murugapoopathy, G. (1967). Diurnal activity of the prawn *Penaeus semisulcatus* De Haan. *J. Mar. Biol. Assoc. India* **10**, 95–98.

Kuttyamma, V. J. (1974). Observations on the food and feeding of some penaeid prawns of Cochin area. *J. Mar. Biol. Assoc. India* **15**, 189–194.

Kuwano, K., and Mitamura, T. (1977). On the Antarctic Krill Protein Concentrate (KPC). 1. Manufacture of the KPC by isopropyl alcohol method and its chemical composition. *Bull. Jpn. Soc. Sci. Fish.* **43**, 559–565.

Kuwano, K., Tsukui, A., and Mitamura, T. (1979). Manufacture of porous Antarctic krill protein concentrate (P-KPC) by a heat denaturation method. *Bull. Jpn. Soc. Sci. Fish.* **45**, 93–97.

Lakshmi, G. J., Venkataramiah, A., and Gunter, G. (1976). Effects of salinity and photoperiod on the burying behavior of brown shrimp *Penaeus aztecus* Ives. *Aquaculture* **8**, 327–336.

Lasker, R. (1966). Feeding, growth, respiration, and carbon utilization of a euphausiid crustacean. *J. Fish. Res. Board Can.* **23**, 1291–1317.

Lee, J. S., and Kolbe, E. (1982). Microbiological profile of Pacific shrimp, *Pandalus jordani*, stowed under refrigerated seawater spray. *Mar. Fish. Rev.* **44**(3), 12–17.

Lee, T.-J., and Matuda, K. (1973). Studies on the shrimp beam trawl used in Taiwan. 1. Field experiments of the beam trawl. *Bull. Jpn. Soc. Sci. Fish.* **39**, 1237–1343.

Leloup, E., and Gilis, Ch. (1967). La crevette grise (*Crangon crangon* L., 1758) dans le sud de la mer du Nord. *Proc. Symp. Crustacea, Ser. 2, Mar. Biol. Assoc. India, Pt. IV*, pp. 1398–1407.

Le Reste, L. (1971). Rythme saisonnier de la reproduction, migration et croissance des postlarves et des jeunes chez la crevette *Penaeus indicus* H. Milne-Edwards de la Baie d'Ambaro. Côte N.O. de Madagascar. *CAH. ORSTOM Ser. Oceanogr.* **9**, 279–292.

Le Reste, L. (1973). Étude du recrutement de la crevette *Penaeus indicus* H. Milne-Edwards dans la zone de Nosy-Bé (Côte Nord-Ouest de Madagascar). *Cah. ORSTOM Ser. Oceanogr.* **11**, 171–178.

Lightner, D. V. (1974). Normal postmortem changes in the brown shrimp, *Penaeus aztecus*. *Fish. Bull.* **72**, 223–236.

Lindner, M. J. (1957). Survey of shrimp fisheries of Central and South America. *U. S. Fish Wildl. Serv., Spec. Sic. Rep. No. 235*, 166 pp.

Lindner, M. J. (1965). Postlarval brown shrimp abundance and fishing success. *Proc. Gulf Caribb. Fish. Inst.* **17**, 19 (Abstr.).

Lindner, M. J. (1966). What we know about shrimp size and the Tortugas fishery. *Proc. Gulf Caribb. Fish. Inst.* **18**, 18–26.

Lindner, M. J., and Anderson, W. W. (1954). Biology of commercial shrimp. *Fish. Bull.* **55**, 457–461.

Lindner, M. J., and Bailey, J. S. (1968). Distribution of brown shrimp (*Penaeus aztecus aztecus* Ives) as related to turbid water photographed from space. *Fish. Bull.* **67**, 289–294.

Lloyd, A. J., and Yonge, C. M. (1947). The biology of *Crangon vulgaris* L. in the Bristol Channel and Severn Estuary. *J. Mar. Biol. Assoc. U. K.* **26**, 626–661.

Loesch, H. (1965). Distribution and growth of penaeid shrimp in Mobile Bay, Alabama. *Publ. Inst. Mar. Sci. Univ. Tex.* **10**, 41–58.

Loesch, H., Bishop, J., Crowe, A., Kuckyr, R., and Wagner, P. (1976). Technique for estimating trawl efficiency in catching brown shrimp (*Penaeus aztecus*), Atlantic croaker (*Micropogon undulatus*) and spot (*Leiostomus xanthurus*). *Gulf Res. Rep.* **5**, 29–33.

Lucas, C. (1974). Preliminary estimates of stocks of the king prawn, *Penaeus plebejus,* in south-east Queensland. *Aust. J. Mar. Freshwater Res.* **25,** 35–47.

Lucas, C., Young, P. C., and Brundritt, J. K. (1972). Preliminary mortality rates of marked king prawns *Penaeus plebejus,* in laboratory tanks. *Aust. J. Mar. Freshwater Res.* **23,** 143–149.

Lucas, C., Kirkwood, G., and Somers, I. (1979). An assessment of the stocks of the banana prawn *Penaeus merguiensis* in the Gulf of Carpentaria. *Aust. J. Mar. Freshwater Res.* **30,** 639–652.

Lusk, G., Karel, M., and Goldblith, S. A. (1964). Astacene pigment loss occurring in freeze-dried shrimp and salmon during storage. *Food Technol.* **18,** 157–158.

Lyubimova, T. G., Naumov, A. G., and Lagunov, L. L. (1973). Prospects of the utilization of krill and other nonconventional resources of the world ocean. *J. Fish. Res. Board Can.* **30,** 2196–2201.

McCoy, E. G. (1968). Migration, growth and mortality of North Carolina pink and brown penaeid shrimp. *N. C. Dep. Conserv. Dev., Raleigh, Spec. Sci. Rep. No. 15.*

McCrary, J. A. (1972). The Kodiak Island Shrimp fishery. *Proc. Natl. Shellfish. Assoc.* **62,** 5–6 (Abstr.).

McElroy, J. (1982). Krill—still an enigma. *Mar. Pol.* **6,** 238–239.

Mackintosh, N. A. (1972). Life cycle of Antarctic krill in relation to ice and water conditions. *Discovery Rep.* **36,** 1–94.

Mackintosh, N. A. (1974). Sizes of krill eaten by whales in the Antarctic. *Discovery Rep.* **36,** 157–178.

McWhinnie, M. A., Denys, C. J., and Angione, P. V. (1981). "Euphausiacea Bibliography—A World Literature Survey." Pergamon, Oxford.

Magill, A. R., and Erho, M. (1963). The developmental status of the pink shrimp fishery of Wahington and Oregon. *Bull. Pac. Mar. Fish. Comm.* **6,** 61–80.

Magoon, C. D., and Tegelberg, H. C. (1969). Washington pink shrimp fishery for the years 1963 through 1968. *State Wash. Dep. Fish., Res. Div. Prog. Rep.,* 27 pp.

Makarov, R. R., Naumov, A. G., and Shevtsov, V. V. (1970). The biology and the distribution of the Antarctic krill. *In* "Antarctic Ecology" (M. W. Holdgate, ed.), Vol. 1, pp. 173–176. Academic Press, New York.

Manar, T. A. (1973). Who eats shrimp? *Mar. Fish. Rev.* **35**(3-4), 3–5.

Marr, J. W. S. (1962). The natural history and geography of the Antarctic krill (*Euphausia superba* Dana). *Discovery Rep.* **32,** 33–464.

Marschall, H.-P. (1983). Sinking speed, density and size of euphausiid eggs. *Meeresforschung* **30,** 1–9.

Martosubroto, P. (1974). Fecundity of pink shrimp *Penaeus duorarum* Burkenroad. *Bull. Mar. Sci.* **24,** 606–627.

Marullo, F. (1973). An automatic pumping device for sampling postlarval shrimp (*Penaeus* spp.). *Mar. Fish. Rev.* **35**(3–4), 24–26.

Marullo, F., Emiliani, D. A., Caillouet, C. W., and Clark, S. H. (1976). A vinyl streamer tag for shrimp (*Penaeus* spp.). *Trans. Am. Fish. Soc.* **105,** 658–663.

Matthews, G. A. (1982). Relative abundance and size distributions of commercially important shrimp during the 1981 Texas closure. *Mar. Fish. Rev.* **44**(9-10), 5–15.

Matylewich, M. A. (1982). Environmental influence on the migratory behavior of the brown shrimp in Pamlico Sound, North Carolina. M.S. Thesis, Old Dominion Univ., Norfolk, Virginia.

Matylewich, M. A., and Mundy, P. R. (1985). An evaluation of the relevance of some environmental factors in the estimation of migratory timing and yield in the brown shrimp of Pamlico Sound, North Carolina. *N. Am. J. Fish. Manage.* **5,** no. 2A, 197–209.

Mauchline, J. (1980). The biology of mysids and euphausiids. *Adv. Mar. Biol.* **18,** 3–623.

Mauchline, J., and Fisher, L. R. (1969). The biology of euphausiids. *Adv. Mar. Biol.* **7**, 1–454.

Meixner, R. (1969). Wachstum, Häutung und Fortpflanzung von *Crangon crangon* (L.) bei Einzelaufzucht. *Meeresforschung* **20**, 93–111.

Meixner, R. (1970). Grössenzunahme bei der Häutung von *Crangon crangon* aus der Nordsee und Ostsee. *Meeresforschung* **21**, 393–398.

Menon, M. K. (1955). Notes on the bionomics and fishery of the prawn *Metapenaeus dobsoni* Miers on the south-west coast of India. *Indian J. Fish.* **2**, 41–56.

Menon, M. K. (1957). Contributions to the biology of penaeid prawns of the south-west coast of India. 1. Sex ratio and movements. Indian J. Fish. **4**, 62–74.

Menon, M. K., and Raman, K. (1961). Observations on the prawn fishery of the Cochin backwaters with special reference to the stake net catches. *Indian J. Fish.* **8**, 1–23.

Menz, A., and Blake, B. F. (1980). Experiments on the growth of *Penaeus vannamei* Boone. *J. Exp. Mar. Biol. Ecol.* **48**, 99–111.

Menz, A., and Bowers, A. B. (1980). Bionomics of *Penaeus vannamei* Boone and *Penaeus stylirostris* Stimpson in a lagoon on the Mexican Pacific coast. *Estuarine Coastal Mar. Sci.* **10**, 685–697.

Menzel, R. W. (1955). Marking of shrimp. *Science* **121**, 446.

Meyer-Waarden, P. F., and Tiews, K. (1965). Further results of the German shrimp research. *Rapp. P.-v. Reun. Cons. Int. Explor. Mer* **156**, 131–137.

Meyers, S. P., and Rutledge, J. E. (1973). Utilization of economically-valuable byproducts from the shrimp processing industry. *In* "Food-Drugs from the Sea, Proceedings 1972" (L. R. Worthen, ed.), pp. 75–85. Marine Technology Society, Washington.

Mezykowski, T., and Ignatowska-Switalska, H. (1981). High levels of prostaglandins PGF_2 and PGE_2 in Antarctic krill *Euphausia superba* Dana. *Meeresforschung* **29**, 64–66.

Mezykowski, T., and Rakusa-Suszewski, S. (1979). The circadian rhythms in *Euphausia superba* Dana and its carbohydrate metabolism. *Meeresforschung* **27**, 124–129.

Moctezuma, M. A., and Blake, B. F. (1981). Burrowing activity in *Penaeus vannamei* Boone from the Caimanero-Huizache lagoon system on the Pacific coast of Mexico. *Bull. Mar. Sci.* **31**, 312–317.

Mohamed, K. H. (1967). Penaeid prawns in the commercial shrimp fisheries of Bombay with notes on species and size fluctuations. *Proc. Symp. Crustacea, Ser. 2, Mar. Biol. Assoc. India, Pt. IV*, pp. 1408–1418.

Mohamed, K. H., and Rao, P. V. (1971). Estuarine phase in the life-history of the commercial prawns of the west coast of India. *J. Mar. Biol. Assoc. India* **13**, 149–161.

Moller, T. H., and Jones, D. A. (1975). Locomotory rhythms and burrowing habits of *Penaeus semisulcatus* (De Haan) and *P. monodon* (Fabricius) (Crustacea: Penaeidae). *J. Exp. Mar. Biol. Ecol.* **18**, 61–77.

Morgan, J. R. (1974). Freezing shrimp in freon without weight loss. *Food Eng.* **46**, 67–68.

Mundy, P. R. (1979). A quantitative measure of migratory timing, illustrated by application of the management of commercial salmon fisheries. Ph.D. Dissertation, Univ. of Washington, Seattle.

Munro, J. L., Jones, A. C., and Dimitriou, D. (1968). Abundance and distribution of the larvae of the pink shrimp (*Penaeus duorarum*) on the Tortugas Shelf of Florida, August 1962–October 1964. *Fish. Bull.* **67**, 165–181.

Nakamura, H., Mohri, Y., Muraoka, I., and Ito, K. (1979). Studies on brewing food containing Antarctic krill. 1. Production of soy sauce with cryo-ground Antarctic krill, *Euphausia superba*. *Bull. Jpn. Soc. Sci. Fish.* **45**, 1389–1393.

Nast, F. (1978). The vertical distribution of larval and adult krill (*Euphausia superba* Dana) on a time station south of Elephant Islands, South Shetlands. *Meeresforschung* **27**, 103–118.

Nast, F. (1982). The assessment of krill (*Euphausia superba* Dana) biomass from a net sampling programme. *Meeresforschung* **29**, 154–165.

Neal, R. A. (1967). An application of the virtual population technique to penaeid shrimp. *Proc. Annu. Conf. Southeast. Assoc. Game Fish Comm.* **21**, 264–272.

Neal, R. A. (1969). Methods of marking shrimp. *FAO Fish. Rep.* **57**, 1149–1165.

Neal, R. A. (1975). The Gulf of Mexico research and fishery on penaeid prawns. *In* "First Australian National Prawn Seminar, Maroochydore, Queensland" (Australian Fisheries Council, ed.), pp. 1–8. Australian Government Publ. Service, Canberra.

Neiva, G. de S. (1969). Observations on the shrimp fisheries of the central and southern coast of Brazil. *FAO Fish. Rep.* **57**, 847–858.

Nemoto, T. (1972). History of research into the food and feeding of euphausiids. *Proc. R. Soc. Edinburgh, Sect. B: Biology* **73**, 259–265.

Nichols, J. P., and Griffin, W. L. (1975). Trends in catch–effort relationships with economic implications: Gulf of Mexico shrimp fishery. *Mar. Fish. Rev.* **37**(2), 1–4.

Nichols, S. (1982). Impacts on shrimp yields of the 1981 fishery conservation zone closure off Texas. *Mar. Fish. Rev.* **44**(9-10), 31–37.

Ogle, J., and Price, W. (1976). Growth of the shrimp, *Penaeus aztecus,* fed a diet of live mysids (Crustacea: Mysidacea). *Gulf Res. Rep.* **5**, 46–47.

Ohshima, T., and Nagayama, F. (1980). Purification and properties of catechol oxidase from the Antarctic krill. *Bull. Jpn. Soc. Sci. Fish.* **46**, 1035–1042.

Omori, M. (1974). The biology of pelagic shrimps in the ocean. *Adv. Mar. Biol.* **12**, 233–324.

Omori, M. (1978). Zooplankton fisheries of the world: A review. *Mar. Biol. (Berlin)* **48**, 199–205.

Omori, M. (1979). Growth, feeding, and mortality of larval and early postlarval stages of the oceanic shrimp *Sergestes similis* Hansen. *Limnol. Oceanogr.* **24**, 273–288.

Omori, M., Konagaya, T., and Noya, K. (1973). History and present status of the fishery of *Sergestes lucens* (Penaeidea, Decapoda, Crustacea) in Suruga Bay, Japan. *J. Cons. Int. Explor. Mer* **35**, 61–77.

Osborn, K. W., Maghan, B. W. and Drummond, S. B. (1969). Gulf of Mexico Shrimp Atlas. *U. S. Bur. Commer. Fish., Fish. Circ. No. 312,* 20 pp.

Oshima. (1974). Studies of the productive effectiveness of prawn stocking. *The Seto Inland Sea Fisheries Aquaculture Society Corporation,* 212 pp. (manuscript).

O'Sullivan, D. (1983a). Fisheries of the Southern Ocean. *Aust. Fish.* **42**(7), 4–11.

O'Sullivan, D. (1983b). Krill research shows large numbers present in AAT waters. *Aust. Fish.* **42**(7), 15–18.

Parrack, M. L. (1979). Aspects of brown shrimp, *Penaeus aztecus,* growth in the northern Gulf of Mexico. *Fish. Bull.* **76**, 827–836.

Parsons, T. R. (1972). Plankton as a food source. *Underwater J.* **4**, 30–37.

Paula, M. A. (1983). The relationship of size class distribution to migratory behavior in the brown shrimp (*Penaeus aztecus*) in Pamlico Sound, North Carolina. M.S. Thesis, Old Dominion Univ., Norfolk, Virginia.

Pearcy, W. G. (1970). Vertical migration of the ocean shrimp, *Pandalus jordani:* A feeding and dispersal mechanism. *Calif. Fish Game* **56**, 125–129.

Pearcy, W. G. (1972). Distribution and diel changes in the behavior of pink shrimp, *Pandalus jordani,* off Oregon. *Proc. Natl. Shellfish. Assoc.* **62**, 15–20.

Pearson, J. C. (1929). Natural history and conservation of the redfish and other commercial sciaenids on the Texas coast. *Bull. U. S. Bur. Fish.* **44**, 129–214.

Pearson, J. C. (1939). The early life history of some American Penaeidae, chiefly the commercial shrimp *Penaeus setiferus* (L.). *Bull. U.S. Bur. Fish.* **49**, 1–73.

Pedraja, R. R. (1970). Change of composition of shrimp and other marine animals during processing. *Food Technol.* **24**(12), 37–39, 42.

Penn, J. W. (1975a). Tagging experiments with western king prawn *Penaeus latisulcatus*

Kishinouye. I. Survival, growth, and reproduction of tagged prawns. *Aust. J. Mar. Freshwater Res.* **26,** 197–211.

Penn, J. W. (1975b). The influence of tidal cycles on the distributional pathway of *Penaeus latisculatus* Kishinouye in Shark Bay, western Australia. *Aust. J. Mar. Freshwater Res.* **26,** 93–102.

Penn, J. W. (1976). Tagging experiments with western king prawn, *Penaeus latisulcatus* Kishinouye. II. Estimation of population parameters. *Aust. J. Mar. Freshwater Res.* **27,** 239–250.

Penn, J. W. (1980). Spawning and fecundity of the western king prawn, *Penaeus latisulcatus* Kishinouye, in Western Australian waters. *Aust. J. Mar. Freshwater Res.* **31,** 21–35.

Penn, J. W., and Stalker, R. W. (1975). A daylight sampling net for juvenile penaeid prawns. *Aust. J. Mar. Freshwater Res.* **26,** 287–291.

Pereyra, W. T., Heyamoto, H., and Simpson, R. R. (1967). Relative catching efficiency of a 70-foot semiballoon shrimp trawl and a 94-foot eastern fish trawl. *Fish. Ind. Res.* **4,** 49–71.

Pérez Farfante, I. (1969). Western Atlantic shrimps of the genus *Penaeus*. *Fish. Bull.* **67,** 461–591.

Perkins, B. E., and Meyers, S. P. (1977). Recovery and application of organic wastes from the Louisiana shrimp canning industry. *Proc. 8th Natl. Symp. Food Process. Wastes* pp. 292–307 (EPA-600/2-77-284).

Permitin, Yu. E. (1970). The consumption of krill by Antarctic fishes. *In* "Antarctic Ecology" (M. W. Holdgate, ed.), Vol. 1, pp. 177–182. Academic Press, New York.

Phillips, G. (1971). Incubation of the English prawn *Palaemon serratus*. *J. Mar. Biol. Assoc. U. K.* **51,** 43–48.

Pike, R. B. (1952). Notes on the growth and biology of the prawn *Pandalus bonnieri* Caullery. *J. Mar. Biol. Assoc. U. K.* **31,** 259–267.

Poffenberger, J. R. (1982). Estimated impacts on ex-vessel brown shrimp prices and value as a result of the Texas closure regulation. *Mar. Fish. Rev.* **44**(9-10), 38–43.

Poleck, T. P., and Denys, C. J. (1982). Effect of temperature on the molting, growth and maturation of the Antarctic krill *Euphausia superba* (Crustacea: Euphausiacea) under laboratory conditions. *Mar. Biol. (Berlin)* **70,** 255–265.

Powell, D. G. (1979). Estimation of mortality and growth parameters from the length frequency of a catch. *Rapp. P.-v. Reun. Cons. Int. Explor. Mer* **175,** 167–169.

Price, A. R. G. (1979). Temporal variations in abundance of penaeid shrimp larvae and oceanographic conditions off Ras Tanura, Western Arabian Gulf. *Estuarine Coastal Mar. Sci.* **9,** 451–465.

Purvis, C. E., and McCoy, E. G. (1978). Population dynamics of brown shrimp in Pamlico Sound. *N. C. Dep. Nat. Econ. Resour., Div. Commer. Sports Fish.,* 26 pp.

Rafail, S. Z. (1973). A simple and precise method for fitting a von Bertalanffy growth curve. *Mar. Biol. (Berlin)* **19,** 354–358.

Rajyalakshmi, T. (1961). Observations on the biology and fishery of *Metapenaeus brevicornis* (M. Edw.) in the Hooghly Estuarine System. *Indian J. Fish.* **8,** 383–402.

Ramamurthy, S. (1967). Studies on the prawn fishery of Kutch. *Proc. Symp. Crustacea, Ser. 2, Mar. Biol. Assoc. India, Pt. IV,* pp. 1424–1436.

Raman, K. (1967). Observations on the fishery and biology of the giant freshwater prawn *Macrobrachium rosenbergii* de Man. *Proc. Symp. Crustacea, Ser. 2, Mar. Biol. Assoc. India, Pt. II.,* pp. 649–669.

Rao, A. V. P. (1967). Some observations on the biology of *Penaeus indicus* H. Milne-Edwards and *Penaeus monodon* Fabricius from the Chilka Lake. *Indian J. Fish.* **14,** 251–270.

Rao, M. R. R., and Novak, A. F. (1975). Thermal and microwave energy for shrimp processing. *Mar. Fish. Rev.* **37**(12), 25–30.

Rao, R. M., and Novak, A. F. (1977). Causes and prevention of weight losses in frozen fishery products during freezing and cold storage. *Proc. Gulf Caribb. Fish. Inst.* **29,** 28–42.

Rasmussen, B. (1951). Some problems in the fishery for deep sea prawns. *Rapp. P.-v. Reun. Cons. Int. Explor. Mer* **128,** 79–81.

Rasmussen, B. (1967a). Note on growth and protandric hermaphroditism in the deep sea prawn, *Pandalus borealis. Proc. Symp. Crustacea, Ser. 2, Mar. Biol. Assoc. India, Pt. II,* pp. 701–706.

Rasmussen, B. (1967b). The fishery for deep sea prawn in Norway. *Proc. Symp. Crustacea, Ser. 2, Mar. Biol. Assoc. India, Pt. IV,* pp. 1437–1441.

Raymont, J. E. G., Srinivasagam, R. T., and Raymont, J. K. B. (1971). Biochemical studies on marine zooplankton. IX. The biochemical composition of *Euphausia superba. J. Mar. Biol. Assoc. U. K.* **51,** 581–588.

Renfro, W. C. (1960). Abundance and distribution of penaeid shrimp larvae. *U. S. Fish Wildl. Serv., Circ. No. 92,* pp. 9–10.

Renfro, W. C., and Pearcy, W. G. (1966). Food and feeding apparatus of two pelagic shrimps. *J. Fish. Res. Board Can.* **23,** 1971–1975.

Robinson, J. G. (1972). Variations in reproductive potential of *Pandalus jordani. Proc. Natl. Shellfish. Assoc.* **62,** 6 (Abstr.).

Rodriguez, G., and Naylor, E. (1972). Behavioral rhythms in littoral prawns. *J. Mar. Biol. Assoc. U. K.* **52,** 81–95.

Roe, R. B. (1969). Distribution of royal red shrimp, *Hymenopenaeus robustus,* on three potential commercial grounds off southeastern United States. *Fish. Ind. Res.* **5,** 151–174.

Roessler, M. A., and Rehrer, R. G. (1971). Relation of catches of postlarval pink shrimp in Everglades National Park, Florida, to the commercial catches on the Tortugas grounds. *Bull. Mar. Sci.* **21,** 790–805.

Ronsivalli, L. J., and Baker, D. W. (1981). Low temperature preservation of seafoods: A review. *Mar. Fish. Rev.* **43**(4), 1–15.

Ross, R. M., and Quetin, L. B. (1983). Spawning frequency and fecundity of the Antarctic krill *Euphausia superba. Mar. Biol. (Berlin)* **77,** 201–205.

Ross, R. M., Daly, K. L., and English, T. S. (1982). Reproductive cycle and fecundity of *Euphausia pacifica* in Puget Sound, Washington. *Limnol. Oceanogr.* **27,** 304–314.

Rothschild, B. J., and J. A. Gulland (conveners). (1982). Interim report of the workshop on the scientific basis for the management of penaeid shrimp. *NOAA Tech. Mem. NMFS-SEFC, No. 98,* 66 pp.

Ruello, N. V. (1969). Prawn behavior in the Hunter Region. *Fisherman* **3**(4), 1–6.

Ruello, N. V. (1973a). Burrowing, feeding, and spacial distribution of the school prawn *Metapenaeus macleayi* (Haswell) in the Hunter River Region, Australia. *J. Exp. Mar. Biol. Ecol.* **13,** 189–206.

Ruello, N. V. (1973b). The influence of rainfall on the distribution and abundance of the school prawn *Metapenaeus macleayi* in the Hunter River Region (Australia). *Mar. Biol. (Berlin)* **23,** 221–228.

Ruello, N. V. (1977). Migration and stock studies on the Australian school prawn *Metapenaeus macleayi. Mar. Biol. (Berlin)* **41,** 185–190.

Savagaon, K. A., Venugopal, V., Kamat, S. V., Kumta, U. S., and Sreenivasan, A. (1972a). Radiation preservation of tropical shrimp for ambient temperature storage. 1. Development of a heat-radiation combination process. *J. Food. Sci.* **37,** 148–150.

Savagaon, K. A., Venugopal, V., Kamat, S. V., Kumta, U. S., and Sreenivasan, A. (1972b). Radiation preservation of tropical shrimp for ambient temperature storage. 2. Storage studies. *J. Food Sci.* **37,** 151–153.

Scattergood, L. W. (1952). The northern shrimp fishery of Maine. *Commer. Fish. Rev.* **14**(1), 1–16.

Scheer, D. (1967). Biology and fishery of the Baltic shrimp (*Leander adspersus* var. *fabricii*) on the coast of the German Democratic Republic. *Proc. Symp. Crustacea, Ser. 2, Mar. Biol. Assoc. India, Pt. II,* pp. 670–675.

Schneppenheim, R. (1980). Concentration of fluoride in Antarctic animals. *Meeresforschung* **28,** 179–182.

Scholz, D. J., Sinnhuber, R. O., East, D. M., and Anderson, A. W. (1962). Radiation-pasteurized shrimp and crabmeat. *Food Technol.* **16,** 118–120.

Schubel, J. R., Carter, H. H., and Wise, W. M. (1979). Shrimping as a source of suspended sediment in Corpus Christi Bay (Texas). *Estuaries* **2,** 201–203.

Schumacher, A., and Tiews, K. (1979). On the population dynamics of the brown shrimp (*Crangon crangon* L.) off the German coast. *Rapp. P.-v. Reun. Cons. Int. Explor. Mer* **175,** 280–286.

Scrivener, J. C., and Butler, T. H. (1971). A bibliography of shrimps of the family Pandalidae, emphasizing economically important species of the genus *Pandalus. Fish. Res. Board Can. Tech. Rep. No. 242,* 42 pp.

Seidel, W. R. (1969). Design, construction, and field testing of the BCF electric shrimp-trawl system. *Fish. Indust. Res.* **4,** 213–231.

Seidel, W. R., and Watson, J. W. (1978). A trawl design: employing electricity to selectively capture shrimp. *Mar. Fish. Rev.* **40**(9), 21–23.

Seki, N., Sakaya, H., and Onozawa, T. (1977). Studies on proteases from Antarctic krill. *Bull. Jpn. Soc. Sci. Fish.* **43,** 955–962.

Shibata, N. (1983a). Change in extractability of muscular protein of Antarctic krill during storage. *Bull. Jpn. Soc. Sci. Fish.* **49,** 745–749.

Shibata, N. (1983b). Extractability of muscular proteins of Antarctic krill in media with different ionic strengths. *Bull. Jpn. Soc. Sci. Fish.* **49,** 739–743.

Shropshire, R. F. (1944). Plankton harvesting. *J. Mar. Res.* **5,** 185–188.

Siegel, V. (1982a). Investigations on krill (*Euphausia superba*) in the southern Weddell Sea. *Meeresforschung* **29,** 244–252.

Siegel, V. (1982b). Relationship of various length measurements of *Euphausia superba* Dana. *Meeresforschung* **29,** 114–117.

Sigurdsson, A., and Hallgrimsson, I. (1965). The deep-sea prawn (*Pandalus borealis*) in Icelandic waters. *Rapp. P.-v. Reun. Cons. Int. Explor. Mer* **156,** 105–108.

Silverman, G. J., Nickerson, J. T. R., Duncan, D. W., Davis, N. S., Schachter, J. S., and Joselow, M. M. (1961). Microbial analysis of frozen raw and cooked shrimp. I. General results. *Food. Technol.* **15,** 455–458.

Simpson, A. C., Howell, B. R., and Warren, P. J. (1970). Synopsis of biological data on the shrimp *Pandalus montagui* Leach, 1814. *FAO Fish. Rep.* **57,** 1225–1250.

Skúladóttir, U. (1979). Comparing several methods of assessing the maximum sustainable yield of *Pandalus borealis* in Arnarfjordur. *Rapp. P.-v. Reun. Cons. Int. Explor. Mer* **175,** 240–252.

Slack-Smith, R. J. (1969). The prawn fishery of Shark Bay, Western Australia. *FAO Fish. Rep.* **57,** 717–734.

Smidt, E. (1965). Deep-sea prawns and the prawn fishery in Greenland waters. *Rapp. P.-v. Reun. Cons. Int. Explor. Mer* **156,** 100–104.

Smidt, E. (1967). Deep sea prawn (*Pandalus borealis* Kr.) in Greenland waters: Biology and fishery. *Proc, Symp. Crustacea, Ser. 2, Mar. Biol. Assoc. India, Pt. IV,* pp. 1448–1453.

Smidt, E. (1969). *Pandalus borealis* in Greenland waters, its fishery and biology. *FAO Fish. Rep.* **57,** 893–902.

Soevik, T., and Braekkan, O. R. (1979). Fluoride in Antarctic Krill (*Euphausia superba*) and Atlantic krill (*Meganyctiphanes norvegica*). *J. Fish. Res. Board Can.* **36,** 1414–1416.

Sofoulis, N. G., Penrose, J. D., Cartledge, D. R., and Fallon, G. R. (1979). Acoustic target strengths of some penaeid prawns. *Aust. J. Mar. Freshwater Res.* **30**, 93–101.

Springer, S., and Bullis, H. R. (1952). Exploratory shrimp fishing in the Gulf of Mexico 1950–1951. *U. S. Fish Wildl. Serv. Fish Leafl. No. 406*, 34 pp.

St. Amant, L. S., Broom, J. G., and Ford, T. B. (1966). Studies of the brown shrimp, *Penaeus aztecus*, in Barataria Bay, Louisiana, 1962–1965. *Proc. Gulf Caribb. Fish. Inst.* **18**, 1–17.

Staples, D. J. (1979). Seasonal migration patterns of postlarval and juvenile banana prawns, *Penaeus merguiensis* de Man, in the major rivers of the Gulf of Carpentaria, Australia. *Aust. J. Mar. Freshwater Res.* **30**, 143–157.

Staples, D. J. (1980a). Ecology of juvenile and adolescent banana prawns, *Penaeus merguiensis*, in a mangrove estuary and adjacent offshore area of the Gulf of Carpentaria. I. Immigration and settlement of postlarvae. *Aust. J. Mar. Freshwater Res.* 31, 635–652.

Staples, D. J. (1980b). Ecology of juvenile and adolescent banana prawns, *Penaeus merguiensis* in a mangrove estuary and adjacent offshore area of the Gulf of Carpentaria. II. Emigration, population structure and growth of juveniles. *Aust. J. Mar. Freshwater Res.* **31**, 653–665.

Staples, D. J., and Vance, D. J. (1979). Effects of changes in catchability on sampling of juvenile and adolescent banana prawns, *Penaeus merguiensis* de Man. *Aust. J. Mar. Freshwater Res.* **30**, 511–519.

Stern, J. A. (1958). The new shrimp industry of Washington. *Proc. Gulf Caribb. Fish. Inst.* **10**, 37–42.

Struhsaker, P., and Aasted, D. C. (1974). Deepwater shrimp trapping in the Hawaiian Islands. *Mar. Fish. Rev.* **36**(10), 24–30.

Subrahmanyam, C. B. (1963). Notes on the bionomics of the penaeid prawn *Metapenaeus affinis* (Milne Edwards) of the Malabar coast. *Indian J. Fish.* **10**, 11–22.

Subrahmanyam, C. B. (1971). The relative abundance and distribution of penaeid shrimp larvae off the Mississippi coast. *Gulf Res. Rep.* **3**, 291–345.

Subrahmanyam, M. (1967). Further observations in lunar periodicity in relation to the prawn abundance in the Godavari Estuarine Systems. *J. Mar. Biol. Assoc. India* **9**, 111–115.

Subrahmanyam, M., and Ganapati, P. N. (1971). Observations on post-larval prawns from the Godavari Estuarine Systems with a note on their role in capture and culture fisheries. *J. Mar. Biol. Assoc. India* **13**, 195–202.

Surkiewicz, B. F., Hyndman, J. B., and Yancey, M. V. (1967). Bacteriological survey of the frozen prepared foods industry. II. Frozen breaded raw shrimp. *Appl. Microbiol.* **15**, 1–9.

Suseelan, C. (1974). Observations on the deep-sea prawn fishery off the south-west coast of India with special reference to pandalids. *J. Mar. Biol. Assoc. India* **16**, 491–511.

Sutter, F. C., and Christmas, J. Y. (1983). Multilinear models for the prediction of brown shrimp harvest in Mississippi waters. *Gulf Res. Rep.* **7**, 205–210.

Suzuki, T. (1981). "Fish and Krill Protein Processing Technology." Applied Science Publ. Ltd., London.

Szewielow, A. (1981). Fluoride in krill (*Euphausia superba* Dana). *Meeresforschung* **28**, 244–246.

Tabb, D. C., Dubrow, D. L., and Jones, A. E. (1962). Studies on the biology of the pink shrimp, *Penaeus duorarum* Burkenroad, in Everglades National Park, Florida. *Fla. Board Conserv. Tech. Ser.* **37**, 5–30.

Tegelberg, H. C. (1972). Sex change in *Pandalus jordani* of the Washington coast. *Proc. Natl. Shellfish. Assoc.* **62**, 7–8 (Abstr.).

Tegelberg, H. C., and Smith, J. M. (1957). Observations on the distribution and biology of the pink shrimp. (*Pandalus jordani*) off the Washington coast. *Wash. Dep. Fish. Fish. Res. Pap.* **2**, 25–34.

Temple, R. F., and Fischer, C. C. (1965). Vertical distribution of the planktonic stages of penaeid shrimp. *Publ. Inst. Mar. Sci. Univ. Tex.* **10,** 59–67.

Thompson, B. G. (1983). Fisheries of the United States, 1982. National Marine Fisheries Service. *Crr. Fish. Statist. No. 8300,* 118 pp.

Thompson, J. R. (1967). Development of a commercial fishery for the penaeid shrimp *Hymenopenaeus robustus* Smith on the continental slope of the southeastern United States. *Proc. Symp. Crustacea, Ser. 2, Mar. Biol. Assoc. India, Pt. IV,* pp. 1454–1459.

Thorson, G. (1946). Reproduction and larval development of Danish marine bottom invertebrates, with special reference to the planktonic larvae in the Sound (Øresund). *Medd. Dan. Fisk. Havunders. Bind 4, Nr. 1,* 523 pp.

Thorson, G. (1950). Reproduction and larval ecology of marine bottom invertebrates. *Biol. Rev. Cambridge Philos. Soc.* **25,** 1–45.

Tiews, K. (1970). Synopsis of biological data on the common shrimp *Crangon crangon* (Linnaeus, 1758). *FAO Fish. Rep.* **57,** 1167–1224.

Tiews, K. (1978a). Non-commercial fish species in the German Bight: Records of by-catches of the brown shrimp fishery. *Rapp. P.-v. Reun. Cons. Int. Explor. Mer* **172,** 259–265.

Tiews, K. (1978b). The predator–prey relationship between fish populations and the stock of brown shrimp (*Crangon crangon* L.) in German coastal waters. *Rapp. P.-v. Reun. Cons. Int. Explor. Mer* **172,** 250–258.

Tokunaga, T., Tida, H., and Nakamura, K. (1977). Formation of dimethyl sulfide in Antarctic krill, *Euphausia superba. Bull. Jpn. Soc. Sci. Fish.* **43,** 1209–1217.

Tom, M., and Lewinsohn, C. (1983). Aspects of the benthic life cycle of *Penaeus (Melicertus) japonicus* Bate (Crustacea Decapoda) along the south-eastern coast of the Mediterranean. *Fish. Res.* **2,** 89–101.

Toyama, K., and Yano, W. (1979). Application of the juice extractor to the Antarctic krill for the elimination of the shell components. *Bull. Jpn. Soc. Sci. Fish.* **45,** 407.

Unar, M. (1974). A review of the Indonesian shrimp fishery and its present developments. *Mar. Fish. Rev.* **36**(1), 61–30.

Vance, D., Staples, D., and Kerr, J. (1983). Banana prawn catches in the Gulf of Carpentaria— trends and predictions. *Aust. Fish.* **42**(6), 33, 35–37.

Van Der Baan, S. M. (1975). Migration of *Crangon crangon* in surface waters near the "Texel" lightship. *Neth. J. Sea Res.* **9,** 287–296.

Velankar, N. K., and Govindan, T. K. (1959). Preservation of prawns in ice and the assessment of their quality by objective standards. *Indian J. Fish.* **6,** 306–321.

Velankar, N. K., Govindan, T. K., Appukuttan, P. N., and Mahadeva Iyer, K. (1961). Spoilage of prawns at 0°C and its assessment by chemical and bacteriological tests. *Indian J. Fish.* **8,** 241–251.

Vinogradov, M. E. (1970). "Vertical Distribution of the Oceanic Zooplankton." Israel Program for Scientific Translations, Jerusalem.

von Bertalanffy, L. (1938). A quantitative theory of organic growth (inquiries on growth laws, II). *Hum. Biol.* **10,** 181–213.

von Bertalanffy, L. (1964). Basic concepts in quantitative biology of metabolism. *Helgol. wiss. Meeresunters.* **9,** 5–37.

Voronina, N. M. (1974). An attempt at a functional analysis of the distributional range of *Euphausia superba. Mar. Biol. (Berlin)* **24,** 347–352.

Warren, J. P., and Griffin, W. L. (1980). Costs and returns in the Gulf of Mexico shrimp industry, 1971–78. *Mar. Fish. Rev.* **42**(2), 1–7.

Warren, P. J., and Sheldon, R. W. (1967). Feeding and migration patterns of the pink shrimp, *Pandalus montagui,* in the estuary of River Crouch, Essex, England. *J. Fish. Res. Board Can.* **24,** 569–580.

Watson, J. W. (1976). Electrical shrimp trawl catch efficiency for *Penaeus duorarum* and *Penaeus aztecus. Trans. Am. Fish. Soc.* **105,** 135–148.

Watson, J. W., and McVea, C. (1977). Development of a selective shrimp trawl for the southeastern United States penaeid shrimp fisheries. *Mar. Fish. Rev.* **39**(10), 18–24.

Welker, B. D., Clark, S. H., Fontaine, C. T., and Benton, R. C. (1975). A comparison of Petersen trawls and biological stains used with internal tags as marks for shrimp. *Gulf Res. Rep.* **5,** 1–5.

Weymouth, F. W., Lindner, M. J., and Anderson, W. W. (1933). Preliminary report on the life history of the common shrimp, *Penaeus setiferus* (Linn.) *Bull. U. S. Bur. Fish.* **48,** 1–26.

White, C. J. (1975). Effects of 1973 river flood waters on brown shrimp in Louisiana estuaries. *La. Wildl. Fish. Comm., New Orleans, Tech. Bull. No. 16,* 24 pp.

White, C. J., and Boudreaux, C. J. (1977). Development of an areal management concept for Gulf penaeid shrimp. *La. Wildl. Fish. Comm., New Orleans, Tech. Bull. No. 22,* 77 pp.

Wiborg, K. F. (1976). Fishery and commercial exploitation of *Calanus finmarchicus* in Norway. *J. Cons. Int. Explor. Mer* **36,** 251–258.

Wickham, D. A. (1967). Observations on the activity patterns in juveniles of the pink shrimp, *Penaeus duorarum. Bull. Mar. Sci.* **17,** 769–786.

Wickins, J. F. (1976). Prawn biology and culture. *Mar. Biol. Annu. Rev.* **14,** 435–507.

Wickstead, J. H. (1967). Pelagic copepods as food organisms. *Proc. Symp. Crustacea, Ser. 2, Mar. Biol. Assoc. India, Pt. IV,* pp. 1460–1465.

Wiesepape, L. M., Aldrich, D. V., and Strawn, K. (1972). Effects of temperature and salinity on thermal death in postlarval brown shrimp, *Penaeus aztecus. Physiol. Zool.* **45,** 22–33.

Wigley, R. L. (1973). Fishery for northern shrimp, *Pandalus borealis,* in the Gulf of Maine. *Mar. Fish. Rev.* **35**(3-4), 9–14.

Williams, A. B. (1955a). A contribution to the life histories of commercial shrimps (Penaeidae) in North Carolina. *Bull. Mar. Sci. Gulf Caribb.* **5,** 116–146.

Williams, A. B. (1955b). A survey of North Carolina shrimp nursery grounds. *J. Elisha Mitchell Sci. Soc.* **71,** 200–207.

Williams, A. B. (1958). Substrates as a factor in shrimp distribution. *Limnol. Oceanogr.* **3,** 283–290.

Williams, A. B. (1959). Spotted and brown shrimp postlarvae (*Penaeus*) in North Carolina. *Bull. Mar. Sci. Gulf Caribb.* **9,** 281–290.

Williams, A. B. (1969). A ten-year study of meroplankton in North Carolina estuaries: Cycles of occurrence among penaeidean shrimps. *Chesapeake Sci.* **10,** 36–47.

Williams, A. B., and Deubler, E. E. (1968). A ten-year study of meroplankton in North Carolina estuaries: Assessment of environmental factors and sampling success among bothid flounders and penaeid shrimps. *Chesapeake Sci.* **9,** 27–41.

Witek, Z., Kalinowski, J., Grelowski, A., and Wolnomiejski, N. (1981). Studies of aggregations of krill (*Euphausia superba*). *Meeresforschung* **28,** 228–243.

Yaldwyn, J. C. (1966). Protandrous hermaphroditism in decapod prawns of the families Hippolytidae and Campylonotidae. *Nature (London)* **209,** 1366.

Yamanaka, H., Kikuchi, T., and Amano, K. (1977). Studies on the residue of sulfur dioxide and the production of formaldehyde in the prawn treated with sodium bisulfite. *Bull. Jpn. Soc. Sci. Fish.* **43,** 115–120.

Yanagimoto, M., Yokoyama, Y., and Kobayashi, T. (1979). An application of water jet to deviscera methods of Antarctic krill (*Euphausia superba*). *Bull. Jpn. Soc. Sci. Fish.* **45,** 375–378.

Yanagimoto, M., Kobayashi, T., and Shiba, M. (1982). De-carapace treatment of Antarctic krills *Euphausia superba* by turbulence. *Bull. Jpn. Soc. Sci. Fish.* **48,** 467–473.

Young, P. C. (1978). Moreton Bay, Queensland: A nursery area for juvenile penaeid prawns. *Aust. J. Mar. Freshwater Res.* **29,** 55–75.

Young, P. C., and Carpenter, S. M. (1977). Recruitment of postlarval penaeid prawns to nursery areas in Moreton Bay, Queensland. *Aust. J. Mar. Freshwater Res.* **28,** 745–773.

Zein-Eldin, Z. P. (1963). Effect of salinity on growth of postlarval penaeid shrimp. *Biol. Bull. (Woods Hole, Mass.)* **125,** 188–196.

Zein-Eldin, Z. P., and Aldrich, D. V. (1965). Growth and survival of postlarval *Penaeus aztecus* under controlled conditions of temperature and salinity. *Biol. Bull. (Woods Hole, Mass.)* **129,** 199–216.

Zein-Eldin, Z. P., and Griffith, G. W. (1966). The effect of temperature upon the growth of laboratory-held postlarval *Penaeus aztecus. Biol. Bull. (Woods Hole, Mass.)* **131,** 186–196.

Zupanovic, S., and Mohiuddin, S. Q. (1973). A survey of the fishery resources in the northeastern part of the Arabian Sea. *J. Mar. Biol. Assoc. India* **15,** 496–537.

2

The Biology and Exploitation of Crabs

PAUL A. HAEFNER, JR.

I. CRAB DEFINED

Description

The term *crab* is most often used in reference to the more than 4500 extant species of decapod crustaceans included in the infraorder Brachyura (Table I). The crab body form, in which the abdomen is reduced, flattened, and carried folded against the ventral surface of the cephalothorax (Figs. 1 and 2), has actually evolved several times in the Crustacea, and it is not confined to the Brachyura. The aberrant infraorder Anomura includes species that resemble the brachyuran body plan to varying degrees (Fig. 3). The re-

THE BIOLOGY OF CRUSTACEA, VOL. 10
Copyright © 1985 by Academic Press, Inc.
All rights of reproduction in any form reserved.

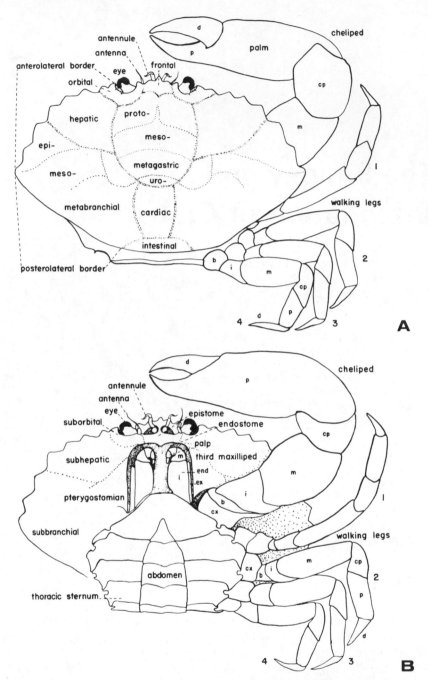

Fig. 1. Schematic drawings of brachyuran crab. (A) Dorsal view, (B) ventral view, areas of carapace indicated. Legs are shown on one side only. b, Basis immovably united with ischium; cp, carpus; d, dactyl; i, ischium; m, merus; p, propodus. (From Williams, 1965.)

TABLE I

Number of Recent Species of Brachyura in Relation to Other Groups of Crustacea[a]

Taxa		Percentage of		
		Reptantia	Decapoda	Crustacea
Section Brachyura	>4,500	66.9	51.5	14.8
Section Anomura	>1,400	20.8	16.0	4.6
Section Palinura	130	1.9	1.5	0.4
Section Astacura	> 700	10.4	8.0	2.3
Suborder Reptantia	>6,730	100.0	77.0	22.1
Suborder Natantia	>2,000		23.0	6.6
Order Decapoda	>8,730		100.0	28.7
Nondecapod Crustaceans	21,770			71.3
Class Crustacea	>30,500			100.0

[a] Data from Kaestner (1970). Courtesy of John Wiley and Sons, Inc., New York.

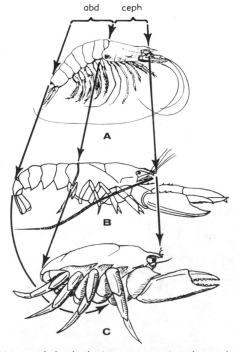

Fig. 2. Main divisions of the body in representative decapods. (A) *Penaeopsis;* (B) *Homarus,* (C) *Carcinus.* abd, Abdomen; ceph, cephalothorax. (From Glaessner, 1969, courtesy of the Geological Society of America and University of Kansas.)

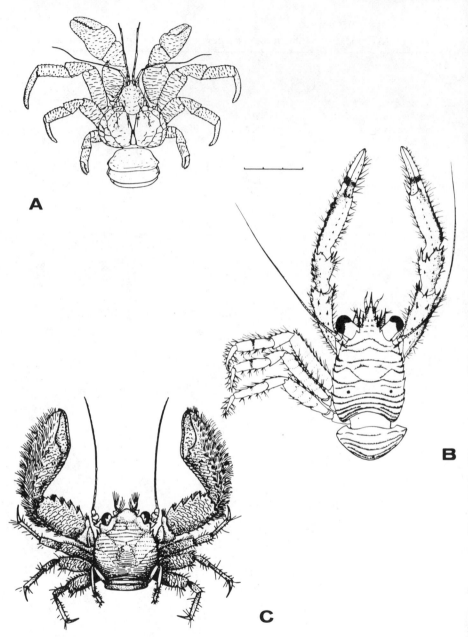

Fig. 3. Anomura. (A) *Birgus latro*, (B) *Galathea rostrata*, (C) *Petrolisthes galathinus*. [(A) and (C) from Kaestner, 1970, courtesy of John Wiley and Sons, Inc., New York; (B) from Williams, 1965.]

semblance is confined principally to the apparent symmetry of the abdomen and to the ability to fold the abdomen under the cephalothorax. Features that distinguish these anomuran crabs from the "true" crabs of the Brachyura are described in Williams (1965) and Kaestner (1970) (see also Volumes 1 and 5 of this treatise).

II. UNIQUENESS OF THE CRAB FORM

Although the phyletic origins and associations of the many diverse crustacean taxa are conjectural (Whittington and Rolfe, 1963; Brooks, 1969; Bowman and Abele, chapter 1, Volume 1), it is generally accepted that of the highly successful Decapoda, the highly specialized crab form represents the most diverse and successful group (Whittington and Rolfe, 1963). The genesis of the brachyuran organization involved changes in shape and structure that were accompanied by maximum diversification of organs and their functions. This new level of organization led to great biological improvement presently manifest by taxonomic diversity and wide ecological distribution (Stĕvcic, 1971).

Adaptations of Importance to Fisheries

Crabs have developed highly successful relationships between the environment and those biological mechanisms involved in evolutionary processes (Warner, 1977). There are filter feeders, sand cleaners, mud, plant, and carrion feeders, predators, commensals, and parasites (Patton, 1967; Caine, 1974, 1975a,b; Virnstein, 1977; Castro, 1978; Telford, 1978). Some are temporary swimmers, but most are walking bottom-dwellers. Some even climb trees and construct burrows on land (Waterman and Chace, 1960; see also Volume 8, Chapter 6, of this treatise).

The Brachyura show extreme versatility in distribution. They are found from the shore to the deep sea, in marine, brackish, and freshwater environments, in all climates, and in terrestrial and semiterrestrial conditions. The Podotremata (Guinot, 1977) are exclusively marine; most other taxa include brackish water species as well as marine forms, which invade estuarine habitats. The family Pseudothelphusidae contains most of the typically freshwater species that complete their life cycles in that habitat (Chace and Hobbs, 1969), but there are also freshwater representatives in the Potamidae (Rathbun, 1901) and Hymenosomatidae (Griffen, 1970). There are various degrees of terrestrialism in crabs, with representatives from families Gecarcinidae, Grapsidae, Ocypodidae, Trichodactylidae, and Myctyridae (Rathbun, 1901, 1930, 1933; Bliss, 1968; Chace and Hobbs, 1969; Griffen,

1970). Some crabs are also known to invade cave habitats occasionally (Chace and Hobbs, 1969). Today the marine littoral zone is conspicuously rich in crabs, although there is paleontological evidence that suggests that the number of species of Brachyura in the littoral zone gradually increased through geological time (Beurlen, 1931, cited by Glaessner, 1969).

The decapod series of abdominal structure (Fig. 2) is an excellent example of what Manton (1968, 1969) described for terrestrial arthropods, that is, conspicuous diagnostic characters are correlated with some all-important habit or habits. Glaessner (1969) suggested that the reduction in size and calcification and the concurrent changes in the function of the abdomen are the most important events in the evolution of decapods. Many of the morphological features of Recent crabs are testimony to the success of the evolved crab form. The crab form has developed directly through the brachyuran series and secondarily in the Anomura.

In crustacean fisheries, the yield and quality of product depend on the anatomy and degree of development of the musculature. These, in turn, are related to the habits of the species in question. For example, lobsters are primarily processed for their tails, which contain well-developed musculature that effects the sudden flexion in backward flight movements. Crabs, on the other hand, are processed for their body meat, the enlarged musculature (Cochran, 1935) that develops with modifications of the pereopods associated with substrate, water current, agility, and body elevation. Portunid crabs display a flattening and expansion of the dactyli of the fifth pereopods, which function as paddles for propulsion (Hartnoll, 1971; Spirito, 1972) and burrowing (Caine, 1974). The muscular development associated with such swimming behavior is responsible for the high unit value of "back fin lump" meat of portunids, such as *Callinectes sapidus.*

Chelipeds are not normally used in locomotion, but are usually highly developed for acquiring food (Elner, 1978, 1980) and for agonistic behavior (Jachowski, 1974). Claw meat is not as highly prized as is meat from other parts of the body, but it comprises a significant portion of the total meat yield, particularly in species such as the stone crab, *Menippe mercenaria,* which has disproportionately oversized chelipeds (Savage and Sullivan, 1978).

With reduction of the abdomen, the center of gravity shifted forward to a point beneath the cephalothorax, and a change in body shape occurred (Fig. 2). The cylindrical cephalothorax of the macruran body plan gave way to the conical shape of the Brachyura. The change in body shape and function of the abdomen is reflected in the internal morphology, as the organs and muscles conform to changes in the endophragmal skeleton (Pearson, 1908; Cochran, 1935; Pyle and Cronin, 1950; Lochhead, 1950a,b; Young, 1959;

Kaestner, 1970; Barnes, 1974; Ingle, 1983; Warner, 1977). The fusion of the internal skeletal matrix in crabs has, in turn, resulted in other modifications unique to Brachyura, such as the restriction of premolt mineral resorption to specific sites (Dennell, 1960; Passano, 1960) and adaptations of configuration of the sternum in burrowing forms.

Configuration of the sternum, shape of the body, and type of chela are interrelated (Schäfer, 1954). The more typical crablike forms (e.g., Menippe, Cancer) are heterochelic and heterodontic. The chelipeds can be folded against the anterolateral margin of the condensed carapace. The sternites are arranged parallel to the transverse plane. Such crabs live primarily in places of strong current or wave action and normally move sideways. Chelipeds of triangularly shaped spider crabs (e.g., Chionoecetes) are neither heterochelic nor heterodontic, and are seldom folded against the carapace. These animals, whose sternites are radially arranged, are able to move in all directions relative to the longitudinal axis of the body. Such crabs are most commonly found in places where water motion is reduced.

Crabs have striking adaptations related to defense and concealment. Properties of the exoskeleton analogous to those of prestressed concrete (Hepburn et al., 1975) are significant to the crab processing industry. Various degrees of calcification are associated with a defensive life style. Cancer crabs are relatively slow-moving, ponderous animals that rely on a heavy shell and mechanically strong chelae for protection (Jeffries, 1966; Warner and Jones, 1976). The less heavily calcified portunids, in contrast, are extremely agile and very fast swimmers. They will normally use their chelipeds in agonistic display; but when the encounter demands it, they will defend themselves with their chelae (Teytaud, 1971; Jachowski, 1974), which have a lower mechanical advantage, but "faster" muscle fibers than those in Cancer (Warner and Jones, 1976). Heterochelic and heterodontic crabs, such as portunids, also have the advantage of possessing a stronger, major chela (crusher) on one side and a faster, minor chela (cutter) on the opposite side (Stěvcic, 1971; Williams, 1974; Vermeij, 1977). Distinctive spines and ridges, which provide some protection, are also widespread among the Brachyura.

The adaptations related to body shape, cheliped development, skeletal configuration, and shell hardness have a direct bearing on the manner in which the crab is processed for its meat, for example, by hand picking, machine butchering, fish press, or brine flotation. These factors, as well as the relative ease of capture and handling, help determine whether any given species will support a fishery.

Crustacean mechanisms for solving basic physiological problems have been discussed to varying degrees in earlier reviews (Waterman, 1960) and are treated elsewhere in this series. Crabs are not totally unique in the ways in which they cope with their environment, but there are certain features that set them off from other crustaceans and other decapods.

Few crabs are specialized feeders; most are opportunistic omnivores (Marshall and Orr, 1960) and, hence, are highly vulnerable to the baited-trap form of capture. Studies of feeding, digestion, and assimilation (e.g., Caine 1975a,b. 1976; Klein Breteler, 1975a; Fuzessery and Childress, 1975; Hill, 1976; Vermij, 1977; Barker and Gibson, 1978; Elner, 1978; Williams, 1978; Wolcott, 1978) continue to support the following generalizations set forth by Yonge (1928), Vonk (1960), and Jennings (1972): (1) the relative dependence of feeding on specific physical and/or chemical mechanisms is associated with the ecology and the mode of existence of the species, (2) the mechanism for capture and digestion of food can be correlated with the niche an animal occupies, and (3) the classification of the feeding mechanism cannot be based on phylogenetic relationships; rather it is influenced by the nature of the food.

The capacity of a crab to survive in any given environment depends not merely on one physiological system (Robertson, 1960b; Lockwood, 1962) but on the interplay of several systems responding to interacting environmental factors (Spaargaren, 1975; Dorgelo, 1976). The blue crab, *Callinectes sapidus,* is an excellent example of how the integration of several regulatory mechanisms (osmoregulation, respiration, and nitrogen metabolism) supports its existence in, and migrations through, an environment that is unstable in both ions and dissolved oxygen (Mangum and Towle, 1976). This model becomes even more complex when the physiological fluctuations of molting (Robertson, 1960a) are superimposed on the integrated system.

Considerable study has been devoted to capacity and resistance adaptations of crabs to varying environmental conditions alone or in combination (Florkin, 1960; Edney, 1960; Bliss, 1968, 1979; Bliss and Mantel, 1968). Responses of crabs have been assayed in terms of survival (Miller and Vernberg, 1968; Diwan and Nagabhushanam, 1976), behavioral modifications (Rao, 1968; Taylor and Butler, 1973; Taylor et al., 1973), osmotic and ionic conformity or regulation (Copeland, 1968; Mantel, 1968; Tagatz, 1969; MacMillen and Greenway, 1978), cardiovascular and ventilatory activity (Redmond, 1968; Horn, 1968; Uglow, 1973; Young, 1973a,b; Spaargaren, 1974; Dimock and Groves, 1975; Coyer, 1977; Diaz and Rodriguez, 1977; McDonald et al., 1977; McMahon and Wilkens, 1977; Batterton and Cameron, 1978), and metabolism (Vernberg and Vernberg, 1968; Veerannan 1972, 1974; Venkatachari and Kadam, 1974; VanWinkle and Mangum, 1975; Ambore and Venkatachari, 1978; Burke, 1979). The significance of such adaptations to survival in the natural habitat has been summarized by Vernberg and Vernberg (1972), and in Vol. 8, page 335.

Growth in crustaceans is a complex, integrated process that dominates the life of the individual. Metabolism, behavior, reproduction, and neurosensory acuity are all affected by the required periodic replacement of the integument and by the involved physiological cycles (Passano 1960, 1961). The more unique adaptations of the molting process in crabs are those involving the localization of mineral resorption sites and ecdysial suture lines (Drach, 1933; Drach and Tchernigovtzeff, 1967), the modifications in the function of the pericardial sacs, and the reduction of the soft-shell stage in terrestrial forms (Bliss, 1968, 1979; Rao, 1968), the behavioral patterns associated with molting and mating (Hartnoll, 1969; Henning, 1975; Bliss et al., 1978), the physiological accommodations during molting (Bliss, 1951, 1953, 1979; Lewis and Haefner, 1976), and the integration of the reproductive and molting cycles (Adiyodi and Adiyodi, 1970; Swartz, 1978). As with most other physiological processes, molting and growth in crabs are influenced by the quality of food consumed as well as by environmental factors (Skinner and Graham, 1970, 1972; Klein Breteler, 1975a,b,c). Studies of this type are the key to the planning and development of mariculture projects, which seek to rear crabs from the egg stage to market size (Reed, 1969; Escritor, 1970, 1972; Bardach et al., 1972; Pagcatipunan, 1972; Welsh, 1974; Millikin, 1978) or which deal with the production of specialty products, such as the soft-shell crab (Haefner and Garten, 1974; Haefner and Van Engel, 1975).

Various aspects of crustacean reproduction have been extensively reviewed (Adiyodi and Adiyodi, 1970; Charniaux-Cotton, 1960; see also Chapter 5, Volume 8, and Chapter 3, Volume 9, of this treatise). The most pertinent studies that describe reproductive adaptations of Brachyura are those of Hartnoll (1968, 1969). Hartnoll (1968) distinguished two basic patterns of vaginal structure in Brachyura (simple and concave) and later correlated the structure with copulatory behavior patterns (Hartnoll, 1969). All crabs that possess simple ducts undergo a pre-ecdysial courtship and mate immediately after ecdysis. A majority of crabs with concave ducts copulate during intermolt after a brief courtship; some, however, do mate in the soft-shell stage. It appears to be of adaptive significance that the Grapsidae, Gecarcinidae, and Ocypodidae, which have terrestrial and semiterrestrial representatives, mate in the intermolt condition. Thus, they avoid the danger of water loss, which would occur during an extended mating period in the soft-shell condition. It is not known why mating is restricted to the period when the female has just molted in certain crabs, such as the Cancridae and the Portunidae, but the associated courtship is of considerable survival value. The hard-shell intermolt male offers considerable protection for the vulnerable female prior to, during, and following mating.

Mating and molting alternate during the reproductive cycle of most adult crabs. Molting and reproduction are major metabolic events involving cyclic mobilization of organic reserves from storage areas such as the midgut glands (hepatopancreas). Although these events are temporally separated, they are inseparably integrated through endocrine function (Adiyodi and Adiyodi, 1970; see also Chapter 3, Volume 9, of this treatise) and are commonly responsive to seasonal changes (Knudsen, 1964; Pereyra, 1966; Rahaman, 1967; Griffin, 1969; Hill, 1975), which influence fishery activity.

III. EXPLOITATION

Knowledge of optimal ranges of environmental factors for any given species and the capacity to predict the effects of environmental perturbations on the crab stocks are basic to the construction of the principles of fishery management and conservation of fishery stocks (Copeland and Bechtel, 1974).

A. General Concerns and Problems of Fisheries

Two concerns involving the conservation and management of sea fisheries of the world have been addressed by FAO (1968), Rothschild (1972), and Gulland (1976); world fish catches have been expanding rapidly, but the potential expansion of the familiar types is limited. Gulland (1976) predicted that the period of expansion would end by 1985.

Three major problems face world fisheries during this period of change (Jackson, 1972; Gulland, 1976). First, fishing pressure on many of the traditional, well-established, high-unit-value fisheries has been so high that only prompt adoption of proper control measures will prevent the drastic decline and near obliteration of stocks. With proper management, stocks should remain at a highly productive level as they are harvested in an efficient manner. Second, because continued local depletion of the more familiar stocks appears likely, it will be necessary to harvest less familiar stocks, the potential of which is very high. Some regions of the world have not yet reached maximum potential. For example, Panikkar (1969) predicted a substantial increase in yield for the Indian Ocean because offshore species were not being exploited and inshore stocks were underfished. More recent reports (Longhurst, 1971; Alverson and Patterson, 1974) have reemphasized that region's potential. Third, increased production of fish protein will depend on cultivation, particularly of species that feed directly on plants. McKernan (1972) and Rounsefell (1975), however, stressed the need for concern among fishery scientists. As the data clearly indicate, one species after another has been declining in abundance under increased fishing pressure.

The principal problems that must be solved in the development of a modern fishery for new species are technical and economic (McHugh, 1968). Methods of catching and processing must be developed at the same time that markets are being established, and the new species must be caught at a cost that will provide a profit in the market. For example, the tanner (snow) crabs (*Chionoecetes* spp.) enjoy nationwide recognition and popularity. Their future in the U.S. market appears limited by the uncertainty of supply and a price too high to maintain their competitiveness with king crabs (*Paralithodes camtschatica*) nationwide and with regional crab favorites in coastal areas (Anderson et al., 1977). One of the major deterrents to expanding the U.S. harvest of tanner crab resources from the Bering Sea has been the size of *C. opilio,* a smaller crab that is more expensive to process for a market requiring a product largely in meat form. Until domestic processor demand increases, the optimum yield of *C. opilio* may not be realized.

The application of technology has contributed to the impressive growth of fisheries, especially among the poorer countries (Gulland, 1976). Expansion of a fishery occurs as local fishing for a species gradually increases. This is followed by extension into more distant waters. Commercial landings can increase as (1) increased market demand parallels distribution and growth of an increasingly prosperous population, (2) catching methods improve, and (3) a manpower supply capable of using new methods exists, but without competition from well-paid alternative jobs ashore that make fishermen reluctant to tolerate the discomfort of going to sea. Indeed, it is the degree of economic and nutritional dependence of communities and geographic regions on fishing that will determine, to a large extent, how a fishery may or may not expand (Turvey and Wiseman, 1957; FAO, 1968).

B. Crab Fisheries

Fisheries exploitation of crab species has not reached anticipated levels because (1) crabs are often unavailable to most types of gear used by the fin fisheries industry and (2) a variety of cultural, social, political, and financial reasons have hindered the development of crab fisheries in regions in which potential resources exist (Longhurst, 1971).

1. LANDINGS REPORTS

The dilemma of world supply and demand for fishery products has been one of the principal on-going concerns of the fishing industry (FAO, 1968, 1971, 1972). The problem of demand exceeding world production of crab resources has been addressed by the Food and Agriculture Organization (FAO) of the United Nations. Summary reports, such as those by Longhurst (1971) and Alverson and Patterson (1974), have examined crab landings. They have indicated those species of crab and geographic areas that could contribute to increased supply and have estimated the potential maximum potential. Crab resources constituted 23–30% of of the world crustacean production from 1958 to 1969 (Table II). Longhurst estimated that untapped resources could increase world crab landings by 72%, providing that effective development of fisheries takes place (Table III). There was some indication of increased landings for certain species from 1970 to 1975 (Table III), and some evidence that world crab production was increasing (Tables IV and V). The irony is that the major deterrent to increased market demand has been a continual shortage of supply (Alverson and Patterson, 1974).

The degree of exploitation has been difficult to assess because of the nebulous quality or actual lack of data for certain fisheries throughout the world. Crab resources either have been ignored or have been considered as so small in importance that they are recorded in statistical "catch-alls," such as "miscellaneous Crustacea" or some similar category. For example, commercial fishermen, buyers, and shippers of crabs in the Chesapeake Bay region have misnamed the three following species of edible crab: the rock crab (*Cancer irroratus*), the Jonah crab (*C. borealis*), and the stone crab (*Menippe mercenaria*). This confusion has resulted in incorrect documentation in fishery statistics (Haefner, 1976). The *Cancer* spp. have only recently been separated in landing statistics (Table IV).

In California, *Cancer magister* is landed as the "market crab," whereas it is referred to as the "dungeness crab" in other markets. Three other crabs of the genus *Cancer* (*C. anthonyi, C. antennarius, C. productus*) are collectively referred to as "rock crabs" in California fisheries and are reported as such in the landing statistics (Table IV). This obviously simplifies the task of those concerned with buying and selling the product, but it places fishery biologists at a distinct disadvantage as they attempt to derive relationships between available stocks and landings of individual species (Gotshall, 1978b). This situation is improving as increased attention is paid to gathering accurate records. The number of individual species recognized in the landing statistics is generally increasing.

The extent of landings in the sport fisheries for crabs is virtually unknown throughout the world. It has been difficult to obtain accurate, commercial landing data (Gotshall, 1978b), but surveys of sport fisheries have been nonexistent or unpublished. Yet, all of the inshore, shallow water species are vulnerable to the sport fishermen, who are under no mandate to report their catch (Smith, 1978; Williams, 1979). Unpublished surveys have revealed large catches by sport fishermen. In 1958, for example, 199,725

TABLE II

Catches of Crustaceans (Tons × 10³, Live Weight)[a]

Crustacean group	1958[b]	1963	1964	1965	1966[c]	1967	1968	1969
Sea spiders, crabs	197	272	288	318	365	368	388	397
Lobsters, rock lobsters, spiny lobsters	79	90	92	92	92	88	96	. . .[d]
Squat lobsters, nephrops	32	38	41	43	45	51	50	59
Shrimps, prawns	502	671	670	663	697	757	771	. . .[d]
Miscellaneous marine crustaceans	36	53	67	69	77	102	106	. . .[d]
Total, all crustaceans	860	1140	1170	1190	1280	1360	1400	. . .[d]

[a] Modified from Longhurst (1971) and FAO Yearbooks of Fishery Statistics, 1973 and 1975. Courtesy of Fishing News Books Ltd., Surrey, England, and FAO, United Nations, Rome, Italy.

[b] 1958, "prawns and shrimps" includes "freshwater crustaceans."

[c] Up to 1966, "prawns and shrimps" included "other marine crabs."

[d] No data.

TABLE III

Summarized Average Annual Tabulation of 1958–1969 and Potential World Production of Crustacea (Tons × 10³)[a]

FAO region	1958–1969[b]			Potential[b]		
	Lobsters	Crabs	Prawns	Lobsters	Crabs	Prawns
I Northwest Atlantic	31.7	2.2	8.6	32.0	10.0	27.0
II Northeast Atlantic	29.9	13.2	65.4	65.3	20.5	82.8
III Mediterranean	4.0	+	23.0	4.0	+	50.0
IV North Pacific	+	208.0	41.1	+	250.0	131.2
V Western central Atlantic	13.3	76.9	125.0	21.0	80.0	160.0
VI Eastern central Atlantic	2.5	+	4.6	4.0	10.0	63.0
VII Western Indian Ocean	0.4	+	21.3	3.1	+	144.0
VIII Eastern Indian Ocean	8.6	+	58.9	10.0	+	98.1
IX Western central Pacific	1.7	39.3	304.1	5.0	150.0	544.1
X Eastern central Pacific	+	<1.5	65.0	5.0	32.5	75.0
XI Southwest Atlantic	3.5	15.8	37.1	4.0	21.0	86.0
XII Southeast Atlantic	10.7	+	2.0	11.5	+	7.0
XIII Southwest Pacific	12.3	1.0	4.5	14.5	1.0	4.9
XIV Southeast Pacific	0.1	14.2	11.5	0.1	66.5	20.0
Total	118.7	372.2	772.1	179.5	641.5	1492.1
		(1.26 × 10⁶)			(2.31 × 10⁶)	

[a] Modified from Longhurst (1971). Courtesy of FAO, United Nations, Rome, Italy, and Fishing News Books Ltd., Surrey, England.

[b] +, Recorded landings less than 0.1 × 10 metric tons.

blue crabs were captured in 3 months by 18,394 daytime crabbers in Great South Bay, Long Island, New York. This represents a small portion of the overall sport crab fishery in New York, where there is a more intense nighttime private boat and bank fishery (P. T. Briggs, Dept. of Environmental Conservation, Stonybrook, New York, United States, personal communication).

Minor bait fisheries exist for which there are few data. On Long Island, New York, all species of fiddler crabs (*Uca* spp.) are captured by some fishing stations operators; others trap the green crab (*Carcinus maenas*). Lobstermen keep the hermit crabs *Pagurus acadianus* and *P. pollicaris* and sell them to fishing station owners. In addition, fiddler crabs are imported from Florida. All are used as bait for tautog (*Tautoga onitis*) (P. T. Briggs, personal communication).

2. THE DOMINANT FISHERIES

Four fisheries (seven species) dominate the world crab landings, and all, at least in part, are captured in the coastal waters of North America (Table

IV). Three of the crabs are brachyurans: blue, *Callinectes sapidus* (Portunidae); dungeness, *Cancer magister* (Cancridae); queen, snow, or tanner, *Chionoecetes* spp. (Majidae). The anomuran king crab, *Paralithodes camtschatica* (Lithodidae), constituted the largest and most valuable single species of crustacean fishery until 1970. Blue crab landings equaled or surpassed those of the king crab from 1971 to 1975. Dungeness crab landings fell dramatically during the same period, whereas landing of snow crabs peaked and then trailed off (Table IV). "Other marine crabs" accounted for roughly one-third of the total crab production during the period 1963–1975 (Longhurst, 1971; Table IV). The overall importance of the top four fisheries has declined over the 1963–1975 period (Alverson and Patterson, 1974). They constituted 71% of the total landings in 1963–1964 and only 58% from 1970–1973, indicative of declining stocks of the major species (Peterson, 1973; Alverson and Patterson, 1974; Fisher and Wickham, 1976) as well as of increased catch of underutilized species. Note, for example, the recent increase in reported landings of *Maia squinado*, *Portunus* spp., *Portunus (= Neptunus) trituberculatus*, *Scylla serrata* and "other marine crabs" (Table IV).

3. PRODUCTIVE GEOGRAPHIC REGIONS

The most productive geographic region in terms of crab fisheries is FAO Region IV (FAO, 1972; Tables III and V), which encompasses all of the North Pacific Ocean from the cold waters north of Hokkaido to the California Current at the level of the California–Mexican border (Longhurst, 1971). The crab resource of the region is composed principally of the red king crab, *Paralithodes camtschatica*, which is taken by fleets from the United States, the Soviet Union, Japan, and Korea. Centers of productivity include the Okhotsk Sea, the Bering Sea, and the Gulf of Alaska. A decline in king crab catches (1967–1975) was attributed to year class fluctuation and overfishing (McMullen et al., 1972; Alverson and Patterson, 1974). This decline, however, forced attention upon the tanner (snow) crabs of the region (*Chionoecetes tanneri, C. opilio, C. bairdii, C. japonicus*), which account for about 50% of the Japanese crab landings. Snow crabs (C. bairdii) are now one of Alaska's most valuable shellfish resources (Bakkala et al., 1976; Bartlett, 1976b). As with the North Atlantic's C. opilio, the C. bairdii fishery expanded more rapidly than biological information could be gathered (Brown, 1971; Hilsinger, 1976; Bakkala et al., 1976). The dungeness crab, *Cancer magister*, is also taken in this region but to a far lesser extent (15–25% of king crab landings). Longhurst (1971) estimated a high potential productivity for this region (Table III), based on potential increased fishing effort on the stocks of dungeness and tanner crabs, as well as on the utilization of potentially commercial species, which include deep water species *Lithodes aequispina* (Anomura: Lithodidae) and *Geryon quinquedens*

TABLE IV

Catches (Metric Tons × 10³) of Marine Crabs by Species and by FAO Regions[a]

Species	FAO region	1970[c,d]	1971[c,d]	1973[c,d]	1975[c,d]
King crab, *Paralithodes camtschatica*	Northwestern Pacific	37.2	29.2	24.9	17.7
	Northwestern Pacific	35.6	39.6	34.9	45.4
	Total	72.8	68.8	59.8	63.1
Mud crab, *Scylla serrata*	Asian	0.2	0.3	0.2	1.3
	East Indian	0	0	0	1.2
	Western central Pacific	4.3	4.6	4.0	4.6
	Total	4.5	4.9	4.2	6.1
Queen crab, *Chionoecetes opilio*	Northwest Atlantic	7.7	6.8	9.9	7.0
Pacific snow crab, *Chionoecetes bairdii, C. japonicus, C. opilio,* and *C. tanneri*	Northwest Pacific	36.0	28.2	20.8	16.1
	Northeast Pacific	23.9	20.0	40.4	29.0
	Total	59.9	48.2	61.2	45.1
Atlantic rock crab, *Cancer irroratus*[b]	Northwest Atlantic	0.8	0.4	0.7	0.7
Dungeness crab, *Cancer magister*	Northeast Pacific	24.5	17.3	6.4	8.2
	Eastern central Pacific	3.1	2.5	0.2	0.2
	Total	27.6	19.8	6.6	8.4
Edible crab, *Cancer pagurus*	Africa	–	–	–	+
	Northeast Atlantic	21.2	23.2	19.6	21.1
	Eastern central Atlantic	0	0	–	+
	Mediterranean	0.1	0.3	0.1	0.1
	Total	21.3	23.5	19.7	21.3
Jonah crab, *Cancer borealis*[b]	Northwest Atlantic	–	–	–	0.2
Pacific rock crab, *Cancer productus*	Eastern central Pacific	0.3	0.2	0.5	0.6
Stone crab, *Menippe mercenaria*	Western central Atlantic	1.3	1.0	1.3	1.3
Red crab, *Geryon quinquedens*	Northwest Atlantic	–	–	0.1	0.3
Swimming crab, *Portunus puber*	Northeast Atlantic	0.4	0.4	0	0.6
Swimming crabs, *Portunus* spp.	Asia	–	–	0	+
	Eastern central Indian	0	0	0	0.2
	Western central Pacific	14.3	15.6	17.6	16.3
	Total	14.3	15.6	17.6	16.5

Species	Region				
Blue crab, *Callinectes sapidus*	Northwest Atlantic	38.2	39.8	31.2	30.9
	Western central Atlantic	29.8	29.7	32.4	34.3
	Total	68.0	69.5	63.6	65.2
Siri crab, *Callinectes danae*	Southwest Atlantic	3.2	3.1	3.4	3.6
Gazami crab, *Neptunus trituberculatus*	Northwestern Pacific	3.7	5.2	12.4	17.9
Green crab, *Carcinus maenas*	Northwestern Atlantic	0	0	0	0
	Northeast Atlantic	0.6	0.6	0	+
	Eastern central Atlantic	–	–	–	–
	Southeast Atlantic	–	–	–	+
	Total	0.6	0.6	0	+
Med. green crab, *Carcinus aestuarii*	Mediterranean	0.1	0.1	–	0.2
Southern King crab, *Lithodes antarcticus*	Southwest Atlantic	0.2	0.3	0.2	0.4
Spinous spider crab, *Maia squinado*	Northeast Atlantic	0.2	0.2	6.0	7.1
	Eastern central Atlantic	–	–	–	+
	Mediterranean	0.1	0.1	0.1	0.1
	Total	0.3	0.3	6.1	7.2
Other marine crabs	Northwest Atlantic	0.2	0.2	0.3	+
	Northeast Atlantic	–	–	–	+
	Western central Atlantic	0.4	0.6	1.8	1.8
	Eastern central Atlantic	1.1	0.9	0.2	0.2
	Mediterranean	0.6	0.5	0.5	0.5
	Southwest Atlantic	13.9	18.1	17.3	18.3
	Southeast Atlantic	–	–	–	2.2
	Western Indian	0.3	0.3	0.2	0.4
	Eastern Indian	0	0.1	0.1	0.1
	Northwest Pacific	29.3	32.4	59.2	59.6
	Western central Pacific	27.3	26.1	24.7	28.2
	Eastern central Pacific	0.2	0.3	0.1	0.4
	Southwest Pacific	0.1	0.1	0.1	0.2
	Southeast Pacific	1.5	1.1	1.3	2.2
	Total	74.9	80.7	105.8	114.9
Total crabs		362.3	349.8	373.5	381.2

[a] Courtesy of FAO, United Nations, Rome, Italy. Data from FAO Yearbook of Fishery Statistics, 1973 and 1975. Compare with Table III.
[b] Rock crab designation ambiguous in early 1970s; included *C. borealis* in some years.
[c] –, No data.
[d] +, Recorded landings less than 0.1 × 10³ metric tons.

TABLE V

Average Biennial Production of Crabs (Metric Tons × 10³) from 1970–1975[a]

FAO region	1970–1971	1972–1973	1974–1975
I Northwest Atlantic[b]	47.1	44.6	44.0
II Northeast Atlantic	23.4	25.1	27.3
III Mediterranean	1.0	0.8	1.1
IV North Pacific	181.1	198.0	199.4
V Western central Atlantic[b]	31.4	33.8	37.3
VI Eastern central Atlantic	1.0	0.2	0.2
VII Western Indian Ocean	0.3	0.2	0.4
VIII Eastern Indian Ocean	0.1	0.1	0.4
IX Western central Pacific[c]	46.1	43.2	53.3
X Eastern central Pacific	3.3	0.8	0.9
XI Southwest Atlantic	19.4	19.9	22.3
XII Southeast Atlantic	0	0	5.9
XIII Southwest Pacific	0.1	0.1	0.3
XIV Southeast Pacific	1.7	1.8	2.4
Total	356.1	368.9	396.6

[a] Courtesy of FAO, United Nations, Rome, Italy. Data from FAO Yearbook of fishery statistics, 1973 and 1975. Compare with Table III.

[b] The discrepancy between Tables III and V for the landings in these regions are related to the reported landings for blue crab (See Table IV). Note that the total combined landings (I,V) are similar to that in Table III.

[c] Increased landings due to "other marine crabs" (see Table IV).

(Geryonidae) and western Pacific species (*Erimacrus isenbeckii, Telmessus cheiragonus, Paralithodes brevipes*). Longhurst (1971) failed to mention the potential for the blue king crab, *Paralithodes platypus,* which is second in abundance to *P. camtschatica* in Alaskan waters, particularly the Bering Sea (Bartlett, 1976a; Kaimmer et al., 1976).

Longhurst (1971) identified the northwest Atlantic region (Tables III–V) as one in which the major crab resource has been the rock crab, *Cancer irroratus* (Caddy et al., 1974; Krouse, 1978). He predicted a fivefold increase in the crab landings based on the untapped resources of the green crab (*Carcinus maenas*), the snow crab (*Chionoecetes opilio*), and the deep-sea red crab (*Geryon quinquedens*), as well as the underdeveloped fishery for rock crabs. He made no mention of the Jonah crab, *Cancer borealis,* which also exists in harvestable stocks and has been landed, sometimes under the guise of "rock crab" or "miscellaneous" (Holmsen, 1973; Rathjen, 1974; Haefner, 1976; Krouse, 1980).

The crab fishery in the northeast Atlantic is based on the European edible crab *Cancer pagurus,* which supplies the bulk of the landings (Tables II–V). Longhurst's (1971) estimated 50 to 100% increase in landings has materialized as the fishery for *C. pagurus* expanded (Brown and Bennett, 1978) and as other species (*C. maenas, Maia squinado,* and portunids) were utilized. Earlier Graham (1956) cited *M. squinado* as an unsaleable crab, which invaded *C. pagurus* pots in large numbers.

Longhurst (1971) estimated that the fishery for blue crabs, *Callinectes sapidus,* in the western central Atlantic had reached a point near its maximum sustainable production. The increase in landings of this species in the western Gulf of Mexico was the only reason he estimated a potential increase in the resource. However, there are a number of other species of crabs, which could increase yield in this region, if fisheries were developed. *Cancer borealis* and *Geryon quinquedens* are captured as incidental species in the Virginia and Maryland offshore lobster fishery. *Cancer borealis* is retained during the last few days of the offshore trips only if vessel holding space is available.

Geryon has been landed, but in negligible quantities in the "miscellaneous" category. *Cancer irroratus* is a by-catch in the Virginia sea bass (Shotton, 1973) and blue crab fisheries (Terrettà, 1973; Haefner and Van Engel, 1975). Efforts to generate interest in these species in the Chesapeake region have met with moderate success. The xanthid crab *Menippe mercenaria* and several species of portunids could also support expandable fisheries in the more southerly reaches of the region (Savage and Sullivan, 1978).

4. REGIONS OF LOW PRODUCTIVITY

Several regions have had very low (in some cases only subjective, undocumented) crab productivity (Tables III–V). High potential productivity has been estimated for those subtropical regions (Indian Ocean, central Pacific, central and southwest Atlantic) known to support sizeable stocks of potentially marketable portunid crabs (Olsen *et al.,* 1978). Alverson and Patterson (1974) suggest that the importance of two tropical portunids, the blue swimmer [*Portunus (= Neptunus) pelagicus*] and mangrove crab (*Scylla serrata*), has been hidden by inadequate catch statistics. These species dominate the crab landings in India, Burma, Pakistan, Bangladesh, Malaysia, Vietnam, and China. The 1972 landings of each of these species from the Indo-Pacific region was estimated to be in excess of 35,000 metric tons (MT) or 10% of the world's production of crabs. In India alone from 1956–1965, portunids constituted an annual average catch of 3800 MT (Rao, 1969). Unfortunately, these values were not specifically recorded in FAO catch statistics at that time. Crabs have been underexploited in India and rank as the least important component (<3%) of the crustacean fisheries (Panikkar, 1969; Rao, 1969). In Taiwan, Indonesia, and the Philippines, efforts have been made to culture *Scylla serrata* as a subsidiary crop in milkfish (*Chanos chanos*) ponds (Bardach *et al.,* 1972; Pagcatipunan, 1972;

Escritor, 1970, 1972). In Queensland, Australia, the fishery for these species amounts to about 400 MT annually (B. J. Hill, C.S.I.R.O. Australia, personal communication). Large estuarine populations of *Scylla serrata* also exist on the eastern coast of South Africa (Hill, 1975, 1979). The expanding fishery is homologous to that of *Callinectes sapidus* on the east coast of North America. Longhurst (1971) reasoned that if the relatively short coastline of the eastern United States produces an average of 75×10^3 tons of portunids (blue crabs) per year at close to maximum sustainable levels, it is not unreasonable to suppose that the subtropical regions, with a much greater shelf area and length of coastline, could produce twice this quantity.

Longhurst (1971) found it impossible to predict potential landings for the low productive regions of the Mediterranean and the southeast Atlantic. The former region is known to support small stocks of crabs (Alverson and Patterson, 1974).

The production figure for the northwest Atlantic (Longhurst, 1971) is potentially quite low. The values presented in Table V for this region are misleading in that a change in reporting has occurred for northwest and western central Atlantic. What appears to be an increase in catch for one region is actually due to landings formerly reported for the other region. All of the species mentioned above (rock, green, snow, red crabs) are taken as incidental catches in the lucrative inshore and offshore American lobster (*Homarus americanus*) fishery. Many of the crabs are processed on board and may not enter the wholesale market, being distributed among local consumers.

On the other hand, large quantities of certain species, such as *Cancer borealis* and *C. irroratus,* are discarded at sea because holding facilities on board the catch vessels are inadequate to support an incidental catch or because processors are not equipped to handle them. There is also a reluctance on the part of many lobstermen to venture into a diversified fishery when the market is questionable. There have been cases in which commercial fishermen have rejected offers to bring in the incidental catches, even when they have been assured of an open market. Some intentionally fish to avoid catching crabs (Massachusetts Lobstermen's Association, 1974).

Considerable effort has been expended to promote commercial and consumer interest in the underutilized resources in the northwest Atlantic. The effort has included descriptive lists of underexploited species (Rees, 1963; Wiler, 1966; Holmsen, 1973; Pinhorn, 1976), the creation of regional surveys (Caddy *et al.,* 1974; Massachusetts Lobstermen's Association, 1974) and national programs (Rathjen, 1974), exploratory fishing and attempts to estimate stock size (Schroeder, 1959; McRae, 1961; Powles, 1966, 1968a,b; Fullenbaum, 1970; Haefner and Musick, 1974; Ganz and Herrmann, 1975; Stasko, 1975; Wigley *et al.,* 1975; Haefner, 1978), and recommendations for handling, processing, and marketing (Varga et al., 1969; Haefner et al., 1973; Holmsen and McAllister, 1974; Marchant and Holmsen, 1975; Van Engel and Haefner, 1975).

Similar efforts are undoubtedly occurring worldwide as other potential fisheries are investigated (for example, see Clarke, 1972; Fukuhara, 1974; LeLoueff et al., 1974; Anderson et al., 1977; Cayre et al., 1979), but such efforts are likely to go unnoticed as many of the data are relegated to regional manuscript reports, which usually receive limited distribution.

The Food and Agricultural Organization (1977) has assembled a Code of Practice for Crabs, which contains technological and hygienic requirements based on good commercial practice and on the FAO/WHO recommended International Code of Practice—General Principles of Food Hygiene. It includes guidelines for handling crabs at sea as well as on shore.

Unfortunately, development of new fisheries is an extremely slow process. Even organized fisheries development programs, such as that described by Rathjen (1974), encounter reluctance on the part of fishermen to enter into new fisheries. It is recognized that exploitation of underdeveloped resources requires a comparatively high input of risk, fishing, and processing technology. Only a well-coordinated effort with financial incentives and social rationales (Miller, 1979) will show positive results.

In some cases, a fishery will develop fortuitously. A good example is the fishery for snow crab (Chionoecetes opilio) in the Canadian western Atlantic. In 1960, very little biological information was available on this species when a small incidental fishery (12,000 lbs) began (Powles, 1966, 1968a,b; Wilder, 1966). When it was recognized that this species was an additional source of crab meat for a depleted market, fishing intensity increased. In Newfoundland alone, landings increased from 93 MT in 1968 to 3400 MT in 1974 (Pinhorn, 1976). The provincial and federal fisheries agencies of Canada were suddenly confronted with a fishery for which limited biological data were available in 1967 (Watson, 1969). A concerted research effort has since provided some basis for fishery management practices. Recognition that controlled entry is the most important management need of the fishery (Pinhorn, 1976) testifies to the rapid expansion of the fishery and demand for the product on the eastern coast of the United States, if not on the western coast (Anderson et al., 1977).

5. GEAR

Historically, a wide variety of methods has been employed to catch crabs but, with few exceptions, gear relying on the use of bait has been most effective, both from the standpoint of catch efficiency and of fishery management. The effectiveness of the baited gear is directly related to the omnivorous, scavenging, feeding behavior of crabs. The fact that crabs spend most of their time crawling on the bottom also simplifies catch efforts by deploying gear that will rest on the bottom and, when baited, lure crabs (Bennett, 1974b; Fuzessery and Childress, 1975; Hipkins, 1972; Pearson and Olla, 1977) into a trap from which escape is controlled. The myriad types of crab fishing gear in use throughout the world are reviewed in Firth (1969), Rao (1969), Sainsbury (1971), Von Brandt (1972), Nedelec (1975), and Rounsefell (1975).

A sampling of gear used in the United States is presented in Table VI. Other types not listed, but which are used locally or have been discontinued, include dip nets and peeler pots (Van Engel et al, 1973), entangling nets (Morris, 1969; Idyll and Sisson, 1971), spears (Warner, 1977), land traps, and hand capture (Feliciano, 1962).

Trawl nets, so common in finfish and shrimp fisheries, are unsatisfactory for crab fisheries. A synoptic comparison of three bottom quantitative sampling techniques by Uzmann et al. (1977) revealed a marked inefficiency of trawl gear in capturing benthic reptant decapods. Trawls are also indiscriminate as to the stage of animals caught. Size, sex, and molt stage selection is difficult to achieve in trawling operations, and continued use of such gear can result in overfishing. Small-size and soft-shelled crabs often form a greater proportion of trawl catches as opposed to trap catches. On-deck sorting of the catch is an inefficient use of crew time, and return of weakened specimens to the sea most likely represents a potential loss to the stock (Powles, 1968b). It has also been demonstrated that the quality of crabs collected by trawl is inferior to that of crabs taken in pots (Haefner and Musick, 1974). Trawls, then, have become the exception in U.S. crab fisheries as most regions have banned their use (Table VI).

Tangle nets are another form of indiscriminate capture, but they are very efficient at capturing their intended prey—the tanner and king crabs (Idyll and Sisson, 1971; Morris, 1969). They were so effective that the crab stocks were in danger of depletion. Outlawing this gear permitted the stocks to rebuild to biologically safe levels (Morris, 1969).

The blue crab (*Callinectes sapidus*) may be subject to the most diverse kinds of fishing gear for any single species (Van Engel, 1962; Isaacson, 1963; Tagatz, 1965; Carley and Frisbie, 1968; More, 1969; Adkins, 1972; Van Engel et al., 1973; Perry, 1975; Warner, 1976; Eldridge and Waltz, 1977; Eldridge et al., 1979; Bishop et al., 1983). Essentially, the various types of fishing gear have been developed specifically to harvest each of several different life history stages of the blue crab. In the more logical situations, regulations governing the use of such fishing gear are based on knowledge of the biology of the species, in particular, the rate of growth and the reproductive potential.

The baited crab pot, or trap, is fished throughout the estuarine area and in coastal embayments during the warmer part of the year when the hard-shell

TABLE VI

Crab Fisheries of the United States Arranged by Method of Capture for 1967[a]

Gear, species, area	Landings (1000 lb)
Otter trawl	
Blue crab	
North Carolina–Georgia	12,038
Pot	
Green crab	
Maine–Rhode Island	65
King, tanner crabs	
Alaska	127,833
Dungeness crab	
Alaska–California	42,562
Blue crab	
New Jersey–Texas	100,204
Stone crab	
South Carolina–Florida	924
Pound net	
Blue crab	
Virginia	2,379
Crab dredge	
Blue crab	
New Jersey–Virginia	15,062
Crab scrape	
Blue crab	
Chesapeake Bay	1900
Dip net	
Blue crab	
Maryland–North Carolina	213
Brush trap	
Blue crab	
Louisiana	58
Drop net	
Blue crab	
Louisiana	605
Trotlines	
Blue crab	
Chesapeake–North Carolina	16,615

[a] Courtesy of C. V. Mosby Co., St. Louis, Missouri. Modified from Rounsefell (1975).

crabs (intermolt stage C; Drach and Tchernigovtzeff, 1967) are actively feeding and thus are attracted to the bait in the trap (Bennett, 1974b). The pots are constructed of a wire mesh large enough to allow undersized crabs to escape or are outfitted with escape ports (Eldridge et al., 1979). In regions where two different species are present but where sound management dictates the exclusion of one species or the other, entrances and escape vents of varying sizes and shapes may be employed (Stasko, 1975; Krouse, 1978).

The trotline consists of a baited line deployed on the bottom in shallow-water reaches of the bay. As the line is slowly brought on board the catch boat, the crabs are retrieved by a dip net just prior to the moment the bait breaks the surface of the water. Because the crabs are sorted by hand, all discards may be conveniently returned to the water.

The crab dredge is used in Virginia during the winter months in those higher salinity reaches near the mouth of the Chesapeake Bay, where the adult female crabs overwinter in a dormant state. These females have spawned during the previous summer and normally will die prior to the next spring. Dredging results in extensive damage to the individual crabs, but the catch is processed for its crab meat. Few, if any, dredged crabs reach market as a whole crab product.

The crab pound net is a stationary device constructed along the shoreline. It is designed to capture peeler blue crabs (intermolt stage D_1-D_3; Drach and Tchernigovtzeff, 1967), which migrate into protective shallow water eel grass (Zostera sp.) beds prior to ecdysis. Because crabs "follow" barriers, they are guided along a leader fence and through a funnel into the pound from which they seldom escape. These peelers are then taken to holding floats and retained until they molt. At such time, they produce the soft-shell crab (intermolt stage A_1; Drach and Tchernigovtzeff, 1967), a highly regarded gourmet item in the Chesapeake Bay region.

The crab scrape is a toothless dredge, which is dragged through shallow-water eel grass beds for the express purpose of capturing crabs that have molted but that have not progressed beyond the soft-shell stage. It is commonly used along the bayside of the eastern shore peninsula of the Chesapeake Bay, but is seldom used elsewhere.

The peeler pot is a unique device designed to capture penultimate instar female blue crabs. These crabs are peelers, which will undergo a terminal molt resulting in a mature crab capable of mating. Female peeler crabs are not actively feeding and so are not attracted to food as bait in the peeler pot. Rather, they are attracted to the live adult male blue crab, which is placed in the upper chamber as bait. The principle of pheromone activity is most likely functioning here, as it does in the attraction of males to premolt penultimate females (Ryan, 1966; Kittredge et al., 1971; Gleeson, 1980).

Dip nets, brush traps, and drop nets are most frequently used in the recreational fishery for blue crab, both hard-shell and soft-shell (Smith, 1978). Accurate figures on recreational yield are not available, but such yield is suspected to be significant even when compared to commercial landings (P. T. Briggs, Dept. Environmental Conservation, Stonybrook, New York, United States, personal communication).

As with finfisheries, most crab fisheries are subject to seasonal efforts, which reflect the biological cycles of the crab as well as control regulations, which should be based on the knowledge of life history of the species in question. Certain types of gear are restricted to specific seasons for good reasons. The crab dredge will work in winter when the crab is sluggish; it is a very poor choice in summer when the blue crabs can avoid the gear. Pots function well in summer when crabs are actively feeding; they do not attract crabs during the inactive winter condition (Bennett, 1974b).

Other conditions might be involved in seasonal restrictions of fishing effort. Certain regions may be placed offlimit, such as the spawning sanctuary for blue crabs in the lower Chesapeake Bay, where fishing for ovigerous females is prohibited during a 4-month period each year. An existing protective regulation, the sanctuary continues to be debated even among fishery scientists. Some are of the opinion that it is ineffective and unnecessary (Van Engel et al., 1973).

Almost all crab fisheries follow a seasonal pattern of effort, which is imposed by choice or by regulation to protect the stocks. In the dungeness crab fishery, a closed season corresponds to a peak incidence of molting. This serves to protect the crabs, which are in a highly vulnerable condition (there is little public demand for soft dungeness crabs as compared to that for the soft blue crab); but it has also been recognized that molting crabs do not actively feed and are thus not attracted to baited traps, and the meat in recently molted crabs (papershell stage B_1; Drach and Tschernigovtzeff 1967), is low in yield and of poor quality (Dassow, 1969). Similar regulations exist for the king crab, edible crab, and snow crab industries (Graham, 1956; Morris, 1969; Fukuhara, 1974).

There are some fisheries in which cessation or reduction of effort is due to conditions unrelated to the biology of the target species. Weather often dictates closure, but other problems such as the heavy growth of filamentous blue-green algae on the fishing gear in the stone crab (Menippe mercenaria) fishery (Savage et al., 1975) hamper operations to the point of closure.

Chapman and Smith (1977) provide evidence that crab catch is affected not only by the bait in the trap but by the presence of dead individuals of the target species. Reduction of catch of Cancer pagurus by 54% occurred when crushed whole crab was added to the bait. This tends to support Hancock's

(1974) suggestion that chemically induced intraspecific avoidance responses are well developed in some crustaceans.

6. PROCESSING

The crab industry generates three marketable products: crab meat, whole crab, and crab waste. Because almost all crabs caught are processed for crab meat, a substantial quantity of crab waste is produced. Until recently, almost all the wastes were discarded or disposed of in the most convenient fashion, that is, dumped into the water adjacent to the processing plants. Waste-water control practices and the search for ways to utilize and dispose of the wastes has resulted in a market for the product. The use of the whole crab as a product depends on the species, size, sex, and season.

Acceptance of a species into the crab meat market depends on its size, hardness of the shell, meat yield, and flavor. Most processing operations have been single species operations. That is, their machines were built to process crabs of a particular size range and shell thickness. Thus, there is an understandable reluctance to expend capital funds for refitting old machinery or for purchase of new machinery to process an unproven product. The use of existing equipment on a markedly different product does not, however, always produce favorable results. King crab butchering machines proved to be quite unsatisfactory for processing tanner crabs (Brown, 1971). Similar problems exist in hand-picking operations. Pickers become proficient with one species and balk at having to work with a totally different crab. Unless machines are available, the processor refuses to buy the crab, and the doors to the fishery are closed at the dock.

In most blue crab meat processing plants, the steps are essentially the same. The steps include weighing, cooking, cooling, picking, packaging, and disposing of the waste (Lee et al., 1963). There are, however, various handling patterns of 28 possible steps depending on the product mix (e.g., backfin lump pick, regular meat pick, claw pick, ice pack, pasteurized) (Lee and Sanford, 1964).

Other species, such as red crab and dungeness crab, may be butchered prior to cooking (Hoopes, 1973; Holmsen and McAllister, 1974). Butchering removes the back shell, viscera, and gills. The shoulder sections with legs and claws attached are then cooked and chilled. After chilling, the sections may be picked (shucked) for canning or frozen for shipment. Picking is still performed by hand, although various types of mechanical devices have been tried with varying success. Picked meat in virtually all crab processing plants is floated through a brine solution to facilitate separation of shell fragments. The meat may then be sold fresh, pasteurized, or frozen.

Specialty products are especially appealing to the customer, and they are more economical to the processor because they usually require less handling in the plant (Holmsen and McAllister, 1974). These products include individually quick frozen claws, and whole, cooked, frozen glazed crabs. Even less handling is required if the whole live crab can be shipped considerable distances to satisfy a strong market demand (Hoopes, 1973; Barnett et al., 1969; Angel et al., 1974). One of the most valuable specialty products is the soft-shell blue crab, a high demand item that has supported its own industry in the Chesapeake Bay (Lee and Sanford, 1962; Haefner and Garten, 1974). Very few, if any, other areas of the world deal in soft-shell crabs.

Meat yield is one of the most serious concerns to the industry. If the crab has a low meat yield, it will never reach the processing plants. Research in the development of machines that will economically increase meat recovery from presently utilized species is a continuing effort. Meat recovery from commercial operations, regardless of the method, seldom achieves the yield derived during semilaboratory conditions (Holmsen and McAllister, 1974).

By-products of crab processing plants include crab waste meal, whole shells, chitin, protein concentrates, recovered meat, and products from viscera (Mendenhall, 1971). Waste meal is the oldest and most widely used commercial by-product of the shellfish industry. It has been used as a fertilizer and in feed for cattle, swine, and poultry, the principal benefit being in the mineral and protein content. Demand for waste meal remains low and is confined to the existing markets, but recent studies suggest their use as fish food supplement (Anonymous, 1974, Simpson, 1978). Whole shells are used for serving stuffed crab and as ashtrays, but a sound market would require bulk use. Their use as abrasives has been suggested, but several studies have failed to demonstrate the value of this. Chitin can be isolated from crustacean shells and is sought after for the chitosan content, which has varied uses, such as in dialysers, photographic emulsions, artificial fabric, printing inks, adhesive, cosmetics, and anticoagulents (Mendenhall, 1971; Muzzarelli and Pariser, 1978).

Protein in crab wastes is present as meat and tissue adhering to the shell and as a complex in the shell structure. Extracted protein is of high purity and could be used for vertebrate food (Mendenhall, 1971).

Innovations in the treatment of shellfish wastes has resulted in the recovery of meat scraps remaining after the usual picking operation. The use of continuous flow centrifuge or perforated steel drum flesh separators can recover enough meat to increase the overall yield by up to 20% (Holmsen and McAllister, 1974).

Crab livers (midgut glands, hepatopancreas) have been sold in Japan as a delicacy and for extraction of vitamins. The ovaries from mature adult female blue crabs are used in a regional (Virginia) delicacy called she-crab soup.

C. Fishery Biology

1. PREDICTION AND MODELING

The ability to predict recognized cyclic variations in natural populations has long been the goal of population ecologists (Pearson, 1948; McHugh and Ladd, 1953; Rothschild *et al.*, 1970). The application of mathematical modeling concepts to the problems of population variability has led to the development of a unique methodology, which shows promise as a tool in fishery biology. Ricker (1954) simulated cyclic populations with stock–recruitment relationships and estimated the effect of exploitation on amplitude and period of oscillations of the populations. The principals set forth in Ricker's (1954) study formed the basis for many other simulations of fish populations in subsequent years.

As computers grow in size and in ease of handling, so too does the discipline of mathematical modeling. Admittedly, theoretical models are approximations of natural populations, but they provide insight into the kinds of data and numerical values that must be acquired for theoretically derived results to be applicable to the study (Botsford and Wickham, 1978). Then too, models are flexible and can continually be manipulated toward realism. For example, the age-specific model of Botsford and Wickham (1978) of dungeness crab populations was an improvement over stock-recruitment models in that density-dependent inter-age relationships could be based on the specific behavior of individuals in the populations, and the dynamics of age structure were included (Fig. 4). The study suggested that fishing may cause the population to become unstable and revealed one density-dependent mechanism (cannibalism). Catch records (Fig. 5) indicated an abundance of legal crabs in each of the past 2 years in excess of any previous year. If a mechanism for suppression of survival of young by older age classes exists, such as cannibalism of the young by older crabs, low levels of abundance would be expected in the next few years. Botsford and Wickham (1978) recommended that fishery management policies be reevaluated following further use and development of density dependent models. McKelvey *et al.* (1980) extended this approach by modeling the *Cancer magister* fishery of California. They examined the plausibility of various alternative biological control mechanisms by incorporating them into a multistage dynamic recruitment model. They admitted that an over-compensatory natality function could be responsible for cyclic year class strength and landings in the dungeness crab population, but they concluded that neither natality nor juvenile cannibalism are significant factors in the population dynamics of *C. magister*. The use of iterative modeling in the simulation of management manipulations of crab populations is generally

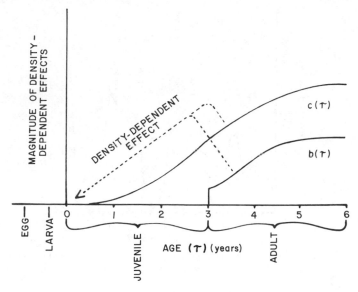

Fig. 4. A schematic example of the age-specific, density-dependent model of the northern California dungeness crab fishery. Age 0 is the time of metamorphosis from the larval stage. The relative effects of older animals through reproduction and the density-dependent mortality on the number at age 0 are indicated by functions b (τ) and c (τ), respectively. These functions indicate the effect per number present at each age. (From Botsford and Wickham, 1978, courtesy of the Department of Fisheries and Oceans, Canada.)

accepted, but questions have been raised concerning the interpretations of the models (Botsford, 1981; McKelvey, 1981).

Sound management practices in any given fishery require that the reproductive biology of the species be described and related to age and growth data for the population. When this has been accomplished, an educated decision can be made on the selection of a minimum legal size, such that stock productivity will not be adversely affected. According to Fukuhara (1974), an ideal management system would include the following: (1) maintenance of the standing stock of mature female crabs at a high level but one that does not depress growth, survival, and fecundity due to density related factors and (2) selection of a retention size in the catch such that the spawning stock contains an adequate number and size of mature males to copulate with all or most of the mature females. His system was based on a model of Bering Sea king crab stocks for which data on size class and sex ratios were available for all four seasons.

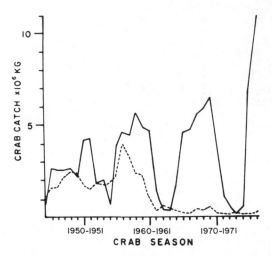

Fig. 5. Total catch of dungeness crab in northern and central California. (———) Northern California; (---) Central California. (From Botsford and Wickham, 1978, courtesy of the Department of Fisheries and Oceans, Canada.

2. AGE AND GROWTH

The analysis of year class structure is a standard procedure in fin-fishery research (Cassie, 1954; Rounsefell, 1975), but it has been difficult to apply to crustacean population investigations because of the inability to directly determine the age of crustaceans (Yano and Kobayashi, 1969). Biologists have tried to estimate age of crustaceans indirectly through probability analysis of population size-frequency distributions and molt increments. Powles (1968a,b), for example, used the method of Harding (1949) to resolve the component normal distributions of *Chionoecetes opilio* from sample data with a polymodal distribution. Thomas (1973) used probability analysis on modal group data for *Homarus americanus* based on Cassie's (1954) method. Because the modal groups represented year classes and/or molt groups (Fig. 6), the method is not universally accepted by crustacean biologists, but similar population analyses have been made throughout the years (MacKay and Weymouth, 1935; Butler, 1961; Farmer, 1973; Klein Breteler, 1975b,c; Haefner, 1978; Reilly and Saila, 1978; Donaldson, 1981). Some have been more successful (conclusive) than others.

Fig. 6. Accumulative percentages of lobster length plotted on probability paper for 1969 and 1970. Solid lines from the length frequencies on the ordinate to the oblique line crossing the 50 accumulative percent line on the abscissa designate the modes, whereas the dash lines represent the standard deviation about this mode. Cumulative percent frequencies of (O) females, and (X) males. (From Thomas, 1973.)

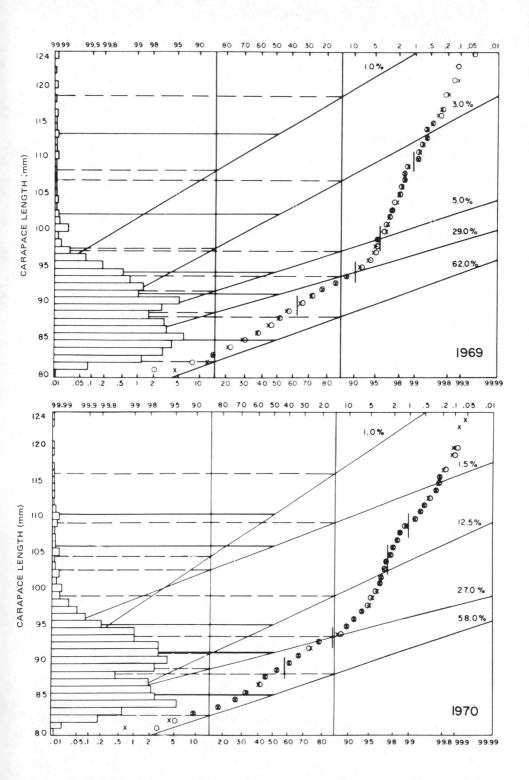

The analysis of discontinuous growth of crustaceans has also provided a basis for mathematically predicting growth and estimating maximum size (Hiatt, 1948; Kurata, 1962; Hancock and Edwards, 1967; Bennett, 1974a; Mauchline, 1976). Such data are vital to the analysis of populations. The Hiatt growth diagram (Figs. 7 and 8) is analogous to the Ford–Walford diagram used in growth analyses of fin fish (Ford, 1933; Walford, 1946), but they are not directly comparable. The intermolt periods of a crustacean increase logarithmically and produce a log time scale within the Hiatt growth diagram. The Ford–Walford time scale is linear, usually measured in calendar years, and it is therefore probably an arbitrary scale relative to that of the growth processes of the fish (Mauchline, 1976).

3. SIZE AT SEXUAL MATURITY

Biometric analysis of the growth of body parts in relation to one another is an important component in crab fishery management investigations. Crustaceans undergo changes in the form of body and parts during growth, which correspond to distinct growth phases (Hartnoll, 1982). During each phase, the different organs grow relative to each other according to the simple relations of allometry. These phases are also separated from one another by critical stages of brief duration when the growth pattern abruptly changes. These changes often correspond to shifts in biochemistry and endocrine function (Teissier, 1960). Because allometric growth can be mathematically described (for reviews, Teissier, 1960 and Hartnoll, 1982, also Chapter 3,

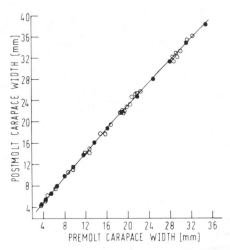

Fig. 7. Hiatt growth diagram for *Pachygrapsus crassipes* males. Open circles, data from Hiatt (1948); closed circles, points calculated by Mauchline (1976). *Marine Biology (Berlin)*, courtesy of Springer-Verlag, New York.

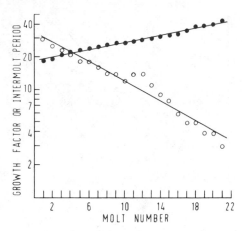

Fig. 8. *Pachygrapsus crassipes.* Growth factor line (open circles) and intermolt period line (closed circles) fitted to data on females from Hiatt (1948). (From Mauchline, 1976, courtesy of Springer-Verlag, New York.)

Volume 2, of this treatise; Weymouth and MacKay, 1936; Newcombe, 1949; Newcombe *et al.*, 1949a,b), the principles can be applied to studies of the variation in the growth of secondary sexual characters and to the determination of sexual maturity (Fig. 9) (Drach, 1933, 1934, 1936; Shotton, 1973; Terretta, 1973; Farmer, 1974; Arnaud and Do-Chi, 1977; Haefner, 1977; Lewis, 1977; Carpenter, 1978; Hartnoll, 1978; Somerton, 1980a,b, 1981; Donaldson *et al.*, 1981; see also Chapter 4, Volume 9, of this treatise). Depending on the species, changes in growth rate that accompany the transition from puberty to maturity may be subtle or exaggerated. Several characters exhibit differences in growth pattern, which can be correlated with adaptive function. Male chelae and the female abdomen both exhibit high positive allometry in the prepuberty phase and considerable size increases at the puberty molt (Hartnoll, 1974). In the course of molts prior to genital maturation of females, the variation of the width of abdominal somites (Y) in relationship to a standard measurement such as carapace width (X), follows a discontinuous growth described by $Y = bX^\alpha$, where α is the regression coefficient and b is the Y-intercept (Tessier, 1960; Hartnoll, 1978; Somerton, 1980a,b). In females approaching genital maturation, this relationship breaks down. As the abdominal somites grow wider at a higher rate than during the prepubertal growth phase, a change in slope of the regression becomes evident (Fig. 9). The body size at which this change in growth rate occurs is one indication of the size range in which sexual maturity occurs (Drach, 1933, 1934; Miller and Watson, 1976; Arnaud and Do-Chi, 1977; Lewis, 1977; Hartnoll, 1978; Somerton, 1980a,b). This phe-

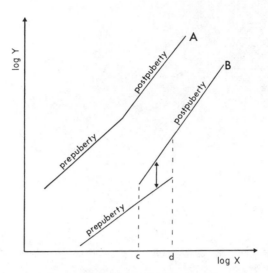

Fig. 9. Log variable (Y) plotted against log body size (X). (A) For species the population of which was analyzed en bloc; the inflection marks the puberty molt. (B) For a species where prepuberty and postpuberty populations were analyzed independently, the discontinuity marks the puberty molt. (c and d) are the minimum and maximum body sizes at which the puberty molt occurs. The arrows indicate the change in size of the variable occurring at the middle of the size range of the puberty molt. [From Hartnoll, 1974, courtesy of N. V. Boekhandel and Drukkerij (Formerly E. J. Brill).]

nomenon can, and should, be confirmed by other observations such as determining the smallest size female that is functionally mature. That is, in studying a population of crabs, one must recognize the size range and molt stage of females that (1) are ovigerous, (2) have mated (spermatophores in the spermathecae; sperm plugs in, or disfigurement and discoloration of, the gonopores; mating marks on the exoskeleton), (3) exhibit advanced gonad development (vitellogenesis) as determined by gonad indices or histological confirmation, and (4) show other secondary sex characteristics such as breeding dress (elaboration of setation on abdomen and pleopods) (Rahaman, 1967; Hartnoll, 1968, 1974; Haley, 1973; Diwan and Nagabhushanan, 1974; Farmer, 1974; Wenner et al., 1974; Haynes et al., 1976; Fielding and Haley, 1976; Haefner, 1977; Lewis, 1977; Carpenter, 1978; Pradeille-Rouquette, 1978; Reilly and Saila, 1978).

Males and females of most crustacean species differ in growth rates, most often postpuberty, that result in a significant difference in the ultimate size of the mature individuals (MacKay and Weymouth, 1935; Weymouth and MacKay, 1936; Newcomb et al., 1949a,b; Butler, 1961; Hancock and Edwards, 1967; Krouse, 1976; Mauchline, 1976; Miller and Watson, 1976;

Reilly and Saila, 1978). This information is also vital to sound evaluations of observed size frequency distributions of populations. Deviations from expected 1 : 1 male-to-female ratios are often explained by the knowledge of differential growth rates (Wenner, 1972; Fielding and Haley, 1976; Swartz, 1976; Haefner, 1978), as well as by observed segregation of crabs by sex and/or size according to depth, temperature, salinity, and other environmental parameters (Pereyra, 1966).

Because the reproductive capacity of the stock is influenced by the sex ratio, it is important to determine the optimal ratio for yielding high recruitment from the spawning population (Rothschild et al., 1970). In most crab species, the male grows to a larger maximum size than does the female, and thus is more vulnerable to the fishery. Assurance of an adequate number of mature males in the spawning stock (Fukuhara, 1974) may not be difficult to achieve, if the promiscuity of male king crabs reported by Powell et al. (1974) is common among all exploited species. Powell and Nickerson (1965) initially found large numbers of new-shelled young adult males engaged in copulation. On the assumption that only anexuviant males contributed to reproduction, the sex ratio of 1 old-shelled male to 4.1 adult females (Gray and Powell, 1966) does not appear to be an undesirable sex ratio. However, if small males produce less reproductive material than large males, the former may be less capable of fertilizing the greater masses of eggs of large females (Powell et al., 1974). Because king crab females normally mate immediately after molting, a shortage of males might result in significant numbers of females failing to mate. McMullen (1969) has demonstrated that females can successfully mate and produce normal clutches of fertilized eggs 0 to 13 days after ecdysis, thus providing some time for them to come in contact with polygamous males. Rothschild et al. (1970) suggested that female crabs can be harvested if utilizable and if they exceed the optimal sex ratio.

4. RECRUITMENT

Also critical to the determination of the annual optimum sustained yield of a crab stock is some measure of the recruitment of young individuals to the population. The age and growth manipulations mentioned above provide some measure as to when a given individual will enter the sampled population and when it will become vulnerable to the fishery.

Of equal importance is information on how environmental parameters influence development and survival of early life history stages (Costlow and Bookhout, 1959; Costlow, 1967; Sastry and McCarthy, 1973; Knowlton, 1974; Sastry, 1977; Vargo and Sastry, 1977) and on what adaptations exist among larval and juvenile forms that support recruitment to populations of adult crabs (Wickham, 1979c; Sulkin et al., 1980; Kelly et al., 1982). Analysis of fishery data for blue crabs indicates that annual fluctuations in commercial abundance are primarily associated with variable rates of survival of crabs in their first year of life (Pearson, 1948; Tagatz, 1965; More, 1969). Allen (1966, 1972) has reviewed the role of rhythms, migrations, and other behavioral manifestations in the population dynamics of decapod crustaceans. More recently, Sandifer (1975) addressed the role of pelagic larvae in the recruitment of young to estuarine decapod populations. He identified the following two mechanisms: (1) retention of larvae by the estuarine system and (2) immigration into the estuary of juveniles and adults, the eggs and larvae of which develop in the coastal oceanic environment. Sulkin *et al.* (1980) demonstrated barokinetic and geotaxic responses of larval blue crabs, *Callinectes sapidus,* which produce a pattern of early stages that move to surface waters and later stages to deeper waters. The coincidence of this pattern with the characteristic estuarine and coastal circulation could provide a mechanism for exchange of larvae between the estuary and coastal waters. Kelly *et al.* (1982) examined behavioral responses of first stage zoeae of *Geryon quinquedens* to gravity, hydrostatic pressure, and thermoclines, and they designed a recruitment model that explains reported depth distribution of postmetamorphic crabs and that can be tested in the field. Such relationships need to be further explored and defined before an accurate estimate of recruitment can be made for commercial species.

5. MORTALITY

The balance sheet on stock size can be completed only if some reasonable measure of mortality can be obtained. In order to determine the relative effect of fishing on the existing stock, the fishery scientist distinguishes between fishing mortality and natural mortality (Ricker, 1975). Population estimates and fishing mortality rates may be calculated from reported landing statistics and tagging studies (Gotshall, 1978b,c). Such estimates provide valuable insights into the condition of crab stocks, but the estimates are still dependent on the data used to calculate them (Gotshall, 1978b). Natural mortality is more difficult to assess (Stevens and Armstrong, 1981), and yet it is critical to the prediction of stock size. Predation, disease, and perturbations of the environment are the principal factors contributing to the natural mortality of the species and can dramatically affect the stock size of a fishery (Williams and Duke, 1979). Fisher and Wickham (1976), for example, have suggested that the fishing pressure on the dungeness crab was not significantly reducing the reproductive capacity of the population. Rather, they provided the following summary of the role of physical factors that have caused population fluctuations: (1) disturbance of larval settlement by shifting ocean currents, (2) the correlation between the coincidence of rainfall during the peak occurrence of salinity-sensitive larval stages and the realized crab catch 4 years later when those larvae would have entered the fishery (Lough, 1975), and (3) the positive correlation between upwelling intensity and crab catch (Peterson, 1973; Botsford and Wickham, 1975).

Hurt et al. (1979) and Love and Westphal (1981) extended the argument for the influence of environmental parameters on realized periodicities in crab populations. Hurt et al. (1979) applied Fourier transform analyses to Chesapeake blue crab (C. sapidus) catches, to Philadelphia air temperature and rainfall, and to earth–moon–sun tidal forces for the years 1922–1976. Substantial agreement was recognized for three periods (17.5, 9.8, 7.4 years) of temperature (average), rainfall (minimal), crab catch, and tidal forces. The agreement was explained on the basis that high tides enhance nutrient concentration of the surface waters of the Chesapeake Bay; minimum rainfall increases salinity of the surface waters; higher temperatures warm the surface waters. All of these conditions enhance blue crab survival and growth. Love and Westphal (1981) studied the relationship between Cancer magister landings and the mean annual sunspot number between 1950 and 1980. Crab catches and sunspot numbers varied in approximately 11-year cycles; the cycle periods for the two were strongly correlated.

Disease can be a major contributor to the decline of a population, particularly if it is aided by environmental perturbations. Fisher and Wickham (1976) attributed decline in dungeness crab population to cyanophytes and to bacteria infesting the egg mass. The major source of the nutrients of such saprophytes is external and most likely originate in waste-water effluents. The epiphytes observed by Fisher and Wickham (1976) are but two of a host of known diseases of crabs (Sindermann and Rosenfield, 1967; Sindermann, 1970; Sprague and Couch, 1971; Baross et al., 1978; Sawyer and MacLean, 1978; Schmidt and MacLean, 1978; Gallagher et al., 1979; Sparks and Hibbits, 1979; Newman and Feng, 1982). Carcinonemertes errans has been identified as a numerically significant predator on the eggs of Cancer magister; epidemic levels in California waters has resulted in the direct mortality of 55% of dungeness crab eggs produced over a 5-year period (Wickham, 1979a, 1980). Incidence of epibiotic fouling and mortality of crab eggs has also been linked to predation by C. errans (Wickham, 1979b). Unfortunately, incidences of infection of the various microbial and parasitic diseases are not as well documented. Usually, only major outbreaks are reported as they attract (and demand) immediate attention when mass mortalities occur. The diseases may result in loss to the fishery through death of the crabs or, as in the case of shell disease, through erosion and discoloration of the shell, which make the product unacceptable to the consumer.

When information such as that described above is assembled and correlated into a seasonal pattern of survival and reproduction for the female crabs and when similar information is documented for the male of the species (see e.g., Brown and Powell, 1972; Powell et al., 1972), one should have met the requirements for a functional management system (Fukuhara, 1974).

D. Underexploited Species

Essentially only seven species dominate the world landings (Longhurst, 1971). This fact generates two related questions, which bear answering. What is there about king crabs, dungeness crabs, blue crabs, and tanner (snow) crabs that enables them to support successful fisheries? Why are other, closely related species relegated to the underexploited category?

Van Engel (1974) described several basic criteria that must be satisfied before a successful fishery can be established. First, the populations of the animal supporting the fishery must be present in relatively consistent (and predictable) levels of high abundance. Second, the species must be reasonably vulnerable to fishing gear. Third, the crab must be relatively large, yet easy to handle, transport, process, and market after capture. Fourth, the product must have consumer acceptance and a potential market.

Although there is nothing unusual among these criteria, it can be demonstrated that many species are underutilized because of failure to fulfill any one of these qualifications. In addition, each undeveloped stock usually has a unique set of problems of varying magnitude generally related to the economics of harvesting, processing, and marketing (Rathjen, 1974).

1. STOCK SIZE

Even utilized species, such as the blue crab (Callinectes sapidus), undergo fluctuations in abundance that defy accurate prediction (Pearson, 1948; McHugh and Ladd, 1953; Millikin, 1978; Miller et al., 1980). Although the stocks have rarely decreased to the point of endangering the fishery, there is continuing effort on the part of most fishery research agencies to design a sampling program that will result in reasonably accurate stock predictions (see e.g., Tagatz, 1965; Adkins, 1972; Chittenden and Van Engel, 1972; Sulkin, 1972; Perry, 1975; Gotshall, 1978a,b; Miller et al., 1980). Significant decreases in stocks of major fisheries of this type have on occasion stimulated high-intensity research programs, which are often short-lived (Mahood et al., 1970). Decreases in stock have also prompted the fishermen to look for new, alternative resources. The Canadian snow crab fishery developed from such pressures (Powles, 1966, 1968a,b). In certain areas, some species, for example, Carcinus maenas, are subject to such dramatic fluctuations in population density that crabbers cannot depend on fishable stocks from year to year (Welch, 1968). The reluctance to enter a risk fishery of this type is understandable.

2. VULNERABILITY TO FISHING GEAR

The size of the red crab (*Geryon quinquedens*) resource is of uncertain magnitude (Wigley et al., 1975; Haefner, 1978; Cayre et al., 1979), but the crab has not been effectively utilized because of its inaccessibility. It is encountered only in deep water (300–1500 m) on the continental shelf and slope. Fishermen have experienced difficulties in maintaining large strings of pots in such areas. Commercial trawling is also of questionable value in this regard because it adversely affects the quality of the catch (Haefner and Musick, 1974).

3. SIZE

The green crab (*Carcinus maenas*), besides being sporadic in abundance, is a relatively small crab, thus making it difficult to extract the meat. Its value in the United States is confined to its desirability as bait in the sport fin fishery and to its undesirability as a predator on soft-shell clams (Welch, 1968). It is common on the coast of Europe and is used there as food (Rees, 1963).

4. HANDLING, TRANSPORTATION, AND PROCESSING

Cancer borealis, C. irroratus, and *Geryon quinquedens* present problems in these categories. Unlike the American lobster, with which these species are often captured, they are difficult to keep alive for extended periods of time (Varga et al., 1969). In addition, the meat is prone to discoloration, unless additional steps are added to the processing (Holmsen and McAllister, 1974; Van Engel and Haefner, 1975). The extreme hardness of the shell of *Cancer* crabs and the adherence of the meat to the shell make it difficult to extract the meat by hand, a method favored in the blue crab industry (Lee et al., 1963). Mechanical extraction is possible but lowers the quality of the meat and reduces the yield (Wilder, 1966). Full utilization of such species may depend on their being marketed as specialty products (Marchant and Holmsen, 1975).

Brown (1971) described two processing conditions that had been deterrents to the growth of the tanner crab fishery in Alaska. The attempt to use equipment and methods designed for the larger king crab resulted in shell fragments in the meat. The hand labor required for shell removal increased the cost of the product. The presence of a black encrustation on the crab shell has also caused processing difficulties.

It should be reemphasized that the lack of any one criterion may limit the marketing potential for a given species, but often it is a combination of factors that brings about a reluctance on the part of crabbers, buyers, and processors to become heavily involved in a new fishery. The fisheries development program (Rathjen, 1974) should produce favorable results if followed through to completion. Plans call for the following efforts for each selected species: (1) review of existing data on resource availability and harvesting technology; (2) development of handling and sorting methods for vessels and plants; (3) study of processing technology and engineering problems to permit efficient preparation of traditional and new products; and (4) determination of ways to increase acceptance of product forms in domestic and export markets.

IV. SUMMARY

The brachyuran level of organization represents the culmination of crustacean evolutionary history. The change in body shape and structure and the accompanying maximum diversification of organs and their functions are manifested by high taxonomic diversity and wide geographic and ecological distribution.

Although the evolution of the crab form is of little concern to potential fishery exploitation, the traits leading to evolutionary success are also those that provide a given species with the adaptive potential to survive, grow, and perpetuate itself in a given set of environmental conditions. An understanding of the total biology and ecology of the crab is basic to the construction of the principles of management and conservation of crab stocks.

The principal problems facing crab fisheries are no different from those of other marine fisheries. Traditional, well-established high-unit value fisheries are near maximum (optimum) sustained yield. Control measures must be brought to bear to prevent future obliteration of stocks.

The ability to predict cyclic variations of fishery stocks and to prevent overfishing are two goals that have eluded fishery scientists for years. Innovations in statistical sampling and analysis have been little match for the biological variability of the populations and their interaction with environmental perturbations, both natural as well as degraded. At the present time, predictions are estimations, at best, and often fall prey to unpredictable natural and man-made catastrophes. This phase of management biology remains the number one challenge. Accurate predictions of stock size and improvements in gear selectivity and management regulations lead to stabilization of the catch and market conditions. These are often offset by variations in recruitment, growth rate, year class failures, and fluctuating product demands.

It is the fishery scientists' role to determine the species that could yield substantial harvests and to provide estimates of the quantity that could be caught. The management of existing fisheries and the development of new fisheries requires a great volume of scientific and, in particular, ecological data (see for example, Pereyra et al., 1976). Scientists know enough about some aspects of fishery production to give useful advice on the technical development and management of fisheries, but sociological aspects are often overlooked (Miller, 1979). Until such time that multidisciplinary studies can keep pace with fishery development, management advice will often be ignored, incomplete, or too late (McHugh, 1968; Gulland, 1976).

Improved technology in catching, handling, and processing should lead to more complete utilization of the crab resource (and less wastage) and a better product. One of the more critical aspects involves the maintenance of the fresh product at sea, in storage, and during transportation. A significant portion of the catch is often lost because of spoilage or simply because facilities are inadequate to retain the catch. Modifications for butchering and processing the catch on board would alleviate the problem of fresh storage.

Advances in food technology would improve the quality of the product, would increase the shelf life of products frozen, pasteurized, boiled, steamed, and/or canned, and would increase the percentage of meat yield.

The potential for increase in crab productivity exists with the development and expansion of fisheries for numerous underexploited species. For those geographic areas subject to potentially expandable crab fisheries, it is essential that an understanding of the crab stocks is acquired, i.e., how they are affected by fishing and by other events in the sea and how changes in stocks will be reflected in the catches. The success of such new ventures depends on well-coordinated efforts among all segments in the fishery—catching, processing, marketing, and managing. Technological advances in fishing methods and processing will be of little value if they are not applied to the biological aspects of the target species to assure its continued survival and proliferation.

REFERENCES

Adiyodi, K. G., and Adiyodi, R. G. (1970). Endocrine control of reproduction in decapod Crustacea. Biol. Rev. **45**, 121–165.

Adkins, G. (1972). A study of the blue crab fishery in Louisiana. La. Wildl. Fish. Comm. Oyster, Water Bottoms Seafoods Div. Tech. Bull. No. 3, 57 pp.

Allen, J. A. (1966). The rhythms and population dynamics of decapod Crustacea. Oceanogr. Mar. Biol. **4**, 247–265.

Allen, J. A. (1972). Recent studies on the rhythms of postlarval decapod Crustacea. Oceanogr. Mar. Biol. **10**, 415–436.

Alverson, F. G., and Patterson, P. H. (1974). International Trade—Crabs. In "International Trade—Tuna, Shrimp, Crab, Fish Meal, Groundfish" pp. 67–102. FAO, UN Indian Ocean Fish, Commission Indian Ocean Programme, IOFC/DEV/74/40, Rome, Italy.

Ambore, N. E., and Venkatachari, S. A. T. (1978). Respiratory metabolism in relation to body size, sex and gill area of the freshwater crab *Barytelphusa guerini* Milne Edwards. *Indian J. Exp. Biol.* **16,** 465–467.

Anderson, E., Gorham, A., Ness, H., Orth, F., Queirolo, L., Richardson, J., and Atkinson, C. (1977). The Bering Sea tanner crab resource: U.S. production capacity and marketing. *Alaska Sea Grant Rep. No. 77-5,* 157 pp.

Angel, N. B., Crow, G. L., Webb, N. B., and Otwell, W. S. (1974). Development of improved handling, holding and transporting techniques for North Carolina blue crab. *N.C. Dep. Admin., Raleigh, N.C.,* 61 pp.

Anonymous (1974). New use found for red crab waste. *Univ. R.I., N. Engl. Mar. Res. Inf. 66, Nov. 1974,* 3 pp.

Arnaud, P. M., and Do-Chi, T. (1977). Données biologique et biométriques sur les lithodes *Lithodes murrayi:* (Crustace: Decapoda: Anomura) des îles Crozet (SW oćean Indien). *Mar. Biol. (Berlin)* **39,** 147–159.

Bakkala, R. G., Kessler, D. W., and MacIntosh, R. A. (1976). History of commercial exploitation of demersal fish and shellfish in the eastern Bering Sea. *In* "Demersal Fish and Shellfish Resources of the Eastern Bering Sea in the Baseline Year 1975" (W. T. Pereyra, J. E. Reeves, and R. G. Bakkala, eds.), pp. 13–36. Natl. Mar. Fish. Serv. Northwest and Alaska Fish. Ctr. Proc. Rept., Seattle, Washington.

Bardach, J. E., Ryther, J. H., and McLarney, W. O. (1972). "Aquaculture: The Farming and Husbandry of Freshwater and Marine Organisms." Wiley, New York.

Barker, P. L., and Gibson, R. (1978). Observations on the structure of the mouthparts, histology of the alimentary tract, and digestive physiology of the mud crab *Scylla serrata* (Forskål) (Decapoda: Portunidae) *J. Exp. Mar. Biol. Ecol.* **32,** 177–196.

Barnes, R. D. (1974). "Invertebrate Zoology." Saunders, Philadelphia, Pennsylvania.

Barnett, H. J., Nelson, R. W., and Hunter, P. J. (1969). Shipping live dungeness crabs by air to retail market. *Commer. Fish. Rev. May 1969,* pp. 21–24.

Baross, J. A., Tester, P. A., and Morita, R. Y. (1978). Incidence, microscopy, and etiology of exoskeleton lesions in the tanner crab, *Chionoecetes tanneri. J. Fish. Res. Board Can.* **35,** 1141–1149.

Bartlett, L. D. (1976a). King crab (family Lithodidae). *In* "Demersal Fish and Shellfish Resources of the Eastern Bering Sea in the Baseline Year 1975" (W. T. Pereyra, J. E. Reeves, and R. G. Bakkala, eds.), pp. 531–544. Natl. Mar. Fish. Serv. Northwest and Alaska Fish. Ctr. Proc. Rept., Seattle, Washington.

Bartlett, L. D. (1976b). Tanner crab (family Majidae). *In* "Demersal Fish and Shellfish Resources of the Eastern Bering Sea in the Baseline Year 1975" (W. T. Pereyra, J. E. Reeves, and R. G. Bakkala, eds.), pp. 545–552. Natl. Mar. Fish. Serv. Northwest and Alaska Fish. Ctr. Proc. Rept., Seattle, Washington.

Batterton, C. V., and Cameron, J. N. (1978). Characteristics of resting ventilation and response to hypoxia, hypercapnia, and emersion in the blue crab *Callinectes sapidus* (Rathbun). *J. Exp. Zool.* **203,** 403–418.

Bauchau, A. (1966). "La Vie des Crabes." Editions Paul Lechavalier, 18, Rue des Ecoles, Paris, France, 138 pp.

Bennett, D. B. (1974a). Growth of the edible crab (*Cancer pagurus* L.) off south-west England. *J. Mar Biol. Assoc. U. K.* **54,** 803–823.

Bennett, D. B. (1974b). The effects of pot immersion time on catches of crabs, *Cancer pagurus* L. and lobsters, *Homarus gammarus* (L.). *J. Cons. Int. Explor. Mer.* **35,** 332–336.

Beurlen, K. (1931). Die Besiedelung der Tiefsee. *Nat. Mus.* **61,** 269–279.

Bishop, J. M., Olmi, E. J., III, Whitaker, J. D., and Yianopoulos, G. M. (1983). Capture of blue crab peelers in South Carolina: An analysis of techniques. *Trans. Am. Fish. Soc.* **112,** 60–70.

Bliss, D. E. (1951). Metabolic effects of sinus gland or eyestalk removal in the land crab, *Gecarcinus lateralis*. *Anat. Rec.* **111**, 502–503.

Bliss, D. E. (1953). Endocrine control of metabolism in the land crab, *Gecarcinus lateralis* (Freminville). I. Differences in the respiratory metabolism of sinus glandless and eye-stalkless crabs. *Biol. Bull. (Woods Hole, Mass.)* **104**, 276–296.

Bliss, D. E. (1968). Transition from water to land in decapod crustaceans. *Am. Zool.* **8**, 355–392.

Bliss, D. E. (1979). From sea to tree: Saga of a land crab. *Am. Zool.* **19**, 385–410.

Bliss, D. E., and Mantel, L. H. (1968). Adaptations of crustaceans to land: A summary and analysis of new findings. *Am. Zool.* **8**, 673–685.

Bliss, D. E., Van Montfrans, J., VanMontfrans, M., and Boyer, J. R. (1978). Behavior and growth of the land crab *Gecarcinus lateralis* (Freminville) in southern Florida. *Bull. Am. Mus. Nat. Hist.* **160**, 111–152.

Botsford, L. W. (1981). Letters and comments: Comment on cycles in the northern California dungeness crab population. *Can. J. Fish. Aquat. Sci.* **38**, 1295–1297.

Botsford, L. W., and Wickham, D. E. (1975). Correlation of upwelling index and dungeness crab catch. *Fish. Bull.* **73**, 901–907.

Botsford, L. W., and Wickham, D. E. (1978). Behavior of age-specific, density-dependent models and the northern California dungeness crab (*Cancer magister*) fishery. *J. Fish. Res. Board Can.* **35**, 833–843.

Brooks, H. K. (1969). Eucarida. *In* "Treatise on Invertebrate Paleontology" (R. C. Moore, ed.), Part (R) Arthropoda 4 (1), pp. R332–R345. Geol. Soc. Am., Boulder, Colorado.

Brown, C. G., and Bennett, D. B. (1978). Population and catch structure of the edible crab (*Cancer pagurus*) in the English Channel. *J. Cons. Int. Explor. Mer.* **39**, 88–100.

Brown, R. B. (1971). The development of the Alaskan fishery for tanner crab, *Chionoecetes species,* with particular reference to the Kodiak area, 1967–1970. *Alaska Dept. Fish. Game. Inf. Lflt. No. 153,* 26 pp.

Brown, R. B., and Powell, G. C. (1972). Size at maturity in the male Alaskan tanner crab, *Chionoecetes bairdi,* as determined by chela allometry, reproductive tract weights, and size of precopulatory males. *J. Fish. Res. Board Can.* **29**, 423–427.

Burke, E. M. (1979). Aerobic and anaerobic metabolism during activity and hypoxia in two species of intertidal crabs. *Biol. Bull. (Woods Hole, Mass.)* **156**, 157–168.

Butler, T. H. (1961). Growth and age determination of the Pacific edible crab *Cancer magister* Dana. *J. Fish. Res. Board Can.* **18**, 873–891.

Caddy, J. F., Chandler, R. A., and Wilder, D. G. (1974). Biology and commercial potential of several underexploited molluscs and crustaceans on the Atlantic coast of Canada. *Proc. Symp. Ind. Dev. Branch Environ. Can., Montreal, Quebec, Febr. 5–7, 1974,* 111 pp.

Caine, E. A. (1974). Feeding of *Ovalipes guadulpensis* (Saussure) (Decapoda: Brachyura: Portunidea), and morphological adaptations to a burrowing existence. *Biol. Bull. (Woods Hole, Mass.)* **147**, 550–559.

Caine, E. A. (1975a). Feeding and masticatory structure of selected Anomura (Crustacea). *J. Exp. Mar. Biol. Ecol.* **18**, 277–301.

Caine, E. A. (1975b) Feeding of *Pinnotheres maculatus* Say (Brachyura: Pinnotheridae). *Forma Funct.* **8**, 395–404.

Caine, E. A. (1976). Relationship between diet and the gland filter of the gastric mill in hermit crabs (Decapoda, Paguroidea). *Crustaceana* **31**, 312–313.

Carley, D. H., and Frisbie, C. M. (1968). The blue crab, oyster and finfish fisheries of Georgia—an economic evaluation. *Georgia Game Fish Comm. Mar. Fish. Div. Contr. Ser. No. 12,* 13 pp.

Carpenter, R. K. (1978). Aspects of growth, reproduction, distribution and abundance of the

Jonah crab, (Cancer borealis) Stimpson, in Norfolk Canyon and the adjacent slope. M.A. Thesis, Univ. of Virginia, Charlottesville.

Cassie, R. M. (1954). Some uses of probability paper in the analysis of size frequency distributions. Aust. J. Mar. Freshwater Res. **5,** 513–522.

Castro, P. (1978). Settlement and habitat selection in the larvae of Echinoecus pentagonus (A. Milne Edwards), a brachyuran crab symbiotic with sea urchins. J. Exp. Mar. Biol. Ecol. **34,** 259–270.

Cayre, P., LeLoeuff, P., and Intes, A. (1979). Geryon quinquedens, le crabe rouge profond. Biologie, pêche, conditionnement, potentialités d'exploitation. La Pêche Marit. No. 1210, 8 pp.

Chace, F. A., Jr., and Hobbs, H. H., Jr. (1969). The freshwater and terrestrial decapod crustaceans of the West Indies with special reference to Dominica. U. S. Natl. Mus. Bull. **292,** 258 pp.

Chapman, C. J., and Smith, G. L. (1977). Creel catches of crab, Cancer pagurus L. using different baits. J. Cons. Int. Explor. Mer. **38,** 226–229.

Charniaux-Cotton, H. (1960). Sex determination. In "The Physiology of Crustacea. Vol. I. Metabolism and Growth" (T. H. Waterman, ed.), pp. 411–448. Academic Press, New York.

Chittenden, M. E., Jr., and Van Engel, W. A. (1972). Effect of a tickler chain and tow duration on trawl catches of the blue crab, Callinectes sapidus. Trans. Am. Fish. Soc. **101,** 732–734.

Clarke, R. A. (1972). Exploration for deep benthic fish and crustacean resources. Hawaii Inst. Mar. Biol. Tech. Rep. No. 29, 20 pp.

Cochran, D. M. (1935). The skeletal musculature of the blue crab, Callinectes sapidus Rathbun. Smithson. Misc. Coll. **92,** 1–76.

Copeland, D. E. (1968). Fine structure of salt and water uptake in the land crab, Gecarcinus lateralis. Am. Zool. **8,** 417–432.

Copeland, B. J., and Bechtel, T. J. (1974). Some environmental limits of six Gulf Coast estuarine organisms. Contr. Mar. Sci. **18,** 169–204.

Costlow, J. D., Jr. (1967). The effect of salinity and temperature on survival and metamorphosis of megalops of the blue crab Callinectes sapidus. Helgol. Wiss. Meeresunters. **15,** 84–97.

Costlow, J. D., Jr., and Bookhout, C. G. (1959). The larval development of Callinectes sapidus Rathbun reared in the laboratory. Biol. Bull. (Woods Hole, Mass.) **116,** 373–396.

Coyer, P. E. (1977). Responses of heart and scaphognathite rates in Cancer borealis and C. irroratus to hypoxia. Comp. Biochem. Physiol. **56**A, 165–167.

Dassow, J. A. (1969). Crab industry. In "The Encyclopedia of Marine Resources" (F. E. Firth, ed.), pp. 150–155. Van Nostrand-Reinhold, Princeton, New Jersey.

Dennell, R. (1960). Integument and exoskeleton. In "The physiology of Crustacea. Vol. I. Metabolism and Growth" (T. H. Waterman, ed.) pp. 449–472. Academic Press, New York.

Diaz, H., and Rodriguez, F. (1977). The branchial chamber in terrestrial crabs: A comparative study. Biol. Bull. (Woods Hole, Mass.) **153,** 485–504.

Dimock, R. V., and Groves, K. H. (1975). Interaction of temperature and salinity on oxygen consumption of the estuarine crab Panopeus herbstii. Mar. Biol. (Berlin) **33,** 301–308.

Diwan, A. D., and Nagabhushanam, R. (1974). Reproductive cycle and biochemical changes in the gonads of the freshwater crab, Barytelphusa cunicularis (Westwood, 1836). Indian J. Fish **21,** 164–176.

Diwan, A. D., and Nagabhushanam, R. (1976). Studies on heat tolerance in the freshwater crab, Barytelphusa cunicularis (Westwood 1836). Hydrobiologia **50,** 65–70.

Donaldson, W. E., Cooney, R. T., and Hilsinger, J. R. (1981). Growth, age and size at maturity of tanner crab, *Chionoecetes bairdi* M. J. Rathbun, in the northern Gulf of Alaska (Decapoda, Brachyura). *Crustaceana* **40**, 286–302.

Dorgelo, J. (1976). Salt tolerance in Crustacea and the influence of temperature upon it. *Biol. Rev. Cambridge Philos. Soc.* **51**, 255–290.

Drach, P. (1933). Sur la croissance de l'abdomen chez les Brachyoures. Cas de *Portunus puber. C. R. Acad. Sceances Paris* **197**, 93–95.

Drach, P. (1934). Sur la croissance de l'abdomen chez les Brachyoures; discontinuities chez *Carcinus maenas* Pennant. *C. R. Sceances Mem. Soc. Biol.* **116**, 138–141.

Drach, P. (1936). Croissance allometrique et dimorphisme chez les Brachyoures. *C. R. Acad. Sceances Paris* **203**, 820–823.

Drach, P., and Tchernigovtzeff, C. (1967). Sur la méthode de détermination des stades d'intermue et son application générale aux crustacés. *Vie Milieu, Ser. A: Biol. Mar. Tome XVIII, Fasc. 3-A*, pp. 595–610.

Edney, E. B. (1960). Terrestrial adaptations. *In* "The Physiology of Crustacea. Vol. I. Metabolism and Growth" (T. A. Waterman, ed.), pp. 367–394. Academic Press, New York.

Eldridge, P. J., and Waltz, W. (1977). Observations on the commercial fishery for blue crabs *Callinectes sapidus* in estuaries in the southern half of South Carolina. *S. C. Mar. Res. Ctr. Tech. Rep. No. 21*, 35 pp.

Eldridge, P. J., Burrell, V. G., Jr., and Steele, F. (1979). Development of a self-culling bluecrab pot. *Mar. Fish. Rev. 1979 (Dec.)*, pp. 21–27.

Elner, R. W. (1978). The mechanics of predation by the shore crab, *Carcinus maenas* (L.), on the edible mussel, *Mytilus edulis* L. *Oecologia* **36**, 333–344.

Elner, R. W. (1980). The influence of temperature, sex and chela size in the foraging strategy of the shore crab, *Carcinus maenas* (L). *Mar. Behav. Physiol.* **7**, 15–24.

Escritor, G. L. (1970). A report on experiments in the culture of the mud crab (*Scylla serrata*). *Proc. Indo-Pac. Fish. Coun.* **14**, 1–11.

Escritor, G. L. (1972). Observations on the culture of the mud crab, *Scylla serrata. In* "Coastal Aquaculture in the Indo-Pacific Region" (T. V. R. Pillay, ed.), pp. 355–361. Fishing News (Books) Ltd., London.

FAO (1968). Fisheries in the food economy. Freedom from Hunger Campaign. Basic Studies No. 19, 79 pp.

FAO (1971). Symposium on investigations and resources of the Caribbean Sea and adjacent waters. *FAO Fish. Rep. No. 71.2*, 347 pp.

FAO (1972). "Atlas of Living Resources of the Seas." Rome, Italy.

FAO (1977). Code of practice for crabs. *FAO Fish. Circ. C349*, 49 pp.

Farmer, A. S. (1973). Age and growth in *Nephrops norvegicus* (Decapoda: Nephropidae) *Mar. Biol. (Berlin)* **23**, 314–325.

Farmer, A. S. (1974). Relative growth in *Nephrops norvegicus* (L.) (Decapoda: Nephropidae) *J. Nat. Hist.* **8**, 605–620.

Feliciano, C. (1962). Notes on the biology and economic importance of the land crab *Cardisoma quanhumi* Latreille of Puerto Rico. *Spec. Contr. Inst. Mar. Biol., Univ. Puerto Rico, Mayaquez*, 29 pp.

Fielding, A., and Haley, S. R. (1976). Sex ratio, size at reproductive maturity, and reproduction of the Hawaiian Kona crab, *Ranina ranina* (Linnaeus) (Brachyura, Gymnopleura, Raninidae). *Pac. Sci.* **30**, 131–145.

Firth, F. E., ed. (1969). "The Encyclopedia of Marine Resources." Van Nostrand-Reinhold, Princeton, New Jersey.

Fisher, W. S., and Wickham, D. E. (1976). Mortalities and epibiotic fouling of eggs from wild populations of the dungeness crab, *Cancer magister. Fish. Bull.* **74**, 201–207.

Florkin, M. (1960). Ecology and metabolism. *In* "The Physiology of Crustacea. Vol. I. Metabolism and Growth" (T. H. Waterman, ed.), pp. 395–410. Academic Press, New York.

Ford, E. (1933). An account of the herring investigations conducted at Plymouth during the years from 1924 to 1933. *J. Mar. Biol. Assoc. U. K.* **19,** 305–384.

Fukuhara, F. M. (1974). Preliminary study of the biological considerations regarding management of the eastern Bering Sea king crab stocks by size, sex and season. *Natl. Mar. Fish. Serv. Northwest Fish. Ctr. Proc. Rep.,* 7 pp.

Fullenbaum, R. F. (1970). A survey of maximum sustainable yield estimates on a world basis for selected fisheries. *Bur. Commer. Fish. Div. Econ. Res. Working Pap. No. 43,* 14 pp.

Fuzessery, Z. M., and Childress, J. J. (1975). Comparative chemosensitivity to amino acids and their role in the feeding activity of bathypelagic and littoral crustaceans. *Biol. Bull. (Woods Hole, Mass.)* **149,** 522–538.

Gallagher, M. L., Rittenburg, J. H., Bayer, R. C., and Leavitt, D. F. (1979). Incidence of *Aerococcus viridans* (var.) *homari* in natural crab (*Cancer irroratus, Cancer borealis*) populations from Maine coastal waters. *Crustaceana* **37,** 316–317.

Ganz, A. R., and Herrmann, J. F. (1975). Investigations into the southern New England red crab fishery. *R.I. Dept. Nat. Res. Div. Fish Wildl. Mar. Fish. Sec. Ms. Rep.* 78 pp.

Glaessner, M. F. (1969). Decapoda. *In* "Treatise on Invertebrate Paleontology" (R. C. Moore, ed.), Part (R) Arthropoda 4 (2), pp. R399–R566. Geol. Soc. Am., Boulder, Colorado.

Gleeson, R. A. (1980). Pheromone communication in the reproductive behavior of the blue crab, *Callinectes sapidus. Mar. Behav. Physiol.* **7,** 119–134.

Gotshall, D. W. (1978a). Relative abundance studies of dungeness crabs, *Cancer magister,* in northern California. *Calif. Fish Game* **64,** 24–37.

Gotshall, D. W. (1978b). Catch-per-unit-of-effort studies of northern California dungeness crabs, *Cancer magister. Calif. Fish Game* **64,** 189–199.

Gotshall, D. W. (1978c). Northern California dungeness crab, *Cancer magister,* movements as shown by tagging. *Calif. Fish Game* **64,** 234–254.

Graham, M. (1956). "Sea Fisheries: Their Investigation in the United Kingdom." Arnold, London.

Gray, G. W., and Powell, G. C. (1966). Sex ratios and distributions of spawning king crabs in Alitak Bay, Kodiak Island, Alaska (Decapoda Anomura, Lithodidae). *Crustaceana* **10,** 303–309.

Griffin, D. J. G. (1969). Breeding and molting cycles of two Tasmanian grapsid crabs (Decapoda, Brachyura). *Crustaceana* **16,** 88–94.

Griffin, D. J. G. (1970). Australian crabs. *Aust. Nat. Hist.* **16,** 304–308.

Guinot, D. (1977). Propositions pour une nouvelle classification des Crustacés Décapodes Brachyoures. *C. R. Acad. Sci., Ser. D* **285,** 1049–1052.

Gulland, J. A. (1976). Production and catches of fish in the sea. *In* "The Ecology of the Seas" (D. H. Cushing and J. J. Walsh, eds.), pp. 283–314. Saunders, Philadelphia, Pennsylvania.

Haefner, P. A., Jr. (1976). *Cancer* crabs: Aids to identification. *Virginia Inst. Mar. Sci. Mar. Res. Adv. No. 10,* 3 pp.

Haefner, P. A., Jr. (1977). Reproductive biology of the female deep-sea red crab, *Geryon quinquedens,* from the Chesapeake Bight. *Fish. Bull.* **75,** 91–102.

Haefner, P. A., Jr. (1978). Seasonal aspects of the biology, distribution and relative abundance of the deep-sea red crab, *Geryon quinquedens* Smith, in the vicinity of the Norfolk Canyon, eastern North Atlantic. *Natl. Shellfish. Assoc. Proc.* **68,** 49–62.

Haefner, P. A., Jr., and Garten, D. (1974). Methods of handling and shedding blue crabs, *Callinectes sapidus, Virginia Inst. Mar. Sci. Mar. Res. Adv. Ser. No. 8,* 14 pp.

Haefner, P. A., Jr., and Musick, J. A. (1974). Observations on distribution and abundance of red crabs in Norfolk Canyon and adjacent continental slope. *Mar. Fish. Rev.* **36**, 31–34.

Haefner, P. A., Jr., and Van Engel, W. A. (1975). Aspects of molting, growth and survival of male rock crabs, *Cancer irroratus*, in Chesapeake Bay. *Chesapeake Sci.* **16**, 253–265.

Haefner, P. A., Jr., Van Engel, W. A., and Garten, D. (1973). Rock crab: A potential new resource. *Virginia Inst. Mar. Sci. Mar. Res. Ser. No. 7*, 3 pp.

Haley, S. R. (1973). On the use of morphometric data as a guide to reproductive maturity in the ghost crab, *Ocypode ceratophthalmus* (Pallas) (Brachyura, Ocypodidae). *Pac. Sci.* **27**, 350–362.

Hancock, D. A. (1974). Attraction and avoidance in marine invertebrates—their possible role in developing an artificial bait. *J. Cons. Int. Explor. Mer.* **34**, 328–331.

Hancock, D. A., and Edwards, E. (1967). Estimation of annual growth in the edible crab (*Cancer pagurus* L.) *J. Cons. Int. Explor. Mer.* **31**, 246–264

Harding, J. F. (1949). The use of probability paper for the graphical analysis of polymodal frequency distributions. *J. Mar. Biol. Ass. U. K.* **28**, 141–153.

Hartnoll, R. G. (1968). Morphology of the genital ducts in female crabs. *J. Linn. Soc. London, Zool.* **47**, 279–300.

Hartnoll, R. G. (1969). Mating in the Brachyura. *Crustaceana* **16**, 162–181.

Hartnoll, R. G. (1971). The occurrence, methods and significance of swimming in the Brachyura. *Anim. Behav.* **19**, 34–50.

Hartnoll, R. G. (1974). Variation in growth pattern between some secondary sexual characters in crabs (Decapoda Brachyura). *Crustaceana* **27**, 131–136.

Hartnoll, R. G. (1978). The determination of relative growth in Crustacea. *Crustaceana* **34**, 281–293.

Hartnoll, R. G. (1982). Growth. *In* "The Biology of Crustacea. Vol. 2. Embryology, Morphology, and Genetics" (L. G. Abele, ed.), pp. 111–196. Academic Press, New York.

Haynes, E., Karinen, J. F., Watson, J., and Hopson, D. J. (1976). Relation of number of eggs and egg length to carapace width in the brachyuran crabs *Chionoecetes bairdi* and *C. opilio* from the southeastern Bering Sea and *C. opilio* from the Gulf of St. Lawrence. *J. Fish. Res. Board Can.* **33**, 2592–2595.

Henning, H. G. (1975). Aggressive, reproductive and molting behavior—growth and maturation of *Cardisoma guanhumi* Latreille (Crustacea, Brachyura). *Forma Funct.* **8**, 463–510.

Hepburn, H. R., Joffe, I., Green, N., and Nelson, K. J. (1975). Mechanical properties of a crab shell. *Comp. Biochem. Physiol.* **50A**, 551–554.

Hiatt, R. W. (1948). The biology of the lined shore crab, *Pachygrapsus crassipes* Randall. *Pac. Sci.* **2**, 135–213.

Hill, B. J. (1975). Abundance, breeding and growth of the crab *Scylla serrata* in two south African estuaries. *Mar. Biol. (Berlin)* **32**, 119–126.

Hill, B. J. (1976). Natural food, foregut clearance-rate and activity of the crab *Scylla serrata*. *Mar. Biol. (Berlin)* **34**, 109–116.

Hill, B. J. (1979). Biology of the crab *Scylla serrata* (Forskål) in the St. Lucia system. *Trans. R. Soc. S. Afr.* **44**, 55–62.

Hilsinger, J. (1976). Snow crab, a life history. *Univ. Alaska Sea Grant Newslett: Alaska Seas Coasts* **4**, 3, 6.

Hipkins, F. W. (1972). Dungeness crab pots. *Natl. Mar. Fish. Serv. Fish. Facts No. 3*, 13 pp.

Holmsen, A. A. (1973). Potential utilization of underexploited species in southern New England. *Univ. R.I. Mar. Adv. Serv. Mar. Memo. Ser. No. 32*, 7 pp.

Holmsen, A. A., and McAllister, H. (1974). Technological and economic aspects of red crab harvesting and processing. *Univ. R.I. Mar. Tech. Rep. No. 28*, 35 pp.

Hoopes, D. T. (1973). Alaska's fishery resources—the dungeness crab. *Natl. Mar. Fish. Serv. Fish. Facts No. 6*, 14 pp.

Horn, M. H. (1968). Observations on the aerating mechanism of the wharf crab, *Sesarma cinereum* (Bosc). *Crustaceana* **15**, 204–208.

Hurt, P. R., Libby, L. M., Pandolfi, L. J., and Levine, L. H. (1979). Periodicities in blue crab population in Chesapeake Bay. *Clim. Change* **2**, 75–78.

Idyll, C. P., and Sisson, R. F. (1971). The crab that shakes hands. *Natl. Geogr. Mag.* **139**, 254–271.

Ingle, R. W. (1983). *In* "Shallow-water Crabs. Synopses of the British Fauna, N.S. 25" (D. M. Kermack and R. S. K. Barnes, eds.). Cambridge Univ. Press, London and New York.

Isaacson, P. A. (1963). Modifications of Chesapeake Bay commercial crab pot. *Commer. Fish. Rev. 25*, 12–16.

Jachowski, R. L. (1974). Agonistic behavior of the blue crab, *Callinectes sapidus* Rathbun. *Behaviour* **50**, 232–253.

Jackson, R. I. (1972). Fisheries and the future world supply. *In* "World Fisheries Policy, Multidisciplinary Views" (B. J. Rothschild, ed.), pp. 3–13. Univ. of Washington Press, Seattle.

Jeffries, H. P. (1966). Partitioning of the estuarine environment by two species of *Cancer*. *Ecology* **47**, 477–481.

Jennings, J. B. (1972). "Feeding, Digestion and Assimilation in Animals," 2nd ed. Macmillan, London.

Kaestner, A. (1970). "Invertebrate Zoology. Vol. III. Crustacea". Wiley (Interscience), New York.

Kaimmer, S. M., Reeves, J. E., Gunderson, D. R., Smith, G. B., and MacIntosh, R. A. (1976). Baseline information from the 1975 OCSEAP survey of the demersal fauna of the eastern Bering Sea. *In* "Demersal Fish and Shellfish Resources of the Eastern Bering Sea in the Baseline Year 1975" (W. T. Pereyra, J. E. Reeves and R. G. Bakkala, eds.) pp. 157–366. Natl. Mar. Fish. Serv. Northwest and Alaska Fish. Ctr. Proc. Rept., Seattle, Washington.

Kelly, P., Sulkin, S. D., and Van Heukelem, W. F. (1982). A dispersal model for larvae of the deep sea red crab *Geryon quinquedens* based upon behavioral regulation of vertical migration in the hatching stage. *Mar. Biol. (Berlin)* **72**, 35–43.

Kittredge, J. S., Terry, M., and Takahashi, F. T. (1971). Sex pheromone activity of the molting hormone, crustecdysone, on male crabs. *Fish. Bull.* **69**, 337–343.

Klein Breteler, W. C. M. (1975a). Food consumption, growth and energy metabolism of juvenile shore crabs, *Carcinus maenas*. *Neth. J. Sea Res.* **9**, 255–272.

Klein Breteler, W. C. M. (1975b). Growth and moulting of juvenile shore crabs, *Carcinus maenas*, in a natural population. *Neth. J. Sea Res.* **9**, 86–99.

Klein Breteler, W. C. M. (1975c). Laboratory experiments on the influence of environmental factors on the frequency of moulting and the increase in size at moulting of juvenile shore crabs, *Carcinus maenas*. *Neth. J. Sea Res.* **9**, 100–120.

Knowlton, R. E. (1974). Larval developmental processes and controlling factors in decapod Crustacea, with emphasis on Caridea. *Thalassia Jugosl.* **10**, 138–158.

Knudsen, J. W. (1964). Observations of the reproductive cycles and ecology of the common Brachyura and crablike Anomura of Puget Sound, Washington. *Pac. Sci.* **18**, 3–33.

Krouse, J. S. (1976). Size composition and growth of young rock crab, *Cancer irroratus*, on a rocky beach in Maine. *Fish. Bull.* **74**, 949–954.

Krouse, J. S. (1978). Effectiveness of escape vent shape in traps for catching legal-sized lobster, *Homarus americanus*, and harvestable-sized crabs, *Cancer borealis* and *Cancer irroratus*. *Fish. Bull.* **76**, 425–432.

Krouse, J. S. (1980). Distribution and catch composition of Jonah crab, *Cancer borealis*, and rock crab, *Cancer irroratus*, near Boothbay Harbor, Maine. *Fish. Bull.* **77**, 685–693.

Kurata, H. (1962). Studies on the age and growth of Crustacea. *Bull. Hokkaido Reg. Fish. Res. Lab.* **22**, 1–48.

Lee, C. F., Knobl, G. M., and Deady, E. F. (1963). Mechanizing the blue crab industry. Part I. Survey of processing plants. *Commer. Fish. Rev.* **25,** 1–10.

Lee, C. F., and Sanford, F. B. (1962). Soft-crab industry. *Commer. Fish. Rev.* **24,** 10–12.

Lee, C. F., and Sanford, F. B. (1964). Crab industry of Chesapeake Bay and the south—an industry in transition. *Commer. Fish. Rev.* **26,** 1–12.

LeLoueff, P., Intes, A., and LeGuen, J. C. (1974). Note sur les premiers essais de capture du crabe profond *Geryon quinquedens* en Cote d'Ivoire. *Doc. Sci. Centre Rech. Oceanogr. Abidjan. V (1–2),* pp. 73–84.

Lewis, E. G. (1977). Relative growth and sexual maturity of *Bathynectes superbus* (Costa) (Decapoda: Portunidae) *J. Nat. Hist.* **11,** 629–643.

Lewis, E. G., and Haefner, P. A., Jr. (1976). Oxygen consumption of the blue crab, *Callinectes sapidus* Rathbun, from proecdysis to postecdysis. *Comp. Biochem. Physiol.* **54A,** 55–60.

Lochhead, J. H. (1950a). Crayfishes (and *Homarus*). *In* "Selected Invertebrate Types" (F. A. Brown, Jr., ed.), pp. 422–447. Wiley, New York.

Lochhead, J. H. (1950b). *Callinectes sapidus, In* "Selected Invertebrate Types" (F. A. Brown, Jr., ed.), pp. 447–462. Wiley, New York.

Lockwood, A. P. M. (1962). The osmoregulation of Crustacea. *Biol. Rev. Cambridge Philos. Soc.* **37,** 257–305.

Longhurst, A. R. (1971). Crustacean resources. *In* "The Fish Resources of the Ocean" (J. A. Gulland, ed.), pp. 206–245. Fishing News (Books) Ltd., Surrey, England.

Lough, R. G. (1975). Dynamics of crab larvae (Anomura, Brachyura) off the central Oregon coast, 1969–1971. Ph.D. Thesis, Oregon St. Univ., Corvallis.

Love, M. S., and Westphal, W. V. (1981). A correlation between annual catches of dungeness crab, *Cancer magister,* along the west coast of North America and mean annual sunspot number. *Fish. Bull.* **79,** 794–796.

McDonald, D. G., McMahon, B. R., and Wood, C. M. (1977). Patterns of heart and scaphognathite activity in the crab *Cancer magister. J. Exp. Zool.* **202,** 33–43.

McHugh, J. L. (1968). The biologist's place in the fishing industry. *BioScience* **18,** 935–939.

McHugh, J. L., and Ladd, E. C. (1953). The unpredictable blue crab fishery. *Natl. Fish. Yearbook, 1953.* pp. 1–3.

MacKay, D. C. G., and Weymouth, F. W. (1935). The growth of the Pacific edible crab, *Cancer magister* Dana. *J. Biol. Board Can.* **1,** 191–212.

McKelvey, R. (1981). Letters and comments: Comment on cycles in the northern California dungeness crab population, Reply. *Can. J. Aquat. Sci.* **38,** 1295–1297.

McKelvey, R., Hankin, D., Yanosko, K., and Snygg, C. (1980). Stable cycles in multistage recruitment models: An application to the northern California dungeness crab (*Cancer magister*) fishery. *Can. J. Fish. Aquat. Sci.* **37,** 2323–2345.

McKernan, D. L. (1972). World fisheries—world concern. *In* "World Fisheries Policy, Multidisciplinary Views" (B. J. Rothschild, ed.), pp. 35–51. Univ. of Washington, Seattle.

McMahon, B. R., and Wilkens, J. L. (1977). Periodic respiratory and circulatory performance in the red rock crab *Cancer productus. J. Exp. Zool.* **202,** 363–374.

MacMillen, R. E., and Greenaway, P. (1978). Adjustments of energy and water metabolism to drought in an Australian arid-zone crab. *Physiol. Zool.* **51,** 230–240.

McMullen, J. C. (1969). Effects of delayed mating on the reproduction of king crab, *Paralithodes camtschatica. J. Fish. Res. Board Can.* **26,** 2737–2740.

McMullen, J. C., Yoshihara, H. T., and Geiger, M. (1972). The king crab fisheries of the Alaska peninsula and Aleutian Islands, 1970–1971. *Alaska Dept. Fish. Game. Inf. Lflt. No. 175,* 21 pp.

McRae, E. D., Jr. (1961). Red crab explorations off the northeastern coast of the United States. *Commer. Fish. Rev.* **23,** 5–10.

Mahood, R. K., McKenzie, M. D., Middaugh, D. P., Bollar, S. J., and Davis, J. R. (1970). A report on the cooperative blue crab study—south Atlantic states. *Georgia Game Fish. Comm. Coastal Fish. Div. Contrib. Ser. No. 19,* 32 pp.

Mangum, C., and Towle, D. (1976). Physiological adaptation to unstable environments. *Am. Sci.* **65,** 67–75.

Mantel, L. H. (1968). The foregut of *Gecarcinus lateralis* as an organ of salt and water balance. *Am. Zool.* **8,** 433–490.

Manton, S. M. (1968). Terrestrial arthropods (2). *In* "Animal Locomotion" (J. Gray, ed.), pp. 333–376. Weidenfield and Nicolson, London.

Manton, S. M. (1969). Evolution and affinities of Onychophora, Myriapoda, Hexapoda, and Crustacea. *In* "Treatise on Invertebrate Paleontology" (R. C. Moore, ed.), Part (R) Arthropoda 4 (1), pp. R15–R56. Geol. Soc. Am. Boulder, Colorado.

Marchant, A., and Holmsen, A. (1975). Harvesting rock and Jonah crabs in Rhode Island: Some technical and economic aspects. *Univ. R.I. Mar. Memo. No. 35,* 15 pp.

Marshall, S. M., and Orr, A. P. (1960). Feeding and nutrition. *In* "The Physiology of Crustacea. Vol. I. Metabolism and Growth" (T. H. Waterman, ed.), pp. 227–258. Academic Press, New York.

Massachusetts Lobstermen's Association (1974). Preliminary investigations into the *Cancer* crab resources of Massachusetts. Ms. Rep. for New England Fish. Dev. Program (Natl. Mar. Fish. Serv.). Mass. Lobstermen's Assoc., Inc., 338 Spring St., Marshfield Hills, Massachusetts, 17 pp.

Mauchline, J. (1976). The Hiatt growth diagram for Crustacea. *Mar. Biol. (Berlin)* **35,** 79–84.

Mendenhall, V. (1971). Utilization and disposal of crab and shrimp wastes. *Univ. Alaska Sea Grant Mar. Adv. Bull. No. 2,* 40 pp.

Miller, D. C., and Vernberg, F. J. (1968). Some thermal requirements of fiddler crabs of the temperate and tropical zones and their influence on geographic distribution. *Am. Zool.* **8,** 459–470.

Miller, R. E., Campbell, D. W., and Lunsford, P. J. (1980). Comparison of sampling devices for the juvenile blue crab, *Callinectes sapidus. Fish. Bull.* **78,** 195–198.

Miller, R. J. (1979). Social rationales for management of a crab fishery. *Fisheries* **4,** 28–31.

Miller, R. J., and Watson, J. (1976). Growth per molt and limb regeneration in the spider crab, *Chionoecetes opilio. J. Fish. Res. Board Can.* **33,** 1644–1649.

Millikin, M. R. (1978). Blue crab larval culture: Methods and management. *Mar. Fish. Rev.* **40,** 10–17.

More, W. R. (1969). A contribution to the biology of the blue crab (*Callinectes sapidus* Rathbun) in Texas, with a description of the fishery. *Texas Parks Wildl. Dep. Tech. Ser. No. 1,* 31 pp.

Morris, M. E. (1969). King-crab industry. *In* "The Encyclopedia of Marine Resources" (F. E. Firth, ed.), pp. 155–159. Van Nostrand-Reinhold, Princeton, New Jersey.

Muzzarelli, R. A. A., and Pariser, E. R., eds. (1978). "Proceedings of the First International Conference on Chitin/Chitosan." MIT Sea Grant Program, Cambridge, Massachusetts.

Nedelec, C., ed. (1975). "Catalogue of Small-Scale Fishing Gear." Whitefriars Press Ltd., London.

Newcombe, C. L. (1949). A method for studying growth in different groups of arthropods. *Science* **109,** 84–85.

Newcombe, C. L., Campbell, F., and Eckstine, A. M. (1949a). A study of the form and growth of the blue crab *Callinectes sapidus* Rathbun. *Growth* **13,** 71–96.

Newcombe, C. L., Sandoz, M. D., and Rogers-Talbert, R. (1949b). Differential growth and moulting characteristics of the blue crab, *Callinectes sapidus* Rathbun. *J. Exp. Zool.* **110,** 113–152.

Newman, M. C., and Feng, S. Y. (1982). Susceptibility and resistance of the rock crab, *Cancer irroratus*, to natural and experimental bacterial infection. *J. Invertebr. Pathol.* **40,** 75–88.

Olsen, D. A., Dammann, A. E., and LaPlace, J. A. (1978). *Portunus spinimanus*, a portunid crab with resource potential in the U.S. Virgin Islands. *Mar. Fish. Rev.* **40,** 12–15.

Pagcatipunan, R. (1972). Observations on the culture of alimango, *Scylla serrata* at Camarines Norte (Phillipines). *In* "Coastal Aquaculture in the Indo-Pacific Region" (T. V. R. Pillay, ed.), pp. 362–365. Fishing News (Books) Ltd., London.

Panikkar, N. K. (1969). Fishery resources of the Indian Ocean. *Bull. Natl. Inst. Sci. India No. 38*, pp. 811–832.

Passano, L. M. (1960). Molting and its control. *In* "The Physiology of Crustacea. Vol. I. Metabolism and Growth" (T. H. Waterman, ed.), pp. 473–536. Academic Press, New York.

Passano, L. M. (1961). The regulation of crustacean metamorphosis. *Am. Zool.* **1,** 89–95.

Patton, W. K. (1967). Commensal Crustacea. *Proc. Symp. Crustacea* **III** 1228–1243.

Pearson, J. (1908). *Cancer. Liverpool Mar. Biol. Comm. Mem.* **XVI,** 209 pp., 13 pl.

Pearson, J. C. (1948). Fluctuations in the abundance of the blue crab in Chesapeake Bay. *U. S. Fish. Wildl. Res. Rep.* **14, 26 pp.**

Pearson, W. H., and Olla, B. L. (1977). Chemoreception in the blue crab, *Callinectes sapidus. Biol. Bull. (Woods Hole, Mass.)* **153,** 346–354.

Pereyra, W. T. (1966). The bathymetric and seasonal distribution and reproduction of adult tanner crabs, *Chionoecetes tanneri* Rathbun (Brachyura: Majidae), off the northern Oregon coast. *Deep-Sea Res.* **13,** 1185–1205.

Pereyra, W. T., Reeves, J. E., and Bakkala, R. G. (1976). Demersal fish and shellfish resources of the eastern Bering Sea in the baseline year 1975. *Natl. Mar. Fish. Serv. Northwest and Alaska Fish. Ctr. Proc. Rep. Seattle, Washington,* 619 pp.

Perry, H. M. (1975). The blue crab fishery in Mississippi. *Gulf Res. Rep.* **5,** 39–57.

Peterson, W. T. (1973). Upwelling indices and annual catches of dungeness crab, *Cancer magister*, along the west coast of the United States. *Fish. Bull.* **71,** 902–910.

Pinhorn, A. T. (1976). Living marine resources of Newfoundland–Labrador: Status and potential. *Bull. Fish. Res. Board Can. No. 194,* 64 pp.

Powell, G. C., and Nickerson, R. B. (1965). Reproduction of king crabs, *Paralithodes camtschatica* (Tilesius), *J. Fish. Res. Bd. Can.* **22,** 101–111.

Powell, G. C., Shafford, B., and Jones, M. (1972). Reproductive biology of young adult king crabs *Paralithodes camtschatica* (Tilesius) at Kodiak, Alaska. *Natl. Shellfish. Assoc. Proc.* **63,** 77–87.

Powell, G. C., James, K. E., and Hurd, C. L. (1974). Ability of male king crab, *Paralithodes camtschatica*, to mate repeatedly, Kodiak, Alaska, 1973. *Fish. Bull.* **72,** 171–179.

Powles, H. W. (1966). Observations on the biology of two species of spider crabs, *Chionoecetes opilio* and *Hyas araneus*, in the Gulf of St. Lawrence. *Fish. Res. Board Can. Manus. Rep. Ser. No. 884,* 15 pp. (14 Figs.).

Powles, H. W. (1968a). Distribution and biology of the spider crab *Chionoecetes opilio* in the Magdalen shallows, Gulf of St. Lawrence. *Fish. Res. Board Can. Manus. Rep. Ser. No. 997,* 106 pp.

Powles, H. W. (1968b). Observations on the distribution and biology of the spider crab, *Chionoecetes opilio. Fish. Res. Board Can. Manus. Rep. Ser. No. 950,* 10 pp. (8 Tables and Figs.).

Pradeille-Rouquette, M. (1978). Physiologie des invertébrés. Évolution ovarienne du Crabe *Pachygrapsus marmoratus* (F.) privé de ses organes Y et de ses pedoncules oculaires. *C. R. Acad. Sci., Ser. D* **287,** 1297–1299.

Pyle, R., and Cronin, E. (1950). The general anatomy of the blue crab *Callinectes sapidus*

Rathbun. *Maryland Bd. Nat. Res. Dept. Res. Educ. Chesapeake Biol. Lab. Solomons, Md. Publ. No. 87,* 38 pp.

Rahaman, A. A. (1967). Reproductive and nutritional cycles of the crab *Portunus pelagicus* (Linnaeus) (Decapoda: Brachyura) of the Madras Coast. *Proc. Indian Acad. Sci.* **65B,** 76–82.

Rao, K. R. (1968). The pericardial sacs of *Ocypode* in relation to the conservation of water, molting, and behavior. *Am. Zool.* **8,** 561–568.

Rao, K. V. (1969). Distributional pattern of the major exploited marine fishery resources of India. *Bull. Cent. Mar. Fish. Res. Inst. No. 6,* 1–69.

Rathbun, M. J. (1901). The Brachyura and Macrura of Porto Rico. *U. S. Fish. Comm. Bull.* **2,** 1–137.

Rathbun, M. J. (1930). The cancroid crabs of America of the families Euryalidae, Portunidae, Atelecyclidae, Cancridae and Xanthidae. *Bull. U. S. Natl. Mus.* **152,** i/xvi, 1–609 (pls. 1–230).

Rathbun, M. J. (1933). Brachyuran crabs of Porto Rico and the Virgin Islands. Scientific survey of Porto Rico and the Virgin Islands, *N. Y. Acad. Sci. 25, Pt. 1,* 121 pp.

Rathjen, W. F. (1974). New England fisheries development program. *Mar. Fish. Rev. 36,* 23–30.

Redmond, J. R. (1968). Transport of oxygen by the blood of the land crab, *Gecarcinus lateralis, Am. Zool.* **8,** 471–480.

Reed, P. H. (1969). Culture methods and effects of temperature and salinity on survival and growth of dungeness crab (*Cancer magister*) larvae in the laboratory. *J. Fish. Res. Board Can.* **26,** 389–397.

Rees, G. H. (1963). Edible crabs of the United States. *Fish. Lflt. No. 550,* 18 pp.

Reilly, P. N., and Saila, S. B. (1978). Biology and ecology of the rock crab, *Cancer irroratus* Say, 1817, in southern New England waters (Decapoda, Brachyura). *Crustaceana* **34,** 121–140.

Ricker, W. E. (1954). Stock and recruitment. *J. Fish. Res. Board Can.* **11,** 559–623.

Ricker, W. E. (1975). Computation and interpretation of biological statistics of fish populations. *Can. Fish. Res. Board Bull.* **191,** 1–382.

Robertson, J. D. (1960a). Ionic regulation in the crab (*Carcinus maenas* (L.) in relation to the moulting cycle. *Comp. Biochem. Physiol.* **1,** 183–212.

Robertson, J. D. (1960b). Osmotic and ionic regulation. *In* "The Physiology of Crustacea. Vol. I. Metabolism and Growth" (T. H. Waterman, ed.), pp. 317–339. Academic Press, New York.

Rothschild, B. J., ed. (1972). "World Fisheries Policy, Multidisciplinary Views." Univ. of Washington Press, Seattle.

Rothschild, B. J., Powell, G., Joseph, J., Abramson, N. J., Buss, J. A., and Eldridge, P. (1970). A survey of the population dynamics of king crab in Alaska with particular reference to the Kodiak area. *Alaska Fish Game Inf. Lflt. No. 147,* 148 pp.

Rounsefell, G. A. (1975). "Ecology, Utilization, and Management of Marine Fisheries." Mosby, St. Louis, Missouri.

Ryan, E. P. (1966). Pheromone: Evidence in a decapod crustacean. *Science* **151,** 340–341.

Sainsbury, J. C. (1971). "Commercial Fishing Methods, an Introduction to Vessels and Gear." Fishing News (Books) Ltd., Surrey, England.

Sandifer, P. A. (1975). The role of pelagic larvae in recruitment to populations of adult decapod crustaceans in the York River estuary and adjacent lower Chesapeake Bay, Virginia. *Estuarine Coastal Mar. Sci.* **3,** 269–279.

Sastry, A. N. (1977). The larval development of the rock crab, *Cancer irroratus* Say, 1817, under laboratory conditions (Decapoda, Brachyura). *Crustaceana* **32,** 155–168.

Sastry, A. N., and McCarthy, J. F. (1973). Diversity in metabolic adaptation of pelagic larval stages of two sympatric species of brachyuran crabs. *Neth. J. Sea. Res.* **7,** 434–446.

Savage, T., and Sullivan, J. R. (1978). Growth and claw regeneration of the stone crab, *Menippe mercenaria. Florida Dep. Nat. Res. Mar. Res. Publ. No. 32,* 23 pp.

Savage, T., Sullivan, J. R., and Kalman, C. E. (1975). An analysis of stone crab (*Menippe mercenaria*) landings on Florida's west coast, with a brief synopsis of the fishery. *Fa. Dep. Nat. Res. Mar. Res. Publ. No. 13,* 37 pp.

Sawyer, T. K., and MacLean, S. A. (1978). Some protozoan diseases of decapod crustaceans. *Mar. Fish. Rev.* **40,** 32–35.

Schäfer, W. (1954). Form und Funktion der Brachyuren-Schere. *Abh. Senckenb. Naturforsch. Ges. No. 489,* pp. 1–65.

Schmidt, G., and MacLean, S. A. (1978). *Polymorphus (Profilicollis) major* Lundstrom 1942 juveniles in rock crabs, *Cancer irroratus,* from Maine. *J. Parasitol.* **64,** 953–954.

Schroeder, W. C. (1959). The lobster, *Homarus americanus,* and the red crab, *Geryon quinquedens* in the offshore waters of the western North Atlantic. *Deep-Sea Res.* **5,** 266–282.

Shotton, L. R. (1973). Biology of the rock crab, *Cancer irroratus* Say, in the coastal waters of Virginia. M.A. Thesis, Univ. of Virginia, Charlottesville.

Simpson, K. L. (1978). Recovery of protein and pigments from shrimp and crab meals and their use in salmonid pigmentation. *In* "Proceeding of the First International Conference on Chitin/Chitosan (R. A. A. Muzzarelli and E. R. Pariser, eds.), pp. 253–262. MIT Sea Grant Program, Cambridge, Massachusetts.

Sindermann, C. J. (1970). "Principal Diseases of Marine Fish and Shellfish." Academic Press, New York.

Sinderman, C. J., and Rosenfield, A. (1967). Principal diseases of commercially important marine bivalve Mollusca and Crustacea. *Fish. Bull.* **66,** 335–385.

Skinner, D. M., and Graham, D. E. (1970). Molting in land crabs: Stimulation by leg removal. *Science* **169,** 383–385.

Skinner, D. M., and Graham, D. E. (1972). Loss of limbs as a stimulus to ecdysis in Brachyura (true crabs). *Biol. Bull. (Woods Hole, Mass.)* **143,** 222–233.

Smith, D. C. (1978). Dipping and picking: A guide to recreational crabbing. *S. C. Mar. Res. Ctr. Sea Grant Mar. Adv. Bull. No. 8,* 10 pp.

Somerton, D. A. (1980a). A computer technique for estimating the size of sexual maturity in crabs. *Can. J. Fish. Aquat. Sci.* **37,** 1488–1494.

Somerton, D. A. (1980b). Fitting straight lines to Hiatt growth diagrams: A re-evaluation. *J. Cons. Int. Explor. Mer.* **39,** 15–19.

Somerton, D. A. (1981). Regional variation in the size and maturity of two species of tanner crab (*Chionoecetes bairdi* and *C. opilio*) in the eastern Bering Sea, and its use in defining management subareas. *Can. J. Aquat. Sci.* **38,** 163–174.

Spaargaren, D. H. (1974). Measurements of relative rate of blood flow in the shore crab, *Carcinus maenas,* at different temperatures and salinities. *Neth. J. Sea. Res.* **8,** 398–406.

Spaargaren, D. H. (1975). Changes in permeability in the shore crab *Carcinus maenas* (L.), as a response to salinity. *Comp. Biochem. Physiol.* **51A,** 549–552.

Sparks, A. K., and Hibbits, J. (1979). Black mat syndrome, an invasive mycotic disease of the tanner crab, *Chionoecetes bairdi. J. Invertebr. Pathol.* **34,** 184–191.

Spirito, C. P. (1972). An analysis of swimming behavior in the portunid crab *Callinectes sapidus. Mar. Behav. Physiol.* **1,** 261–276.

Sprague, V., and Couch, J. (1971). An annotated list of protozoan parasites, hyperparasites and commensals of decapod Crustacea. *J. Protozool.* **18,** 526–537.

Stasko, A. (1975). Modified lobster traps for catching crabs and keeping lobsters out. *J. Fish. Res. Board Can.* **32,** 2515–2520.

Stěvcic, A. (1971). The main features of brachyuran evolution. *Syst. Zool.* **20**, 331–340.

Stevens, B. G., and Armstrong, D. A. (1981). Mass mortality of female dungeness crab, *Cancer magister,* on the southern Washington coast. *Fish. Bull.* **79**, 349–352.

Sulkin, S. D. (1972). Blue crab study in Chesapeake Bay—Maryland. *Univ. Maryland Nat. Res. Inst. Prog. Rep. Ref. No. 72-37,* 17 pp.

Sulkin, S. D., Van Heukelem, W., Kelly, P., and Van Heukelem, L. (1980). The behavioral basis of larval recruitment in the crab *Callinectes sapidus* Rathbun: A laboratory investigation of ontogenetic changes in geotaxis and barokinesis. *Biol. Bull. (Woods Hole, Mass.)* **159**, 402–417.

Swartz, R. C. (1976). Sex ratio as a function of size in xanthid crab, *Neopanope sayi. Am. Nat.* **110**, 898–900.

Swartz, R. C. (1978). Reproductive and molt cycles in the xanthid crab, *Neopanope sayi* (Smith, 1869). *Crustaceana* **34**, 15–32.

Tagatz, M. E. (1965). The fishery for blue crabs in the St. Johns River, Florida, with special reference to fluctuation in yield between 1961 and 1962. *U. S. Fish. Wildl. Serv. Spec. Sci. Rep. - Fish No. 501,* 11 pp.

Tagatz, M. E. (1969). Some relations of temperature acclimation and salinity to thermal tolerance of the blue crab, *Callinectes sapidus. Trans. Am. Fish. Soc.* **98**, 713–716.

Taylor, E. W., and Butler, P. J. (1973). The behaviour and physiological responses of the shore crab *Carcinus maenas* during changes in environmental oxygen tension. *Neth. J. Sea Res.* **7**, 496–505.

Taylor, E. W., Butler, P. J., and Sherlock, P. J. (1973). The respiratory and cardiovascular changes associated with the emersion response of *Carcinus maenas* (L.) during environmental hypoxia, at three different temperatures. *J. Comp. Physiol.* **86**, 95–115.

Teissier, G. (1960). Relative growth. *In* "The Physiology of Crustacea. Vol. I. Metabolism and Growth" (T. H. Waterman, ed.) pp. 537–560. Academic Press, New York.

Telford, M. (1978). Distribution of two species of *Dissodactylus* (Brachyura: Pinnotheridae) among their echinoid host populations in Barbados. *Bull. Mar. Sci.* **28**, 652–658.

Terretta, R. T. (1973). Relative growth, reproduction and distibution of the rock crab, *Cancer irroratus,* in Chesapeake Bay during the winter. M.A. Thesis, College of William and Mary, Williamsburg, Virginia.

Teytaud, A. R. (1971). The laboratory studies of sex recognition in the blue crab *Callinectes sapidus* Rathbun. *Univ. Miami Sea Grant Tech. Bull. No. 15,* 63 pp.

Thomas, J. C. (1973). An analysis of the commercial lobster (*Homarus americanus*) fishery along the coast of Maine, August 1966 through December 1970. *Nat. Mar. Fish. Serv. Spec. Sci. Rep. - Fish. No. 667,* 57 pp.

Turvey, R., and Wiseman, J., eds. (1957). "The Economics of Fisheries. Proceedings of a round table orgnized by the International Economic Assoc." F.A.O., Rome, Italy, Sept. 1956.

Uglow, R. F. (1973). Some effects of acute oxygen changes on heart and scaphognathite activity in some portunid crabs. *Neth. J. Sea Res.* **7**, 447–454.

Uzmann, J. R., Cooper, R. A., Theroux, R. B., and Wigley, R. L. (1977). Synoptic comparison of three sampling techniques for estimating abundance and distribution of selected megafauna: Submersible vs camera sled vs otter trawl. *Mar. Fish. Rev. 39,* 11–19.

Van Engel, W. A. (1962). The blue crab and its fishery in Chesapeake Bay. Part 2. Types of gear for hard crab fishing. *Commer. Fish. Rev. 24,* 1–10.

Van Engel, W. A. (1974). Underutilized crustaceans of the Chesapeake Bay and Chesapeake Bight. Manuscript of presentation at Fish Expo 1974, Norfolk, Virginia, 6 pp. Virginia Inst. Mar. Sci., Gloucester Point, Virginia.

Van Engel, W. A., Cargo, D. G., and Wojcik, F. J. (1973). The edible blue crab . . . abundant crustacean. *Atl. States Mar. Fish. Comm. Mar. Res. Atl. Coast Lflt. No. 15,* 8 pp.

Van Engel, W. A., and Haefner, P. A., Jr. (1975). Discoloration in rock crabs: What to do about it. *Virginia Inst. Mar. Sci. Mar. Resour. Adv. Ser. No. 9*, 1 p.

VanWinkle, W., And Mangum, C. (1975). Oxyconformers and oxyregulators: A quantitative index. *J. Exp. Mar. Biol. Ecol.* **17**, 103–110.

Varga, S., Dewar, A. B., and Anderson, W. E. (1969). Survival of red crabs held on ice and refrigerated air. *Halifax Dept. Fish. For. Tech. Rep. No. 2*, 3 pp.

Vargo, S. L., and Sastry, A. N. (1977). Acute temperature and low dissolved oxygen tolerances of brachyuran crab (*Cancer irroratus*) larvae. *Mar. Biol. (Berlin)* **40**, 165–171.

Veerannan, K. M. (1972). Respiratory metabolism of crabs from marine and estuarine habitats. I. *Scylla serrata. Mar. Biol. (Berlin)* **17**, 284–290.

Veerannan, K. M. (1974). Respiratory metabolism of crabs from marine and estuarine habitats: An interspecific comparison. *Mar. Biol. (Berlin)* **26**, 35–43.

Venkatachari, S. A. T., and Kadam, G. A. (1974). Temperature characteristics of oxygen consumption of the freshwater crab, *Barytelphusa guerini* H. Milne Edwards and their relation to stress. *Monit. Zool. Ital. N.S.* **8**, 19–28.

Vermeij, G. J. (1977). Patterns in crab claw size: the geography of crushing. *Syst. Zool.* **26**, 138–151.

Vernberg, W. B., and Vernberg, F. J. (1968). Physiological diversity in metabolism in marine and terrestrial Crustacea. *Am. Zool.* **8**, 449–458.

Vernberg, W. B., and Vernberg, F. J. (1972). "Environmental Physiology of Marine Animals." Springer-Verlag, Berlin and New York.

Virnstein, R. W. (1977). The importance of predation by crabs and fishes on benthic infauna in Chesapeake Bay. *Ecology* **58**, 1199–1217.

Von Brandt, A. (1972). "Fish Catching Methods of the World." Fishing News (Books) Ltd., London.

Vonk, H. J. (1960). Digestion and metabolism, *In* "The Physiology of Crustacea. Vol. I. Metabolism and Growth" (T. H. Waterman, ed.), pp. 291–316. Academic Press, New York.

Walford, L. A. (1946). A new graphic method of describing the growth of animals. *Biol. Bull. (Woods Hole, Mass.)* **90**, 141–147.

Warner, G. F. (1977). "The Biology of Crabs." Van Nostrand-Reinhold, Princeton, New Jersey.

Warner, G. F., and Jones, A. R. (1976). Leverage and muscle type in crab chelae (Crustacea: Brachyura) *J. Zool.* **180**, 57–68.

Warner, W. W. (1976). "Beautiful Swimmers, Watermen, Crabs and the Chesapeake Bay." Little, Brown, Boston, Massachusetts.

Waterman, T. H., ed. (1960). "The Physiology of Crustacea. Vol. I. Metabolism and Growth." Academic Press, New York.

Waterman, T. H., and Chace, F. A., Jr. (1960). General crustacean biology. *In* "The Physiology of Crustacea. Vol. I. Metabolism and Growth" (T. H. Waterman, ed.), pp. 1–30. Academic Press, New York.

Watson, J. (1969). Biological investigations on the spider crab, *Chionoecetes opilio. Can. Fish. Rep.* **13**, 24–47.

Welch, W. R. (1968). Changes in abundance of the green crab, *Carcinus maenas* (L.), in relation to recent temperature changes. *Fish. Bull.* **67**, 337–345.

Welsh, J. P. (1974). Mariculture of the crab *Cancer magister* (Dana) utilizing fish and crustacean wastes as food. *Humboldt State Univ. Sea Grant Proj. Rep. HSU-SG-4*, 76 pp.

Wenner, A. M. (1972). Sex ratio as a function of size in marine Crustacea. *Am. Nat.* **106**, 321–351.

Wenner, A. M., Fusaro, C., and Oaten, A. (1974). Size at onset of sexual maturity and growth rate in crustacean populations. *Can. J. Zool.* **52**, 1095–1106.

Weymouth, W., and MacKay, D. C. G. (1936). Analysis of the relative growth of the Pacific edible crab, Cancer magister. Proc. Zool. Soc. London **1**, 257–280.

Whittington, H. B., and Rolfe, W. D. I., eds. (1963). "Phylogeny and Evolution of Crustacea." Spec. Publ. Museum of Comparative Zoology, Cambridge, Massachusetts.

Wickham, D. E. (1979a). Predation by the nemertean Carcinonemertes errans on eggs of the dungeness crab Cancer magister. Mar. Biol. (Berlin) **55**, 45–53.

Wickham, D. E. (1979b). Carcinonemertes errans and the fouling and mortality of eggs of the dungeness crab, Cancer magister, J. Fish. Res. Board Can. **36**, 1319–1324.

Wickham, D. E. (1979c). The relationship between megalopae of the dungeness crab, Cancer magister, and the hydroid, Velella velella, and its influence on abundance estimates of C. magister megalopae. Calif. Fish Game **65**, 184–186.

Wickham, D. E. (1980). Aspects of the life history of Carcinonemertes errans (Nemertea; Carcinonemertidae), an egg predator of the crab Cancer magister. Biol. Bull. (Woods Hole, Mass.) **159**, 247–257.

Wigley, R. L., Theroux, R. B., and Murray, H. E. (1975). Deep-sea red crab, Geryon quinquedens, survey off northeastern U.S. Mar. Fish. Rev. 37, 1–21.

Wilder, D. G. (1966). Canadian Atlantic crab resources. Fish. Res. Board Can. Gen. Ser. Circ. No. 50, 6 pp.

Williams, A. B. (1965). Marine decapod crustaceans of the Carolinas. Fish. Bull. **65**, 1–298.

Williams, A. B. (1974). The swimming crabs of the genus Callinectes (Decapoda: Portunidae) Fish. Bull. **72,** 685–798.

Williams, A. B., and Duke, T. W. (1979). Crabs (Arthropoda: Crustacea: Decapoda: Brachyura). In "Pollution Ecology of Estuarine Invertebrates" (C. W. Hart, Jr. and S. L. H. Fuller, eds.), pp. 171–233. Academic Press, New York.

Williams, J. G. (1979). Estimation of intertidal harvest of dungeness crab, Cancer magister, on Puget Sound, Washington, beaches. Fish. Bull. **77,** 287–292.

Williams, M. J. (1978). Opening of bivalve shells by the mud crab Scylla serrata Forskål. Aust. J. Mar. Freshwater Res. **29,** 699–702.

Wolcott, T. G. (1978). Ecological role of ghost crabs, Ocypode quadrata (Fabricius) on an ocean beach: Scavengers or predators? J. Exp. Mar. Biol. Ecol. **31,** 67–82.

Yano, I., and Kobayashi, S. (1969). Calcification and age determination in Crustacea. I. Possiblity of age determination in crabs on the basis of number of lamellae in cuticles. Bull. Jpn. Soc. Sci. Fish. **35,** 34–42.

Yonge, C. M. (1928). Feeding mechanisms in the invertebrates. Biol. Rev. Cambridge Philos. Soc. **3,** 21–76.

Young, J. H. (1959). Morphology of the white shrimp Penaeus setiferus (Linnaeus 1758). Fish. Bull. **59,** 1–168.

Young, R. E. (1973a). Aspects of the physiology and ecology of haemocyanin in some West Indian mangrove crabs. Neth. J. Sea. Res. **7,** 476–481.

Young, R. E. (1973b). Responses to respiratory stress in relation to blood pigment affinity in Goniopsis cruentata (Latreille) and (to a lesser extent) in Cardisoma guanhumi Latreille. J. Exp. Mar. Biol. Ecol. **11,** 91–102.

Fisheries Biology of
Lobsters and Crayfishes

J. STANLEY COBB and DENIS WANG

I. INTRODUCTION

The lobster and crayfishes of the world are a large and diverse group in both a taxonomic and an ecological sense. They are found in marine and fresh waters from the deepest oceans and far recesses of caves to the shallow subtidal and narrow rivulets. The species that we shall discuss in this chapter are found in five of the families of decapods. In so diverse a group, what common bases for discussion can there be? We think that there are at least two bases. First, although found in diverse habitats, many fill the same ecological roles. Second, many species are commercially important, with similar fishing methods and management problems. It is these common aspects that will structure this chapter. In it, we briefly review some of the basic aspects of the biology of lobsters and crayfishes that are important to the fisheries, but the greater proportion of the chapter will be devoted to a discussion of ecology and population dynamics as well as to a description of the various fisheries, some of the management problems that exist, and some of the solutions to those problems.

Lobsters constitute an important part of the world's fisheries. In 1981, the most recent year for which comprehensive statistics are available, lobsters of all sorts made up 5.5% of the total world marine crustacean catch. Clawed lobsters (Nephropidae) were the largest at 89,373 metric tons (MT), whereas the spiny and slipper lobsters (Palinuridae and Scyllaridae) together contributed 65,614 MT. The reported catch of slipper lobsters was less than 500 MT in 1981, which probably does not reflect the true size of the fisheries worldwide. Perhaps more impressive than the total weight of the catch is the value of the landed product. In areas where lobsters are an export item or consumed in quantity in the country of origin, they command a price of U.S. $3.00 to more than $6.00 per kg. The amount and value of the freshwater crayfish catch is more difficult to estimate. In some areas, for example, the southeastern United States and Europe, crayfishes form a significant proportion of the freshwater fisheries. The catch of *Procambarus clarkii* in the United States in some years may reach 20,000 MT, but in most places the fisheries are unregulated, and there is little or no reporting system.

Under the general heading assigned for coverage in this review are included the many diverse types of clawed lobsters, spiny lobsters, and freshwater crayfishes. The vast array of common names given to these animals is bewildering; Farmer (1975) lists 53 for *Nephrops norvegicus* alone, thus we will use the scientific name only. The groups we are concerned with are the Nephropidae (clawed lobsters), the Palinuridae (spiny lobsters), and three families of crayfish, the Astacidae, Parastacidae, and Austroastacidae. Low population densities or inaccessability of habitat of many species means that only a relatively few species in these groups can be commercially important. We have confined the review to consideration of commercially important species with only occasional reference to others.

II. GENERAL BIOLOGY

The lobsters and freshwater crayfishes are similar in many ways. Their external morphology differs mostly in the presence or absence of large claws, and their life history patterns show the same general trends and varied adaptations for specific environments. They are usually the largest benthic crustacean in an ecosystem, they prey on smaller crustaceans, molluscs, and annelids, and their predators generally are large benthic feeding fishes, sharks, or octopus. Almost all are shelter-seeking, nocturnal animals. Spiny lobsters usually shelter communally, whereas clawed lobsters and crayfishes shelter individually. Several of the species undertake long-distance migrations, while many others are relatively sedentary. In this section, we provide a brief overview of several aspects of lobster and crayfish biology that are particularly relevant to fisheries management.

A. General Body Plan

As in other Malacostraca, there are five cephalic and eight thoracic segments fused together to form a cephalothorax covered by the shieldlike carapace. All the segments bear a pair of appendages, such as antennae, mouthparts, and walking legs. The compound eyes and their moveable stalks are not true appendages. The last six segments make up the abdomen or tail. In most groups, the first five abdominal segments bear paired, biramous pleopods. The most anterior of these are modified, in male clawed lobsters and crayfishes, into an intromittent organ for transfer of spermatophores. There are no first pleopods in the Parastacidae, and no intromittent organ. The telson, not a true segment, forms the central portion of the tail fan but carries no appendages. The uropods that flank the telson to make up the remainder of the tail fan are broad, flattened appendages that are modified from the pleopods of the last abdominal segment. The powerful musculature of the abdomen and the bladelike aspect of the tail fan are adaptations for the backward swimming escape response common to all lobsters and crayfishes.

In the clawed lobsters and freshwater crayfishes, the first walking legs are modified into large claws. The two claws are dimorphic in many species. This is most extreme in *Homarus,* with one claw, the "crusher," which is a more massive claw with large, rounded denticles. The smaller "cutter" claw has pointed and smaller denticles. The claws clearly demonstrate allometric growth. In the very earliest *Homarus* juveniles, they make up less than 5% of the total body weight, whereas in the largest adults they are nearly 50% of the total body weight (Lang *et al.*, 1977). In the earliest juvenile stages, both look like cutter claws; it is not until about the eighth or ninth molt after hatching that a difference in external morphology and neuromuscular development can be seen.

The first antennae, or antennules, are the site of long distance chemoreception (Ache, 1977). In the clawed lobsters and crayfishes, they are smaller than in the spiny lobsters, but there does not seem to be a major difference in their use. The long second antennae are large, sturdy, and covered with stout spines in the spiny lobsters, whereas in the clawed lobsters and crayfishes they are slender and whiplike. In all, one function of the second antenna appears to be tactile perception of the environment. In spiny lobsters, an additional function is in aggression against conspecifics and defense against predators.

The adaptions to specific environmental conditions are widely varied. Eyeless or functionally blind forms are found in the deep sea [*Acanthacaris caecus* (Holthius, 1974)] and in caves [*Troglocambarus maclenei* (Hobbs, 1974)]. In burrowing species, the chelae are often flattened and the spines reduced, whereas those species that live in open situations or in dens are more likely to have well-developed spines. Deep sea and cave dwellers are much less robust, have longer legs, reduced chelae, and longer antennae than do their more shallow-dwelling relatives. In the freshwater crayfishes, allometric dimensions of the carapace and abdomen may be very different between lotic and lentic species as adaptations to water flow and oxygen levels (Hobbs *et al.*, 1976).

B. Measurements

Lobsters and crayfishes have been measured along many different dimensions for many purposes. The most accepted linear measurement has become that of carapace length, the distance from the posterior rim of the eye socket to the posterior margin of the carapace, or in spiny lobsters the distance from the anterior margin of the carapace between the horns to the posterior margin of the carapace. Total length or tail length, both of which were used extensively some time ago, and in some areas are still used, have the disadvantage that the lengths change upon relaxation of the muscle tissue, upon preservation, or upon the application of pressure as when a fisherman is trying to "persuade" a lobster to be of legal size. Carapace length should not be measured from the tip of the rostrum or the horns because they often are broken or worn and thus may not accurately reflect the true size of the animal.

The relative development of several external structures have been proposed as measures of sexual maturity. In *H. americanus* (Aiken and Waddy, 1980) and *N. norvegicus* (Farmer, 1974a), the volume of the male crusher claw begins allometric growth some time after sexual maturity. Relative abdominal width measured both inside and outside the pleural spurs of the second abdominal segment has been used as an index of female sexual maturity in *H. americanus* (Templeman, 1935; Aiken and Waddy, 1980), *H. gammarus* (Simpson, 1961), and *N. norvegicus* (Farmer, 1974b). We recommend that the width measurement be taken on the outside rather than on the inside of the abdomen for purposes of standardization.

The maximum size of lobsters and crayfishes varies greatly among species. A terminal molt has not been described for any of the species, so an absolute upper limit in either length or weight cannot be defined. The largest individuals are always male. Wolff (1978) lists the largest *Homarus americanus* as 19.25 kg and 37.9 cm measured from tip of the rostrum to posterior margin of the carapace, and the largest *H. gammarus* as 8.4 kg and 25 cm. The spiny lobster species that grows to the largest size is *Jasus verrauxi*; individuals of up to 23.5 cm carapace length and 11.4 kg have been reported from New Zealand (Kensler, 1967). Freshwater crayfishes range in size from 20 to 30 mm (total length) in some lotic species to about 60 cm (total length) in the Australian *Astacopsis gouldi*, the largest freshwater crayfish known (Frost, 1975).

C. Life History Patterns

Lobsters and crayfish have similar patterns of life history, but there are many variations on the theme. In all, fertilized eggs are carried externally underneath the abdomen. The marine lobsters have pelagic larval stages, whereas the larvae of freshwater crayfish cling to the female. As adults, the animals are usually sexually dimorphic, with the males having a narrower abdomen, larger claws or first walking legs, and modified first pleopods.

1. CLAWED LOBSTERS

Homarus americanus, H. gammarus, and *Nephrops norvegicus* all have very similar life cycles and will be considered together here. Mating usually takes place when the female is soft, often within hours of the time of molting (Hughes and Matthiessen, 1962; Farmer, 1974b; Atema *et al.*, 1979). Sperm are stored in the seminal receptacle and remain viable for many months. It generally has been assumed that fertilization occurs as the eggs pass out of the oviducts and flow over the opening of the seminal receptacle on their way to attachment on the pleopods, but Farmer (1974b) has suggested the possibility of internal fertilization in the Nephropidae. Oviposition generally occurs in the summer and eggs hatch in the following spring. The eggs are carried for 6–10 months in *Nephrops* (Farmer, 1974b) and 10–12 months in *Homarus* (Perkins, 1972; Branford, 1978) before hatching. The length of the incubation period varies with locality and can be shortened in the laboratory by increasing the temperature (Perkins, 1972; Branford, 1978). It has been reported that female *Homarus* spawn every second year (Hughes and Matthiessen, 1962) and *Nephrops* every year (Farmer, 1974b).

Larval release occurs at night over the period of about 1 week in the laboratory (*H. gammarus,* Ennis, 1973a; *H. americanus,* Ennis, 1975; *Nephrops,* Farmer, 1974b) and, presumably, in the field as well. Ennis (1975) has suggested that the risk of predation on the larvae is lowered when they are released in small batches at night over a long period of time. In *Homarus* and *Nephrops,* there is a prelarval stage, three larval stages, and a postlarval stage. The larval stage is identified by the molt preceeding, thus a first-stage lobster has molted once, and the fourth-stage has molted four times. The fourth stage, or postlarva, is similar in appearance to the adult and markedly different from the preceeding zoeal stages. After taking up a benthic existence midway in the fourth stage, juvenile morphology, habitat, and behavior are very similar to those of the adults. Size and age at sexual maturity vary greatly depending both upon species and environmental conditions (Farmer, 1975; Aiken and Waddy, 1980), but maturity occurs after 2–4 years in *Nephrops* and 4–7 years in *Homarus.*

2. FRESHWATER CRAYFISH

The life history adaptations of crayfish species can vary greatly depending upon environment, but they follow the same basic pattern as the clawed lobsters. In many species, males molt into a reproductive stage, form I, during the breeding season. Morphological differences in form I males include modifications of pleopods, ambulatory legs, and chelae (Hobbs, 1974). Crayfishes mate in the spring and the summer after a period of courtship in which visual, chemical, and tactile signals are important (Ameyaw-Akumfi and Hazlett, 1975; Hayes, 1975; Stein, 1976; Pippitt, 1977). Some time later, the female extrudes eggs, which are fertilized by sperm stored in the seminal receptacle. The eggs are carried on the female's pleopods for several months. After hatching, the larvae remain attached to the abdomen of the female by telson threads for several days during which time they molt up to 4 times. After this period, the juveniles wander away. One or two generations, depending on the length of a favorable breeding season, are produced each year. In warm areas, some species may reproduce year round. Life-spans range from 1 to 2 years in relatively dense lentic populations of *Pacifastacus leniusculus* to several years in the burrowing Tasmanian *Parastacoides tasmanicus* and in the troglobitic *Orconectes inermis.*

3. SPINY LOBSTERS

The life cycles of spiny lobsters vary considerably among species. Some, such as *Jasus edwardsii* mate shortly after the female molts; but in most, the receptive period of the female is not linked so closely to molting. The male deposits a sperm packet (the "tar spot") on the sternum of the female. In some species of *Jasus,* insemination and fertilization both are thought to be internal, but in most other spiny lobsters they are external. All spiny lobsters spawn at least once a year after reaching sexual maturity, several spawn biannually, and at least one, *Panulirus homarus* can spawn up to 4 times a year (Berry, 1973). The sperm packet is carried for several months before eggs are extruded, fertilized, and attached to the pleopods underneath the abdomen. In the species that carry an external sperm packet, fertilization is initiated when the female scratches at the protective outer matrix with the fifth legs, exposing the sperm (Berry, 1970). The incubation period ranges from a few weeks (four in *Panulirus argus;* Munro, 1974) to several months (up to six in *Palinurus delagoae;* Berry, 1973). The newly hatched larvae are planktonic and concentrate at the surface. They are oceanic in distribution and spend many months at sea before metamorphosing into the puerulus stage, which then swims inshore from the edge of the continental shelf and settles in shallow water. The juveniles of several species of spiny lobsters remain in shallow inshore nursery areas for 1 to several years before moving into deeper water. In *Panulirus cygnus,* this movement is a synchronus one, occurring in November, and it is associated with molting into a "white phase" coloration that lasts several weeks (George, 1958; George et al., 1979).

D. Larval Ecology

An important part of the life cycle to fisheries biologists is the larval period because this is a time of dispersion and of high mortality. A recent, extensive review of the ecology of larval lobsters has been provided by Phillips and Sastry (1980).

The larvae of *Homarus* are present in the plankton during early and mid-summer. Several studies (Templeman, 1937; Templeman and Tibbo, 1945; Rogers et al., 1968; Scarratt, 1973a) have used nets towed at various depths and found that by far the largest concentration of larvae is at the surface. Squires et al. (1971) and Caddy (1979) suggested that vertical movements may result in position-keeping, particularly with regard to wind-driven turbulence. *Homarus* larvae are patchily distributed (Stasko, 1980; Fair, 1980), and sampling to determine abundance accurately is very difficult. In contrast to *Homarus,* the larvae of *Nephrops* are rarely taken near the surface (Jorgensen, 1925; Fraser, 1965; Hillis, 1972); they seem to be most abundant at depths of 20–150 m (Santucci, 1926; Williamson, 1956; Hillis, 1972).

Predators undoubtedly include all the larger plankton-feeding fishes, ctenophores, and probably gulls and terns. Estimates of mortality have been made from relative proportions of the several stages in the samples (Scarratt, 1964, 1973a; Lund and Stewart, 1970; Caddy, 1979), but variations in development time due to differences in water temperature, larval behavior, and vertical and horizontal distribution make such estimates difficult. Mead and Williams' (1903) suggestion that larval mortality rates are very high is probably as good an estimate as we have had to date.

The freshwater crayfishes do not have a free swimming larval stage. The female carries very few eggs as compared to lobsters. After hatching, the larvae remain on the pleopods of the female, during which time development to an almost adult form occurs. At the end of the brooding period, the juveniles take up a benthic, free-living existence. The lack of a pelagic larval stage presumably is an adaptation to stream life, where larvae could be swept away from parent populations with little chance for return. Brooding and maternal protection of juveniles may also increase offspring survival. Penn (1943) found that *Procambarus clarkii* juveniles remained with their mother for 5 days during which time they molted twice. Similarly, Mason (1970) saw juvenile *P. leniusculus* attached by telson threads to the mother's pleopods for 17 days and two molts. After becoming free-ranging, a maternal brood pheromone may be important in maintaining the protective relationship between offspring and mother. Little (1975, 1976) has suggested that an attractant brood pheromone, the production of which is mediated by a mechanoreceptive feedback system, is present until juveniles reach the fourth stage in the three species that he studied.

Spiny lobster larvae spend 4 to 11 months in the plankton, a lengthy period compared to the usual 20 to 30 day larval life of the clawed lobsters. However, the egg-bearing period of clawed lobsters is long and that of spiny lobsters is short; the total embryonic and larval period of the two groups is roughly comparable. A long larval period allows widespread dispersal, and larvae have been collected at great distance from possible parent populations: Phyllosomas of *Panulirus cygnus* 1500 km from spawning grounds in the Indian Ocean (Phillips *et al.,* 1979) and those of *P. penicillatus* (Olivier, 1791) and *P. gracilis* 4000 km away in the Pacific (Johnson, 1974). It is not known if larvae found so far from the spawning area actually return to it. The current patterns of the areas in which the larvae are distributed are of great importance but are little known.

Phyllosoma larvae of *Panulirus cygnus* spend 9 to 11 months in the plankton, during which time they are carried far from the west coast of Australia before returning to the continental shelf and metamorphosing to the puerulus stage (Chittleborough and Thomas, 1960; Phillips *et al.*, 1978). The larvae undergo a diurnal vertical migration apparently correlated with light intensity (Rimmer and Phillips, 1979). The migration and the vertical distribution of the larvae are important in accounting for their horizontal distribution. Diurnal migrations of the early stages place them at the surface at night when offshore wind-driven currents predominate and below the layer of wind-induced transport during the day, resulting in a net offshore movement. The later stages are found at greater depths during the day and less frequently at the surface at night; this subjects them to a coastward mass transport of water, which underlies the immediate surface layer. As the late stage of phyllosoma larvae approach the edge of the continental shelf, they metamorphose into the puerulus stage (Sweat, 1968; Phillips *et al.*, 1978). The puerulus swims shoreward, arriving at the shallow nursery reefs on nights near the new moon during summer. A review of the interactions between oceanic circulation and the planktonic existence of *P. cygnus* has recently been provided by Phillips (1981).

The feeding habits of phyllosoma larvae are not well known. In the laboratory, *P. interruptus* larvae consume chaetognaths, fish larvae, hydromedusae, and ctenophores (Mitchell, 1971). Many species will accept brine shrimp larvae, but others will not (Phillips and Sastry, 1980). They are probably preyed upon by a variety of fishes, including tuna. No estimates of larval mortality have been made.

E. Distribution

Lobsters and crayfishes are found and fished in aquatic environments throughout the world. Because we are concerned with commercially important species in this chapter, our discussion of distribution will focus on habitats and species accessible to fisherman.

1. THE CLAWED LOBSTERS

The distribution of the two North Atlantic genera, *Homarus* and *Nephrops*, is shown in Fig. 1. *Homarus americanus* is found on the continental shelf and on the upper slope along the western border of the north Atlantic Ocean from North Carolina to Labrador at depths ranging from the intertidal

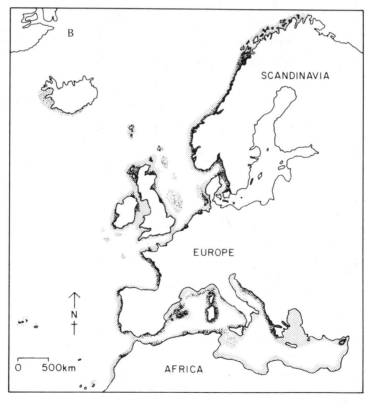

to about 700 m. Inshore, most of the population is found bewteen Rhode Island and Newfoundland. Offshore lobsters are found in highest concentrations on Brown's Bank and near submarine canyons from Corsair Canyon southeast of Georges Bank, to Norfolk Canyon off the coast of Virginia. *Homarus gammarus* ranges from northern Norway to about 30°N off the Moroccan coast. It is found throughout the Mediterranean area, but not in the Baltic Sea (Havinga, 1938; Dybern, 1973; Cooper and Uzmann, 1980). The reported depth range of *H. gammarus* is from the intertidal down to 40 m (Dybern, 1973), but recently "offshore" fishing has developed as far as 65 km from the coast (Bennett, 1980). Both species of *Homarus* typically are found in rocky areas, but when such habitat is not available, as in muddy harbors or in the offshore canyons, *Homarus* will excavate burrows or depressions, often associated with some solid object (Thomas, 1965; Cobb, 1971; Cooper and Uzmann, 1980).

The geographical distribution of *Nephrops norvegicus* is very similar to that of *H. gammarus* (Fig. 1b). It also occurs off Iceland but has not been found in the Black Sea. It makes burrows in soft substrates at depths ranging from 15 to 800 m (Farmer, 1975).

2. SPINY LOBSTERS

The many species of palinurid lobsters are found in temperate and tropical oceans throughout the world (Fig. 2). They are abundant and commercially important in the coastal areas of the north Atlantic, where *Panulirus argus* extends to about 30°N latitude along the United States coast, and *Palinurus elephas* is found as far north as Norway in the eastern North Atlantic. In the Pacific, the northern limits of distribution for both *Panulirus interruptus* along the California coast, and *Panulirus japonicus* in Japan are at about 35°N latitude. The southern distributional limit for species in the Atlantic are at about 35°S latitude for *Palinurus gilchristi* in South Africa and *Jasus tristani* at Tristan da Cunha. *Jasus edwardsii* extends south to about 50°S latitude in the Pacific Ocean near New Zealand. The commercially important palinurid genera do not appear to overlap in distribution when plotted by depth and latitude (Fig. 3). The two shallow-dwelling genera, *Jasus* and *Panulirus,* are found in temperate and tropical waters, respectively. The remaining six genera of palinurids are found in deep water without a great deal of latitudinal overlap (George and Main, 1967). Most of the shallow water species characteristically inhabit reeflike areas of rock or coral that have crevices available for dens (George, 1974; Engle, 1979). Some species in deeper

Fig. 1. (A) The distribution of *Homarus americanus* (shaded) along the coast of North America. (B) The distribution of *Homarus gammarus* (shaded) and *Nephrops norvegicus* (stippled) in the eastern North Atlantic.

water have been reported from trawl samples on mud substrates. This may be an artifact of sampling method, however *Palinurus delagoae* is trawled in areas devoid of rocks, and its behavior in an aquarium additionally suggests that this is not a shelter-seeking species (Berry, 1971a).

3. CRAYFISH

Crayfishes are found in freshwater environments throughout the world but are most abundant in temperate latitudes. Over 300 species and subspecies are recognized on the North American continent. In the Australian region, the only other part of the world where there is a large degree of diversity and speciation, 111 species are known (Bouchard, 1978). Hobbs (1974) gave a synopsis of the families and genera of crayfish. The range and number of species of each genus is shown in Table I. There are only a few species known from South America and the eastern Asiatic region. Other than on the island of Madagascar (one species), there are no crayfishes endemic to Africa.

In North America, three species are fished intensively. *Procambarus clarkii* and some congeners are harvested and cultured for both food and bait in the southern United States, particularly in Louisiana, South Carolina, and Tennessee. *Pacifasticus leniusculus* is fished on the west coast of North America from California to British Columbia. *Orconectes virilis* and congeners are caught in the northcentral and northeast regions of North America.

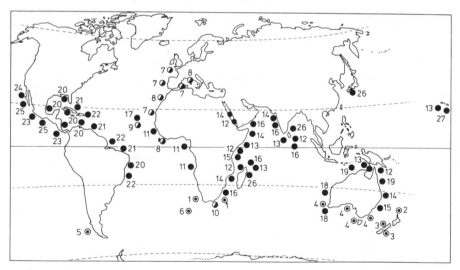

Fig. 2. The geographical distribution of commercially important palinurid species. Genera: (◉) *Jasus*, 1–6; (◑) *Palinurus*, 7–10; (●) *Panulirus*, 11–27. (From Phillips *et al.*, 1980.)

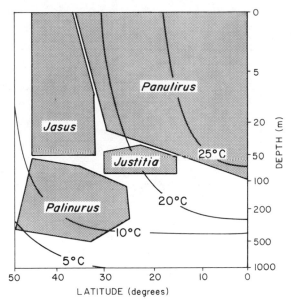

Fig. 3. Distribution by depth and latitude of the palinurid genera *Jasus, Palinurus, Panulirus,* and *Justitia.* Approximate isotherms are also shown. There is no distributional overlap among the genera shown. (Redrawn from George and Main, 1967.)

In Europe, fisheries had existed since the sixteenth century on the genus *Astacus,* primarily *A. astacus* and *A. leptodactylus.* The crayfish fungal plague *Aphanomyces astaci* was introduced in about 1880 from North America and drastically reduced or exterminated many native stocks. Parts of Europe such as the Soviet Ukraine, the Turkmen Caspian, and parts of Norway were not affected and continue to have thriving fisheries on native crayfish. However, most stocks were greatly diminished throughout western and northern Europe (Unestam, 1973). Disease-resistant North American species were introduced to replace native stocks. *Orconectes limosus* was introduced to Poland in 1889 (Kossakowski, 1973) and to France in 1925 (Laurent, 1973), where it is presently successfully established and fished. *Pacifastacus leniusculus* was introduced to Finland and Sweden in the 1960s (Westman, 1975; Brinck, 1977), to Austria in the early 1970s, and to Poland in 1972 (Kossakowski, 1973). *Procambarus clarkii* has been introduced to Europe, Africa, and Central America (Huner and Barr, 1980). Crayfish populations and the fisheries on them are presently growing in these countries.

The fisheries for crayfish in Australia are small and are mostly located in the southeastern portion, and Tasmania. Three genera fished in those loca-

TABLE I

The Range and Number of Species of 27 Crayfish Genera[a]

Genus	Range	Number of species
Astacidae		
Astacus	Middle and eastern Europe	6
Austropotamobius	Western and middle Europe	3
Pacifastacus	Western North America; introduced to Japan, Sweden	5
Cambaridae		
Barbicambarus	Eastern North America	1
Cambarus	Eastern North America	55
Fellicambarus	Central North America	9
Faxonella	South central North America	3
Hobbseus	South central North America	5
Orconectes	Eastern and central North America; introduced to western Europe	63
Procambarus	Guatemala and Cuba to northern United States; introduced to California, Hawaii, Japan	117
Troglocambarus	Florida	1
Cambarellus	Mexico, south central United States	12
Cambaroides	Eastern Asia: Amur basin, Korea, Japan	4
Parastacidae		
Astacoides	Madagascar	1
Astacopsis	Tasmania	4
Cherax	Australia, except desert regions; New Guinea	39
Engaeus	Southeastern Australia	23
Engaewa	Western Australia	3
Euastacoides	Northeastern Australia	2
Euastacus	Eastern Australia	27
Geocherax	Southeastern Australia	2
Gramastacus	Southeastern Australia	2
Paranephrops	New Zealand	2
Parastacoides	Tasmania	6
Parastacus	Chile, Argentina, Uruguay, southern Brazil	6
Samastacus	Chile	2
Tenuibranchiurus	Northwestern Australia	1

[a] From Hobbs, 1974.

tions are *Astacopsis, Euastacus,* and *Cherax*. Additionally, there are small populations of *Cherax destructor* and *C. tenuimanus* in southwestern Australia subject to sports fisheries (Frost, 1975).

Some other crayfish fisheries are found in Africa and South America where *Procambarus clarkii,* the red swamp crayfish, has been introduced within the past 20 years. Huner (1977) reported species introductions in Kenya (1966–1970) and Uganda (1963–1964). He also reported introductions to Costa Rica (1966) and possibly Nicaragua (1976), but no significant harvest occurs in those countries. Such introductions present the danger that the new species may become a pest. For example, Lowery and Mendes (1977) suggested *P. clarkii* may become a predator on the eggs of harvested native fish species in Kenya river systems.

F. Habitat Behavior

1. CLAWED LOBSTERS

Homarus americanus and *H. gammarus* are generally found in rocky areas of the subtidal habitat. Burrows are also found in muddy areas (Thomas, 1968; Cobb, 1971; Cooper et al., 1975). Sand, silt, or mud is excavated from underneath a rock in such a way that the resulting shelter is about one-half as high as it is wide and of variable length (Cobb, 1971). These species generally inhabit burrows solitarily, but large and small lobsters may have closely spaced burrows or occasionally even share the same shelter. *Homarus americanus* is found not only in relatively shallow waters but in deep submarine canyons where shelters are limited. There, depressions are excavated near any solid object, such as a boulder, rock, anemone, or discarded fish net. In areas of steeply sloping embankments where the sediment is cohesive, extensive burrow systems occupied by lobsters, crabs, fishes, and shrimps have been discovered (Warme et al., 1978).

Nephrops norvegicus is found at depths of 15–800 m in areas of fine, cohesive mud that is stable enough to permit construction of unlined burrows (Rice and Chapman, 1971). The burrows are roughly circular in cross-section, extend 20 to 30 cm below the surface, have several side tunnels, and often have more than one entrance. They are usually occupied by a single *Nephrops;* when there is more than one, the occupants are usually of different sizes, each occupying its own tunnel within the burrow system (Chapman, 1980). Densities of 3–5 *Nephrops* burrows per m² were seen in the Sound of Jura with use of both underwater television and a submersible (Chapman, 1979).

The clawed lobsters and freshwater crayfishes modify preexisting shelters or construct new burrows using behavior patterns that are remarkably similar among all the species observed (Cobb, 1971; Rice and Chapman, 1971; Dybern, 1973). The spiny lobsters do not show similar behaviors and do not seem to modify the shelters they inhabit in any major way (Kanciruk, 1980; Cobb, 1981).

Both *Homarus* and *Nephrops* are nocturnal, emerging when the light intensity falls below some critical level and returning to shelter when above it. In *Homarus americanus,* this pattern can be somewhat modified by the presence or absence of conspecifics or shelter (Zeitlin-Hale and Sastry, 1978). Egg-bearing female *H. gammarus* are more active than are non-ovigerous females (Branford, 1979). *Homarus americanus* becomes restless 1 to 2 hours before sunset in the field, moving back and forth in the entrance to their shelter and start to emerge shortly after sunset (Weiss, 1970; Cooper and Uzmann, 1980). Smaller juveniles, less than 40-mm CL spend their entire time within the burrow according to Cooper and Uzmann (1980), but D. Wang (unpublished) has seen animals as small as 25-mm carapace length (CL) out of their burrows immediately after sunset. *Nephrops* smaller than 10 to 15-mm CL probably do not emerge from their burrows after sunset (Chapman, 1980). Small lobsters are highly vulnerable to moderate- and large-size fish predators; it is reasonable to assume that the presence of predators may affect the activity patterns of the animals. Cooper and Uzmann (1980) saw a longhorn sculpin prey upon a small lobster at sunset when the lobster was moving around at the mouth of its burrow. A better knowledge of the activity rhythms of small lobsters may be important to our understanding of natural mortality.

Larger *Nephrops* show peak emergence from burrows at dawn and dusk as shown both by direct field observations (Chapman and Rice, 1971; Chapman *et al.,* 1975; Atkinson and Naylor, 1976; Chapman and Howard, 1979) and by changes in capture rate using various fishing methods (Simpson, 1965; Hillis, 1971; Farmer, 1975). Laboratory studies have confirmed the nocturnal nature of *Nephrops* (Atkinson and Naylor, 1973, 1976; Arechiga and Atkinson, 1975; Naylor and Atkinson, 1976). However, the observation that *Nephrops* appears to be fully nocturnal in shallow water, crepuscular in deeper water, and active during the brightest parts of the day at the deepest part of its range suggests that emergence is keyed to light intensity, with a threshold being approximately 10^{-5} μW/cm^{-2} (Hillis, 1974; Chapman *et al.,* 1975; Chapman, 1980; Moller and Naylor, 1980). The light intensity during peak emergence at 184 m was practically the same as during peak emergence at 10 m (Atkinson and Naylor, 1976). The presence of food and feeding activity plays an important role in the emergence of *Nephrops* (Chapman and Howard, 1979; Moller and Naylor, 1980).

In both *Homarus* and *Nephrops,* it appears that activity is in large part dependent upon light intensity; thus, moon phase and water turbidity, as well as levels of hunger, should affect the time and duration of emergence. Weather, storm surge, and currents all affect water clarity, and a correlation between these factors, food availability, activity, and catch rates is to be expected.

The eye of *Nephrops* is well adapted to functioning at low levels of irradiance (Arechiga and Atkinson, 1975), but when the animal is brought to the surface, retinal degradation occurs after exposure to full sunlight for 2 hours. The degradation appears to be irreversible and thus may result in permanent blindness (Loew, 1976). This has important implications both to laboratory experimentation and to the rapid sorting of the catch on the deck of a fishing vessel if small animals are to be returned to the sea uninjured. No work similar to Loew's has been done with *Homarus* or *Panulirus*.

2. CRAYFISH

Little research has been done on freshwater crayfish habitat behavior, but there are many observations that suggest that several factors may be critical in determining crayfish population density and distribution. Crayfishes have been found to inhabit areas with large amounts of cover: either a rocky substratum, which provides burrow space, or a vegetated area which supplies both food and protective cover. Niemi (1977) found highest densities of *Astacus astacus* in gravel and stony bottom areas whereas Kossakowski (1975) found densities of *Orconectes limosus* as high as 77 animals per m^2 in dense macrophytic zones. Often juveniles and sometimes adults will climb in the vegetation. Flint (1977) compared crayfish densities over several bottom types and found *Pacifastacus leniusculus* density greatest on rocky bottom, less in macrophytic areas, and least in open sandy areas. *Orconectes propinquus* shifted microhabitat choice from a sand to a pebble substrate in the presence of a fish predator (Stein and Magnuson, 1976). In the field, some sizes and life stages (juveniles, females and recently molted animals) appeared to modify their microdistribution similarly to minimize risk of predation (Stein, 1977a). In a laboratory study, Crawshaw (1974) found that, during active periods, crayfishes select warmer temperature ranges. *Pacifastacus leniusculus* moves several hundred meters per month (Flint, 1977).

3. SPINY LOBSTERS

The diversity of habitat use in the spiny lobsters can be illustrated by Berry's (1971a) study of the distribution and general ecology of spiny lobsters found off the coast of South Africa. The five species of *Panulirus* found there all seek shelter in the relatively shallow water of reefs and lagoons in areas ranging from surf to calm conditions. Species are specific to turbid or clear water. On the other hand, *Palinurus delagoae* inhabits the edge of the continental shelf at depths of 180–400 m in areas devoid of rock. *Palinurus gilchristi* is found at depths of 50–100 m, apparently occupying rock shelters. The genera *Puerulus* and *Linuparus* are occasionally trawled in deep (200–300 m) waters, the former on muddy bottoms, the latter apparently in rocky areas. George (1974) reviewed the habitat distribution of reef species of *Panulirus* in the Indo-West Pacific region, and his general conclusions are illustrated in Fig. 4.

Many of the spiny lobsters seek shelter in holes, crevices, or dens formed in rock or coral reef structures. Unlike the clawed lobsters, they shelter communally in groups of two to over one hundred (Lindberg, 1955; Fielder, 1965; Herrnkind *et al.*, 1975; Cobb, 1981). First year juveniles of *Panulirus interruptus* occupy beds of shallow surf grass (*Phyllospadix*) in areas of dense plant cover and narrow bedrock crevices, most sheltering separately. The age 2$^+$ juveniles were found in deeper water, and more frequently were aggregated (Engle, 1979). *Panulirus cygnus* juveniles aged 2 to 4 years are found 'in large groups in shallow limestone patch reefs (Chittleborough, 1970; Cobb, 1981). Engle (1979) and Chittleborough (1974) have shown the home range of juvenile *Panulirus* to be relatively limited. Chittleborough indicated that after foraging the juvenile *P. cygnus* probably return to the same "home" patch reef.

Spiny lobsters are nocturnally active, spending the daylight hours in caves, dens, or grass beds. A majority of *P. argus* left their dens just after

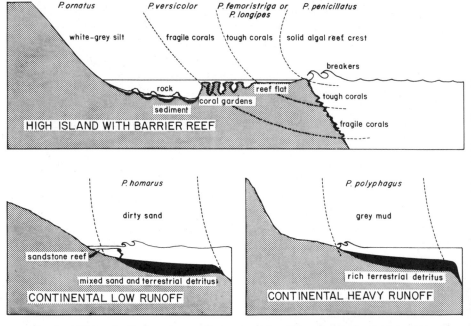

Fig. 4. Habitat distribution of several coastal species of *Panulirus* in the tropics. (Redrawn from George, 1974.)

sunset and returned before sunrise (Cooper and Herrnkind, 1971; Herrnkind et al., 1975). Full moonlight suppressed activity in *P. argus* (Suttcliffe, 1956) and reduced catches in the fishery have been reported at times of bright moonlight for *P. cygnus* (Morgan, 1974b) and *P. japonicus* (Kubo and Ishiwata, 1964; Yoza et al., 1977). Kanciruk and Herrnkind (1978) described different summer and winter patterns of daily activity in *P. argus*. In winter, activity began at sunset, continued a few hours, but did not last throughout the night. In summer, the lobsters were active during the entire period of darkness. Despite sheltering communally during the day, nocturnal foraging is not done in groups. Outside a den at night, spiny lobsters are almost always alone and, if they meet, show aggressive behavior (Herrnkind et al., 1975; Hindley, 1977; J. S. Cobb, unpublished).

The behaviors of burrow emergence, locomotor activity, feeding, and their relationships to temperature and light have a great deal of influence on *catchability*, the probability of capture of any individual. *Homarus americanus* increased its walking rate between 2° and 10°C and between 20° and 25°C but showed no change between 10° and 20°C. Experimental fishing showed that catchability was linearly related to temperature. A proportional relationship between activity and catchability index can be used to correct catch per unit effort data for varying catchability (McLeese and Wilder, 1958). Similarly, locomotor activity of *Panulirus cygnus* increased with increasing temperature between 17° and 25°C but declined at 30°C. Additionally, the activity of animals in late premolt (D_2 or later) was significantly below those in intermolt (Morgan, 1978). Many factors, including water temperature, food availability, hunger, molt state, and egg-bearing, affect movement and thus catchability. Morgan (1974b) additionally used salinity in his catchability model. Variation in catchability of the freshwater crayfish *Cherax tenuimanus* was analyzed by Morrissy and Caputi (1981). They sampled using drop nets, then drained the pond to inventory the entire population. They found that inclusion of water temperature, secci disk values (a measure of illuminance), size rank of the individual, and a coefficient of size variation substantially increased the ability of their model to estimate population size.

G. Migrations

Many tagging experiments with *Homarus gammarus*, *Nephrops*, and inshore *H. americanus* indicate that extensive movements do not occur. Nightly wanderings of *H. americanus* from shelter tend to be no more than several hundred meters, and often they return to the same burrow or one nearby (Cooper, 1970; Stewart, 1972; Cooper et al., 1975). Nondirectional movements on the order of a few kilometers have been reported for inshore

H. americanus by Wilder (1963), Fogarty *et al.* (1980), and Krouse (1981); however, seasonal inshore–offshore movements have been noted in Canada (Bergeron, 1967). In the Gulf of Maine, small-scale movements from shallow into deeper water were associated with strong winds and turbulence rather than season (Cooper *et al.*, 1975). Some exceptions to the generalization of only local movements by inshore *H. americanus* appear. Dow (1974) tagged some large lobsters off the Maine coast that moved 270 km southwesterly along the coast, and Stasko (1980) reviewed accumulating evidence for long distance (> 93 km) movements of lobsters off southwest Nova Scotia. Krouse (1980) and A. Campbell (personal communication) have evidence that inshore mature lobsters, in Maine and the Bay of Fundy, are much more likely to move long distances than are immatures. An earlier, less-conclusive report (Morrissey, 1971) indicated that egg-bearing females tagged in shallow water off Massachusetts moved greater distances than did immatures. In contrast, Fogarty *et al.* (1980) found no differences in distance travelled according to size and sex of Rhode Island lobsters. An intriguing hypothesis (A. Campbell, personal communication) is that the movements of sexually mature females into shoal warmer waters during the summer allow the most rapid egg development rate. Tagging experiments on *H. gammarus* give no indication of extensive migrations (Thomas, 1954; Gibson, 1967; Gunderson, 1969; Watson, 1974). *Nephrops* is similar. In nine tagging experiments summarized by Chapman (1980), there were few recaptures that showed significant movement. Chapman *et al.* (1975) placed ultrasonic tags on *Nephrops* and found that two females moved only short distances and returned to home burrows, while three males travelled farther and did not return to the previously occupied burrow.

Homarus americanus, from the offshore populations, exhibit extensive onshore–offshore movements, as illustrated in Fig. 5. Egg-bearing female lobsters were captured in Veatch Canyon and released 218 km away in Narragansett Bay, Rhode Island, by Saila and Flowers (1968). The lobsters remained in the Bay, released their eggs, and then moved offshore. Three tagged animals were recaptured near the site of original capture. Uzmann *et al.* (1977) and Cooper and Uzmann (1971) have showed that at least 20% and probably 30 to 40% of the offshore lobsters annually engage in directed shoalward migrations in spring and summer. Lobsters tagged on Cox Ledge off the coast of Rhode Island in late summer moved offshore to the outer continental shelf. (Fogarty *et al.*, 1980). Uzmann *et al.* (1977) hypothesized that these inshore–offshore movements, which maintain the migratory lobster within a temperate range of 8°–14°C, allow more rapid growth and development than would be possible for animals either remaining inshore, where temperatures drop as low as 0°C in winter, or remaining offshore, where temperatures rarely go above 12°C. Additionally, they found signifi-

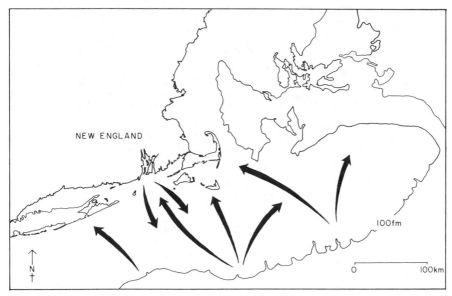

Fig. 5.   General directions of migrations of offshore *Homarus americanus*. Arrows indicate only general trends of data collected, not specific migratory routes. (Compiled from data in Cooper and Uzmann, 1971; Uzmann *et al.*, 1977; and Fogarty *et al.*, 1980.)

cant east–west movement along the edge of the shelf and, hence, believe that there are not discrete populations in each canyon.

Spiny lobsters exhibit a wide variety of movements that have a great diversity of form and apparent adaptive significance. Herrnkind (1980) has recently reviewed the available data for many species. Most of the migrations are seasonal and often involve inshore–offshore movements, but some are extensive along-shore mass movements.

Two species of *Jasus* are known to perform long-distance migrations. Figure 6A shows the springtime migrations of *J. edwardsii* in a southerly direction along the southeast New Zealand coast maintaining a depth of about 40 m (Street, 1971). Juvenile *J. verreauxi* concentrate in the North Cape area of New Zealand's North Island and migrate along shore in a westerly direction to the main fishery area off Cape Reinga (Fig. 6D). This movement apparently is associated, at least in females, with the onset of sexual maturity (Booth, 1979). Neither of these movements change the depth range of the migrants, but they may be an essential feature of the recruitment mechanism, similar to that of *P. ornatus,* described below.

The migration of *Panulirus argus* has been well documented (Crawford and DeSmidt, 1922; Herrnkind and Cummings, 1964; Herrnkind, 1969). It occurs over much of the range of the species, but it is especially striking in

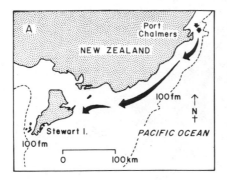
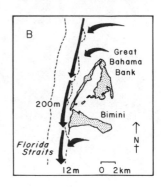
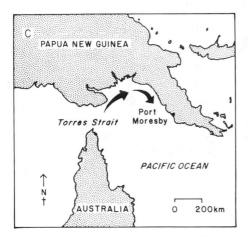
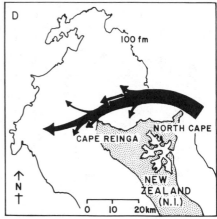

Fig. 6. Long distance movements by *Jasus edwardsii* (A), *Panulirus argus* (B), *Panulirus ornatus* (C), and *J. verrauxi* (D). Each is an along shore movement and in none are return migrations known. It is likely that the movements of *J. edwardsii, J. verrauxi,* and *P. ornatus* are spawning migrations and the larvae return to the area where the migrants started. (From: (A and B) Herrnkind, 1980; (C) data in Moore and MacFarlane, 1977; (D) Booth, 1979 in *The New Zealand Journal of Marine and Freshwater Research.*

the Bahamas (Fig. 6B), where large numbers of migrants move day and night in queues of 2–60 lobsters (Herrnkind, 1969, 1980). It starts suddenly in midautumn (October–November) in the Bahamas and early winter (December–January) off Yucatan in the Caribbean. Movements are associated with periods of storm, often the first strong, frontal storm of the season in the area. The direction of migration is constant from year to year in each area, but directions vary among areas. The total distance, precise origin, and endpoints of the movements are not known (Herrnkind, 1980). Return migrations have not been described. The single file nature of the migrations of *P. argus* significantly reduces the drag upon each individual in the queue

(Bill and Herrnkind, 1976). The adaptive significance of this migration is not entirely clear, but it may be potentially valuable in avoiding the stress of severe winters in shallow areas. The migrants are not in reproductive condition.

Panulirus ornatus, a spiny lobster found in the Gulf of Papua and in the Torres Strait northeast of Australia, has long-distance migrations associated with reproduction (Fig. 6C). Lobsters tagged in the northern Torres Strait are caught in the Gulf of Papua and at Yule Island. Tagging studies have shown no return migration, and substantial densities of juvenile *P. ornatus* are found only in the northern Torres Strait. Spawning occurs during the migration of the 3- and 4-year-old animals. The movement probably is a contranatant spawning migration that assures larval transport to suitable areas (Moore and MacFarlane, 1977). The beginning of the migration is within the jurisdiction of Australia, whereas the terminus and spawning locations are in the waters of Papua New Guinea, making for an interesting political situation.

Movements of spiny lobsters may also be associated with stages of the life cycle. First-year *Panulirus interruptus* in southern California are found in shallow surf grass (*Phyllospadix*) beds. In the autumn, age 2^+ juveniles migrate out of the surf grass beds to deeper water, coincident with recruitment of the new year class of pueruli to the *Phyllospadix* habitat (Engle, 1979). Juvenile (3–5 years) *Panulirus cygnus,* in Western Australia, move from shallow reefs to deeper offshore water. The movement comes immediately after a molt characterized by a change in color from a dark red to a pale color, locally known as "white" lobsters. The molt and movement is coherent and very predictable, starting in November and lasting most of the summer. It forms the basis for an important part of the Western Australian lobster fishery (George, 1958; Chittleborough, 1970; Morgan, 1974a).

Freshwater crayfishes do not appear to move significant distances, although local movements in response to low water, pond drying, or extreme water temperatures may occur. In order to avoid these adverse conditions, crayfishes may burrow and become quiescent (e.g., Caldwell and Bovbjerg, 1969). Young-of-year may be particularly susceptible to environmental extremes, and Williams et al. (1974) found that *Cambarus fodiens* hatchlings remained in their burrows throughout the first winter. Many crayfish emigrate from areas with unsuitable conditions if possible (Momot, 1966). For example, *Cherax destructor,* in Australia, migrates overland in winter when conditions become poor in its original aquatic system (N. M. Morrissy, personal communication).

The adult population of *P. leniusculus* migrates to greater depths to avoid low water temperatures in winter and water turbulence due to storms (Flint, 1977). Crayfish activity is reduced at low temperatures (Roberts, 1944), and storm activity can cause mortality. Goldman (1973) found that interspecific competition for habitat space causes migration. Larger crayfish seem able to migrate to less crowded, deeper water with less rocky cover because predation risks are reduced due to their size relative to that of possible predators. Seasonal stream migration is also common in many species, for instance, moving downstream during spring and upstream in autumn (Henry, 1951).

The size of home range in the crayfish is unclear. Home areas in stream-dwelling *Orconectes juvenalis* were measured (Merkle, 1969), but it was not determined whether or not the movements observed were caused by environmental factors or were due to a behavioral strategy that could be termed a home range. Another stream dweller, *Faxonella clypeata* (Mobberly and Pfrimmer, 1967), did not have a home range but moved about the stream bottom at random. Hazlett *et al.* (1974) found marked crayfish *O. virilis* present in the same areas for varying periods. Some animals moved small distances frequently; some animals remained in the same area for periods ranging up to 9 months and in the same burrows for up to 34 days. The long residence time occurred especially in animals that had just molted or in animals in areas of favorable conditions of food and shelter availability. There were no significant differences in movements between sexes. However, the presence of a home range may be questioned because of 4328 animals captured and presumably marked, less than 7% fit the authors' criteria of exhibiting home range. Generally, there were no discernable patterns in the direction of movement. Mason (1975) found that 70 to 80% of recruits remained in their home pool, which suggests that movements are reduced under favorable conditions.

Juvenile crayfish seem to move to no greater or lesser degree than do adults (Merkle, 1969; Hazlett *et al.*, 1974), frequently moving small distances and infrequently moving larger ones. Habitat selection is probably not very different between juveniles and adults because their functional form and needs are fairly similar. However, habitat distribution may be influenced by food preferences or by competition with adult crayfishes. Crayfishes can be expected to seek habitats with favorable conditions, but this may be prevented by environmental conditions, which form barriers to movement. Dispersal can also be limited by ecological barriers, such as high current or poor water conditions (Fitzpatrick and Hobbs, 1968).

H. Feeding and Predation

Lobsters are omnivorous feeders, preying primarily upon other Crustacea, polychaetes, molluscs, and echinoderms. Less often found upon examination of foreguts are coelenterates, fishes, and plants. Crayfishes show much the same diet but include more plant material. Lobsters and crayfishes may be the most important benthic omnivores in many areas.

Nephrops norvegicus is a varied feeder regardless of sex or size, indiscriminately foraging on the available food organisms occurring on or just within the sea floor (Thomas and Davidson, 1962; Farmer, 1975). *Homarus americanus* appears to be somewhat more selective, although stomach contents often reflect the relative abundance of prey species in the environment (Weiss, 1970; Ennis, 1973c; Elner and Jamieson, 1979). The same seems to be true of *H. gammarus* (Hallback and Warren, 1972). Evans and Mann (1977) and Hirtle and Mann (1978) suggested that *H. americanus* prefers rock crabs (*Cancer*) to sea urchins (*Stronglyocentrotus*) and that sea urchins will be taken only when they are considerably more abundant than crabs. Weiss (1970) has described the lobster's crab-catching behavior, and Elner and Jamieson (1979) have observed how lobsters open sea scallops (*Placopecten*). *Homarus americanus* occasionally will bury the remains of an uneaten meal outside its shelter (Smith, 1976).

The diet of *Homarus* appears to vary with the molt cycle and with a higher proportion of calcium-rich material, such as starfish and sea urchin, which are present in the gut during the molting season (Weiss, 1970; Hallback and Warren, 1972; Ennis, 1973c). Feeding practically stops during the winter months when the temperature is below 5°C and is greatly reduced during the late stages of ecydsis (Weiss, 1970).

The diet of crayfish is variable among species. Crayfish are generally considered omnivores, but they may be more predaceous or herbivorous during particular life stages. In *Pacifastacus leniusculus,* the diet of juveniles included about 62% insect prey, but adults fed primarily on plant material (Mason, 1975). Young-of-year *Astacus leptodactylus* fed heavily on zooplankton and benthic invertebrates, including aquatic insect larvae, amphipods, mysids, cladocerans, and daphnids. Juveniles also fed on plant detritus to some degree. Amphipods in the diet of *A. leptodactylus* changed from 5% in subyearlings to 63% in yearlings. Daily feeding rhythms also changed, 1-month-old animals had four daily peak periods of feeding activity, whereas 4-month-old animals had two daily peaks. Along with this, the theoretical daily ration and net conversion efficiency decreased with age, in the former from 34.5% in subyearlings to 2.9% in yearlings and in the latter from 75 to 21% (Tcherkashina, 1977).

Diet may also change depending on the types of food available or on the physiological diet requirements. Tcherkashina (1975) reported that *Astacus leptodactylus* ate more molluscs after molting and more chironomid larvae and algae during the summer when these foods were abundant. Capelli (1980) found that the diet of *Orconectes propinquus* changed over seasons as well as with respect to depth, apparently due to changing food availability and to food preferences. Cannibalism was also observed in adults feeding on newly hatched juveniles.

The importance of crayfish grazing to the primary production of the periphyton community has been shown by Flint and Goldman (1975). Grazing by *Pacifastacus leniusculus* significantly reduced primary productivity as compared to substrates with ungrazed periphyton populations. Low grazing pressure (*P. leniusculus* biomass less than 131 g/m²) enhanced primary productivity by removing old and senescent periphyton, whereas heavy grazing (density greater than 203 g/m²) reduced primary productivity as compared to ungrazed tests with macrophytes. In contrast, Mason (1975) has suggested that the major role of *Pacifastacus leniusculus* in stream ecosystems is the biological conditioning of allochthonous detritus.

Covich (1977) discovered in feeding experiments with *Procambarus clarkii* that selection and ingestion of food items may be positively related. As has been found in other omnivores, feeding on plant food leads to feeding on animal food as well. In his study, snails and plant foods had just such a positive selective relationship, and Covich (1977) predicted that an increased availability and feeding on plant food might increase the amount of animal food ingested even if the latter did not increase in availability.

Spiny lobsters are similar to clawed lobsters and crayfishes in dietary habits. Although *Homarus* can use either the mandibles or the large claws for crushing and opening hard-shelled prey, the massive mandibles of spiny lobsters are the primary tool for prey opening (Munro, 1974; Pollock, 1979). The list of groups preyed upon include sponges, mollusks, echinoderms, polychaetes, crustaceans, fishes, and plants (summarized by Kanciruk, 1980). A recent study of the stomach contents of *Panulirus cygnus* at two locations in Western Australia showed that plants (algae and seagrass), mollusks, and crustaceans made up the largest porportion of a wide range of identifiable material (Fig. 7). The higher proportion of plant material in the stomach of lobsters from Dongara may reflect high lobster population density and consequent competition for scarce "quality" foods (L. M. Joll, personal communication). In contrast, some species of spiny lobsters appear to have a very limited diet. *Panulirus homarus* feeds primarily upon the mussel *Perna perna*. Studies by Newman and Pollock (1974a) suggested that growth rates of *J. lalandii* were limited by food availability in some areas, a surprising conclusion because mussel biomass can be very high. Pollock (1979) showed that large mussels cannot be opened by any but the largest lobsters and that the smaller mussels are in short supply relative to the population of *J. lalandii,* leading to competition for food. He concluded that growth rates of these lobsters could be limited by food supply even in areas where mussel biomass is comparatively large. Chittleborough (1975) suggested that growth rates of juvenile *P. cygnus* in densely populated juvenile nursery areas at Dongara, Western Australia, is limited by food supply.

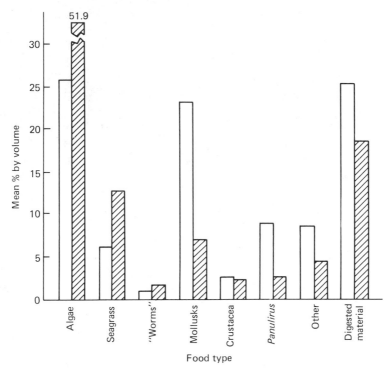

Fig. 7. The estimated percentage by volume of foregut contents in juvenile *Panulirus cygnus* from two locations on the Western Australian coast. Cliff Head has a lower population density of lobsters and a greater availability of "quality foods" than Dongara. (☐) Cliff Head; (▨) Dongara. Data kindly supplied by Lindsay Joll, C.S.I.R.O., Marine Laboratories, Perth, Australia.

Pollock's (1979) study illustrates not only the effect of prey species availability on predator growth but the impact that large numbers of lobsters have on the population of small and medium-size mussels. *Homarus americanus* was hypothesized to be a keystone predator in the nearshore environment, exerting control on the sea urchin (*Stronglyocentrotus droebrachiensis*) population. Removal of lobsters may allow an increase of sea urchins, which would lead to higher grazing pressure on kelp and a resultant decrease in the kelp population (Breen and Mann, 1976; Mann, 1977; Hirtle and Mann, 1978). This hypothesis may or may not be supported by further observations, but it is undeniable that lobsters and crayfishes, as major predators in the benthic environment, can alter density and availability of prey species.

III. POPULATION DYNAMICS

A. Stock Identity

The problem of stock identity is an important one to lobster population biology because many of the vital statistics differ with geographic location and because there often are many different fishery regulations throughout the range of a species. A few studies have addressed the problem of stock identity in *H. americanus,* but very little is known of spiny lobsters or crayfishes. In general, the three following types of approaches have been made: (1) analysis of genetic variation, (2) analysis of morphometric variation, and (3) estimation of dispersal among areas.

The data gathered concerning stock identity for *H. americanus* do not permit strong conclusions. Morphological differences between inshore and offshore groups were suggested by Saila and Flowers (1969). Isozyme polymorphisms studied by Barlow and Ridgeway (1971) showed no difference between the inshore and offshore populations, but Tracey *et al.* (1975) found a difference at one of the 44 loci they examined. Despite the extreme genetic homogeneity at the other 43 loci, Tracey *et al.* (1975) concluded that the difference indicated an absence of gene flow among some local inshore groups and also between inshore and offshore populations. Uzmann (1970) noted a difference in parasite infestation between lobsters caught inshore and those caught offshore. These studies suggest a reduced gene flow between the inshore and the offshore populations of *H. americanus.* However, tagging studies and consideration of larval biology dispute this. Inshore–offshore migrations have been demonstrated convincingly by Cooper and Uzmann (1971), Uzmann *et al.* (1977), and Fogarty *et al.* (1980). Briggs and Mushacke (1980) suggested that a large proportion of the lobsters taken off the south shore of Long Island, New York, in July and August are migrants from offshore. Many tagging studies have indicated that lobsters from the inshore areas move very little (Wilder, 1963; Cooper, 1970; Cooper *et al.,* 1975), but others (e.g., Morrissey, 1971; Dow, 1974) give evidence for long distance movement of larger lobsters. Larval *H. americanus* spend 2 to 5 weeks in the plankton, supposedly at the mercy of currents during the dispersal stage. If the larvae are truly planktonic, there is an excellent chance that larvae released in one area may become the juvenile recruits in another. Thus, there is a diversity of opinion concerning the question of genetic homogeneity in populations of *Homarus americanus.* Perhaps more important than gene flow between populations is the fact that vital statistics such as growth rate, size at maturity, and fecundity vary markedly over the range of *H. americanus* (New England Fishery Management Council, 1982). It is clear from a biologist's point of view that lobsters should therefore be managed as separate stocks. However, social and economic factors must also weigh heavily in decisions concerning management. In the United States, management recommendations are based on the assumption of a single stock. In Canada, on the other hand, there are several management areas, each with somewhat different regulations.

Tagging studies of *Homarus gammarus* have shown only local movements of 12 km or less (Thomas, 1954; Gibson, 1967; Gunderson, 1969; Watson, 1974). The nature of the distribution of *H. gammarus* larvae off the northeast coast of England is very similar to that of *H. americanus* larvae in the northwest Atlantic (Nichols and Lawton, 1978). No genetic or morphometric studies on geographic distribution of *H. gammarus* are available.

Tagging studies of *Nephrops norvegicus* that were summarized by Chapman (1980) have indicated that there is very little movement of individuals from the locality of release. Farmer (1974b) found differences in relative proportions of some body measurements of *Nephrops* from various regions, and he suggested that there may be partial reproductive isolation among neighboring populations. There is a geographic variation in the infestation rate of the parasitic trematode *Stichocotyle nephropsis* (Symonds, 1972). Unlike *Homarus*, *Nephrops* is very selective of bottom type and is found only on fine, cohesive muds, resulting in a discontinuous distribution (Chapman, 1980). It appears that the larval period is the only dispersal stage of *Nephrops*. We suspect that the geographically isolated populations are also genetically isolated and thus should be dealt with as separate stocks.

The long pelagic period of the larval stage of spiny lobsters allows wide dispersion and presumably a genetic coherence of the population over wide geographic ranges. The larvae are carried far offshore by ocean currents, but the maximum distance from which they can return is not known for any species. Larvae that do not return may be a source of recruitment for other areas. There is a possibility of larval exchange between New Zealand and Australian populations of *Jasus verrauxi* (R. W. George, personal communicated cited by Phillips and Sastry, 1980) and between African and Madagascar populations of *Panulirus homarus* (Berry, 1974). Lindberg (1955) suggested that the southern California population of *P. interruptus* is maintained by larvae carried northward on deep countercurrents. *Panulirus cygnus* may form a genetically coherent stock along the western Australian coast (Morgan, 1980a). The population of *Panulirus argus* in Florida and the Caribbean was considered by Sims and Ingle (1966) to be of a single genetic stock, despite the great distances involved. They proposed that the recruits to the Florida fishery were derived from the Caribbean. However, recent work by Menzies and Kerrigan (1979; and personal communication) suggested possible genetic heterogeneity between widely separated populations of *Panulirus argus*. The situation may be different in New Zealand where Smith *et al.* (1980) found very low levels of genetic variation as determined by gel electrophoresis of proteins in both *Jasus edwardsii* and *J. novaehollandiae.* Samples of *J. edwardsii* were taken from three separate locations, and no differences were seen, suggesting that they were drawn from the same stock. Additionally, Smith *et al.* (1980) found very little genetic variation between *J. edwardsii* and *J. novaehollandiae;* it is doubtful whether they should be considered separate species.

Some population parameters vary considerably over the range of a species. The reported growth rates of *P. argus,* as summarized by Munro (1974), vary considerably from Florida to Jamaica, but there is less variation between Jamaica and the Lesser Antilles. Chittleborough (1975) has shown variations in growth rate of *Panulirus cygnus* with several environmental factors, and Smale (1978) discussed reported variations in *P. homarus* growth rates. Growth rates of *Jasus lalandii* vary with the availability of suitable food according to locality off the South African west coast (Newman and Pollock, 1974a). Size at maturity and time of spawning also vary geographically in several species (Morgan, 1980a). These population parameters are most important to decisions concerning fishery management. Thus, as with *H. americanus,* it would seem reasonable that wherever practical management practices be based upon them rather than upon the precept of genetic continuity.

Unlike lobsters, crayfish populations tend to be reproductively isolated in separate drainage systems often with little opportunity for exchange between one population and another. Hybridization due to environmental, morphological, and behavioral factors is rare among crayfish species (Smith, 1981). Additionally, there is no larval dispersal stage, because the larvae are not pelagic. Thus population parameters can be obtained simply by ordinary sampling procedures, including mark–recapture techniques, netting, baited traps, and electric shock. Estimates of sex ratio, length–frequency distribution, age structure, fecundity, and recruitment can be readily obtained. We have found no specific references to stock identity in crayfish. Brown (1981), using electrophoretic techniques, found low genetic variation among six species of crayfish (*Cambarus* and *Procambarus*) in South Carolina. She did not sample separate populations of the same species so statements about variation within species could not be made.

B. Stock Abundance

Estimates of population size can be made either as relative abundance by use of catch and effort data usually available for most developed lobster fisheries or as absolute abundance by use of either census or mark–recapture techniques. Measurements of relative abundance are useful, particularly to provide an index of stock abundance within a fishery over a series of years. Absolute estimates may be more useful because they allow comparisons among species or seasons. Appropriate tags for mark–recapture studies have been developed (Scarratt and Elson, 1965; Cooper, 1970) that will last through the molt and do not induce high mortality or significantly change behavior. Morgan (1974a) addressed some of the problems of mark–recapture work in a study of spiny lobster abundance in western Australia.

There are very few data on the density of any of the clawed lobster populations. *Homarus americanus* populations in the inshore waters from Northumberland Strait to Connecticut surveyed in suitable habitat by SCUBA divers ranged from 70 to 32,500 lobsters per ha. Where available, biomass estimates were 56 to 1780 kg/ha (Scarratt, 1968a, 1972, 1973b; Cobb, 1971; Stewart, 1972; Cooper *et al.*, 1975). Cooper and Uzmann (1977) reported densities of first-year juveniles in mud–rock substrate to be up to 20×10^4/ha. Estimates of *H. americanus* densities in the offshore areas of 10 lobsters/ha on the open shelf and upper slope and 50 lobsters/ha (42.5 kg/ha) in the canyons were made by Cooper and Uzmann (1980) during summer months. At this time, a proportion of the lobsters have moved into shoaler waters (Uzmann *et al.*, 1977), thus these figures probably were underestimates of the total population. Data on stock densities of *H. gammarus* are unavailable. Densities of *Nephrops norvegicus* that were estimated from SCUBA diving and submersible observations range from 1×10^3 to 5×10^4/ha (Farmer, 1975) in appropriate habitats. Because the fishery for *Nephrops* is largely a trawl operation, diel changes in emergence from the burrows change the apparent density on a daily basis (Chapman and Howard, 1979).

Few estimates of absolute population density have been made for spiny lobsters. At the Abrolhos Islands in Western Australia, Morgan (1974a) estimated the density of *Panulirus cygnus* as 329 to 2065 animals per ha, depending upon the season. In shallow-water nursery reefs, juvenile *P. cygnus* densities were measured from 2×10^3 to 14×10^3/ha (Chittleborough and Phillips, 1975). Engle (1979) measured densities of 1^+ year class juveniles of *P. interruptus* as up to 4.9 ± 1.8 animals per m² in surfgrass habitat. In marked contrast, estimates of absolute density of *P. argus* in the Caribbean range from 3.9 to 7.3 lobsters/ha (Peacock, 1974) to 4.8 to 34.0 lobsters/ha (Herrnkind *et al.*, 1975). As Morgan (1980a) suggested, this difference in density between temperate (*P. cygnus*) and tropical (*P. argus*) species may be a typical feature in spiny lobster population estimates.

Crayfish populations often are large components of aquatic system biomass perhaps due to short food webs. Momot *et al.* (1978) calculated crayfish as being 30% of biomass in their stream system. Reported densities of populations in natural as well as man-made ponds have ranged as high as 77×10^4 crayfish/ha (Kossakowski, 1975), whereas standing crop estimates have ranged up to 1345 kg/ha (Momot et al., 1978).

C. Growth

The growth of lobsters and crayfishes is a complex phenomenon that is affected by many environmental factors. Growth proceeds almost continually, but size increase occurs only at molting. Thus, the measured growth rate includes two components—the size increase at molt and the length of time between successive molts. In general, size increase at molt is proportionately large when the animal is small and decreases with increasing size, whereas intermolt period is short at first but increases with age, as Mauchline's (1977) curves for Homarus americanus illustrate (Fig. 8; see also Chapter 3, Volume 2, of this treatise). This generalization seems to hold true for all lobsters and crayfishes (Fielder, 1964; Hewett, 1974; Munro, 1974; Chittleborough, 1975; Farmer, 1975; Mason, 1975; Hobbs, 1977; Pollock and Roscoe, 1977; Smale, 1978; Engle, 1979). Intermolt period seems to be the factor that responds first to environmental conditions and thus is most likely to determine growth rate (Hewett, 1974; Chittleborough, 1975; Aiken, 1980).

The time between molts can be divided into several stages that are identifiable morphologically and physiologically (Drach, 1939; Drach and Tchernigovtzeff, 1967; see also Chapters 1, 2, Volume 9, of this treatise). Aiken (1973) presented a classification scheme based on morphological development of the setae in Homarus americanus; similar schemes have been used for Panulirus cygnus by Dall and Barclay (1977) and for crayfish (Stevenson, 1972; Van Herp and Bellon-Humbert, 1978; Vranckx and Durliat, 1978). The four major stages are as follows: (1) the period when the animal is soft just after ecdysis, (2) the period during which the shell is hardening, (3) the "intermolt" stage, a period of rapid tissue growth, and (4) the premolt period. The proportion of time spent in each stage is quite variable but generally can be characterized as: (1) and (2) together, 5% or less; (3) 40–60%; and (4) 40–60%.

Many of the body parts can be regenerated after loss through autotomy or breakage (see Chapter 4, Volume 4, and Chapter 2, Volume 9, of this treatise). Generally, the new appendage appears fully formed but smaller after the succeeding molt. It increases to full size or nearly so after two molts. Regeneration generally results in a reduction in the overall growth rate of the individual. Lost eyestalks and mouthparts will not regenerate, but a heteromorphic appendage (antenna or leg) occasionally may appear in place of the lost part. The regeneration of limbs lost through autotomy

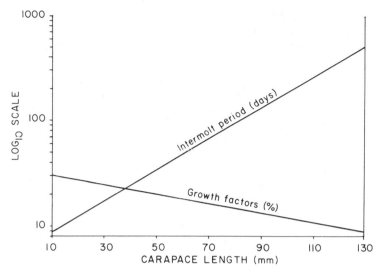

Fig. 8. An illustration of the change with size in the two components of crustacean growth rate, intermolt period, and size increase for *Homarus americanus*. (From Machline, 1977.)

affects growth rate. In *P. cygnus* and *P. argus,* injury, such as the loss of legs, significantly decreased the mean size increment per molt (Chittleborough, 1975; Davis, 1979). In *P. interruptus,* two lost legs decreased size increment at molt (Engle, 1979). Accelerated molting after limb loss can be seen in *P. interruptus* (Lindberg, 1955) and *Procambarus clarkii* (Bittner and Kopanda, 1973). In *Homarus americanus,* regeneration of lost appendages decreases both size increment and molt frequency (Emmel, 1906).

Growth rate is difficult to estimate from the natural population not only because of the discontinuous nature of size increase but also because of the lack of any way to determine age directly. The three following methods have been used (Ennis, 1980b): (1) length–frequency analysis (Cassie, 1954) of the population to establish cohorts, which presumably coincide with some age group, (2) average molt increment and proportion of the population molting from tagged animals and close examination of shell condition or pleopod stage, (3) tag–recapture data (size at release, size at recapture, and time at large) to determine the parameters of the von Bertalanffy equation. Caddy (1977) has suggested that the method of length–frequency analysis may not be appropriate for lobster populations because the modes may not represent single age groups for a species in which there is considerable variability in growth rate. An exception may be intensively exploited species, and the method has been applied with apparently good results to Newfoundland stocks of *H. americanus* (Ennis, 1980b). Size–frequency data have been used successfully to estimate growth in juvenile *Jasus edwardsii* (McKoy and Esterman, 1981). In the crayfish *Orconectes virilis*, cohorts were recognized and used to estimate growth rates (Momot and Gowing, 1977a). The von Bertalanffy (1938) growth model has been used frequently, and a multitude of research reports document lobster growth in terms of the parameters for that equation. Additionally, growth rates for several species of lobsters and crayfishes have been estimated from laboratory experiments in which growth increment and intermolt period were measured.

The parameters of the von Bertalanffy equation allow comparison of growth patterns between areas. Russell (1980) suggested that it is possible to provide standard errors for the estimates of the growth parameters using the least squares procedures described by Tomlinson and Abramson (1961). We present here for purposes of illustration data for *Homarus americanus* from several locations in the northwest Atlantic (Table II). Russell's (1980) analysis of data from the United States indicated there were regional differences among inshore *H. americanus* in the catabolic growth coefficient (k) but not in the asymptotic length (L_∞). It is difficult to compare the offshore populations, but it is apparent that values of L_∞ are substantially larger than for most inshore populations. McKoy and Esterman (1981) have shown in a similar fashion that growth rates of male *Jasus edwardsii* vary considerably with location on the North Island of New Zealand.

Despite the utility and frequent use of the von Bertalanffy equation for

TABLE II

Parameters for the von Bertalanffy Growth Equation for *Homarus americanus* at Selected Locations[a]

Location	Sex	k	L_∞	t_0	Reference
Bay of Fundy	Male	0.065	281	+0.76	Campbell (1983)
	Female	0.089	207	+0.42	
Newfoundland	Male	0.3896	105	−0.7964	Ennis (1980b)
	Female	0.2396	112	−0.6894	
Offshore	Male	0.096	270	+0.5	Cooper and Uzmann
	Female	0.074	240	+0.3	(1980)
Maine	Male	0.048	267	−0.773	Thomas (1973)
	Female	0.087	241	−0.096	Krouse (1977)
Rhode Island	Male	0.0936	190	−0.290	Russell *et al.* (1978)
	Female	0.0966	185	−0.198	
New Jersey	Male	0.127	190	+0.653	Halgren (1976)

[a] Equation: $l_t = L_\infty[1-e^{-k(t - t_0)}]$ where l_t = carapace length at time t, L_∞ = mean asymptotic carapace length, k = growth coefficient, t_0 = hypothetical age at length zero.

determining growth rates, there are some reservations about its validity when applied to lobsters. Growth rate is incremental, not constant, violating an assumption of the model. Intermolt period is not constant over year or animal size. The variation around the von Bertalanffy curve is high, probably due to the degree to which molt frequency and increment are influenced by environmental conditions, making precise definition of the growth curve difficult. Nonetheless, Morgan (1980a) feels that the von Bertalanffy model adequately describes growth in *Panulirus cygnus,* giving results very similar to an empirical model. However, differences between an empirical growth model and the von Bertalanffy curve were suggested for *Jasus edwardsii* in New Zealand (Saila *et al.,* 1979) and for *Homarus americanus* (Saila and Marchessault, 1980). A problem arises when different growth models are applied to the same data with varying results because the growth parameters used can substantially affect inferences derived from yield-per-recruit models. Where does this leave the fishery biologist who needs a way of determining growth rate that provides suitable parameters for yield models? We agree with Anthony and Caddy (1980) that the von Bertalanffy model is acceptable for yield-per-recruit assessments over a limited size range when molt frequency is constant, but not for describing lobster growth over the entire size range.

The foregoing discussion should illustrate the need for further research into the estimation of growth rates. In particular, the problem of crustacean age determination is an important one to fishery biologists.

A number of factors can markedly influence the growth rates of lobsters and crayfishes. These include age, nutrition, temperature, photoperiod, season length, and reproductive condition. The effects of temperature on growth rate can be seen from laboratory experimentation, from seasonal changes, and along latitudinal gradients. Slowing of growth rates during winter months is well documented for lobsters and crayfishes inhabiting temperate and boreal areas (Hopkins, 1967; Brown and Browler, 1977; Phillips *et al.,* 1977; Davis, 1979; Engle, 1979; Aiken, 1980; Pratten, 1980). In *Homarus,* temperatures below 5°C block molt induction; thus, all the individuals in a population will progress to the first stage of proecdysis (D_o) and will remain there until the temperature increases, thereby explaining the synchrony of the spring molt (Aiken and Waddy, 1976). Many laboratory experiments attest to the relationship between temperature and growth rate (Morrissy, 1974, 1976; Flint, 1975; LaCaze, 1976; Momot and Gowing, 1977a; Tcherkashina, 1977; Aiken, 1980). The seasonal effect probably is not due entirely to temperature. Photoperiod affects growth in *Orconectes* (Aiken, 1969) and at low temperatures in *Homarus* (Aiken and Waddy, 1976). Secondary effects of seasonal changes in pond productivity such as population density and food availibility affect crayfish growth rates (Abrahamsson, 1973; Morrissy, 1980).

The amount and, probably, the quality of food available affect growth rates. In the laboratory, *Panulirus cygnus* grows more slowly when food is limited (Chittleborough, 1975). Newman and Pollock (1974a) and Pollock (1979) demonstrated a correlation between growth rates of *Jasus lalandii* and abundance of benthic food off South Africa. On crowded reefs, the growth of *P. cygnus* may be slowed due to food shortage (Chittleborough, 1975). Tcherkashina (1977) found growth rates reduced in natural populations of crayfish as compared to the theoretical optimal growth rates obtained from animals in pond culture. Momot and Jones (1977) suggested that quality of food and temperature regime are more important factors influencing growth in young-of-year than is population density.

Social interactions slow growth rates in larval and juvenile *H. americanus* (Cobb, 1970; Cobb and Tamm, 1974; Cobb et al., 1982). The opposite is true for *Panulirus cygnus* (Chittleborough, 1975). The difference may be due to social structure; *P. cygnus* shelters communally and *H. americanus* solitarily.

In general, lakes and ponds are usually more productive for crayfish growth than are rivers and streams (Mason, 1975). Factors that can limit growth include resource limitations in type and abundance. For instance, some crayfish species will have greater growth on a primarily animal diet, whereas others have a greater assimilation efficiency on allochthonous leaf material. Water conditions also influence crayfish growth. For example, moderate eutrophication may increase crayfish growth, but extreme eutrophication increases turbidity, which will decrease dissolved oxygen concentration (Morrissy, 1974). The balance between higher aquatic plants and phytoplankton can also affect crayfish growth. Phytoplankton at high densities reduces light levels in the water column and compete for nutrients with macrophytes. This in turn reduces the abundance of food resources for crayfish, directly in plant material and indirectly in epiphytes and epizoans. Konikoff (1977) found high growth rates for *Procambarus clarkii* in areas with dense water hyacinth. Other environmental factors can also affect growth rate. Chaismartin (in Laurent, 1973) has suggested that calcium can limit growth in *A. astacus* and *O. limosus* if it is below 2.8 mg/liter. Magnuson et al. (1975) found no *O. virilis* in lakes with calcium less than 2 ppm. Shelter space, water depth (Flint and Goldman, 1977), constancy of conditions, water quality (J. F. Huner and R. P. Romaire, personal communication; Morrissy, 1980) also are important to crayfish growth.

D. Sexual Maturity

The measurement of the size or age at attainment of sexual maturity has several difficulties, not the least of which is what measure to use. Van Engel

(1980) listed the following criteria for *Homarus americanus,* and they can be generalized or modified for all lobsters and crayfishes: stage of ovarian development, contents of male testes and female seminal receptacle, presence of eggs on the abdomen, and morphometry of male claw and female abdomen. Additionally in females, the presence of engorged cement glands and the presence of nonplumose setae on the pleopods may be used (Aiken and Waddy, 1980, 1982). Enlarged first, second, and third legs of male *Panulirus* (Berry, 1970; George and Morgan, 1979; Grey, 1979) and female fifth legs (George and Morgan, 1979; Grey, 1979) have been used as indicators of sexual maturity. Each expresses in some way the degree of readiness of a lobster to reproduce. The most widely used criterion of female sexual maturity is the presence of eggs on the abdomen, and of male maturity the presence of mature sperm. Considerably more attention has been paid to the onset of female than to male maturity.

In consideration of maturity, both morphological and behavioral aspects should be taken into account. Males of *Homarus americanus* have mature sperm at 40–50 mm (Templeman and Tibbo, 1945; Krouse, 1973). However, small males appear "not to be interested in mating" (Aiken and Waddy, 1980) and may have to compete with other males for females or mating sites (Atema and Cobb, 1980). The increase in relative claw size after maturity that is typical of mature male *Homarus* (Aiken and Waddy, 1980) and freshwater crayfishes (Stein, 1976) does not occur until the animal is larger, leading Ennis (1980a) and Van Engel (1980) to suggest the visceral changes precede external modifications. So, despite the physical ability to inseminate a female, a small male may not be able to function as a mature male until it is interested in mating and able to compete successfully for females, which may happen at a considerably larger size. A further complication in clawed lobsters is that the female can mate before ovarian maturity and store the sperm for up to 2 years. It is because of these several complexities in the consideration of sexual maturity that we recommend with Ennis (1980a) that the presence of fertilized eggs on the abdomen or the presence of well developed cement glands, which indicate egg extrusion to be imminent be used as criteria for female sexual maturity.

A listing of size at female sexual maturity and time of spawning for representative species is found in Table III. Perusal of the table should convince the reader of the variety of breeding adaptations. Where populations have been closely examined over a wide geographical range, great variation in size at maturity has been found. Female *Panulirus argus* first bear eggs at about 69 mm CL in Panama (Butler and Pease, 1965), about 80 mm in Florida (Davis, 1975) and Bimini (Kanciruk and Herrnkind, 1976), 85–90 mm at Bermuda (Sutcliffe, 1952), and 95 mm in Jamaica (Munro, 1974). A correlation with water temperature as suggested by Bradstock (1950) for

TABLE III

Size at Female Sexual Maturity and Season of Spawning Period in Several Lobster and Crayfish Species

Species	Range of sizes at 50% female sexual maturity (mm)	Season of spawning period	Reference
Homarus americanus	70–120[a]	Summer	Ennis (1980a); Krouse (1973); Skud and Perkins (1969); Smith (1977); Templeman (1936b)
H. gammarus	77–90[a]	Summer	Simpson (1961)
Nephrops norvegicus	20–25[a]	Summer	Farmer (1974b, 1975); Hillis (1979)
Panulirus argus	69–95[a]	Year round, spring	Munro (1974); Kanciruk and Herrnkind (1976)
		Spring and fall, depending upon area	Butler and Pease (1965); Sutcliffe (1952)
P. cygnus	80–100[a]	Winter	George et al. (1979); Grey (1979)
P. delagoae	70[a]	Spring and summer	Berry (1973)
P. homarus	50[a]	Spring and summer	Heydorn (1969a); Berry (1971b)
P. gracilis	85[a]	Year around	Loesch and Lopez (1966)
P. versicolor	65[a]		George and Morgan (1979)
Jasus lalandii	45–59[a]	Winter	Heydorn (1965); Matthews (1962)
J. edwardsii	70–120[a]	Winter	Bradstock (1950); Street (1969); Annala et al. (1983)
Procambarus clarkii	31[b]	Summer	Penn (1943)
Pacifastacus leniusculus leniusculus	65–80[b]	Spring	Abrahamsson and Goldman (1970); Abrahamsson (1971)
Astacus astacus	76[b]	Fall	Abrahamsson (1971)
O. virilis	25[a]	Fall, spring	Momot (1967)

[a] Carapace length.
[b] Total length.

Jasus lalandii (now *J. edwardsii* and *J. lalandii*) and Annala *et al.* (1983) for *J. edwardsii* is not obvious for *P. argus,* but further research may change this opinion. There is a wide variation in size at sexual maturity in *Homarus americanus.* Higher summer temperatures in some parts of the Canadian Maritimes than in others may lead to variation in size at maturity (Templeman, 1936a), and there is evidence for a similar relationship around Long Island, New York (Smith, 1977; Briggs and Mushacke, 1980). Aiken and Waddy (1980) cite laboratory experiments that support the relationship between small size at maturity and higher summer temperatures. It is also possible that an intense fishery may select for reduced size at maturity.

E. Sex Ratios

Prior to maturity, the sex ratio of lobsters and crayfishes approximates 1.0. The local sex ratio of spiny and clawed lobsters in commercial or experimental catches may vary considerably from 1 : 1. Females or males may be predominant depending upon season, location and method of capture (e.g., Davis, 1974; Briggs and Mushacke, 1980; Ennis, 1980a; Lyons *et al.,* 1981). In many cases, the sex ratio of both spiny and clawed lobsters over the size at maturity taken from catch statistics shows a systematic variation from a preponderance of females in the smaller size classes to a complete dominance of males of the largest sizes (Skud, 1969; Farmer, 1975; Morgan, 1984). Different mortality and growth rates between males and females, due in large part to the female reproductive cycle, probably explain the shape of the curves shown in Fig. 9. A model that assumed differential mortality rates for berried and unberried female *H. americanus* showed that different sex ratios resulted from different mortality rates (Saila and Flowers, 1965). Using data presented by Skud (1969) on the offshore *H. americanus* fishery, D. Campbell (personal communication) showed that the observed changes in sex ratio can be explained simply by a linear increase in fishing mortality rate as the females increase in size, assuming male mortality is constant and higher at the time of entry into the fishery. An additional factor in establishing the variation in sex ratios may be the slower growth rate of females after maturity. The preponderance of females in the size classes just above maturity may be a result of more of the fishing pressure directed towards males because of protection of egg-bearing females by law or by behavioral attributes. We would predict that the heavier the fishing pressure the more the sex ratio will be skewed toward females in this size range.

Sex ratios in crayfish populations appear to be close to 1 : 1 if differential distributions are taken into account. For instance, Aiken (1965) reported a sex ratio of 3.89 : 1 ($\eth$: $\mathfrak{P}$) for *Orconectes virilis* in New Hampshire but fished only in shallow water. Momot (1967) showed that adult female *O.*

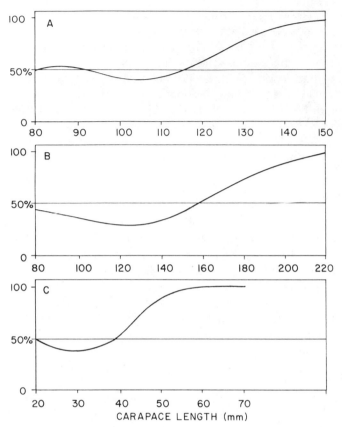

Fig. 9. Relationship between sex ratio (expressed as percent males) and size in three species of lobsters. [Data were taken from catch records by (A) Morgan (1984) for *Panulirus cygnus,* by (B) Skud (1969) for *Homarus americanus,* and by (C) Figueiredo and Thomas (1967) for *Nephrops norvegicus.*]

virilis migrate into deeper water in the summer, perhaps explaining Aiken's data. Abrahamson and Goldman (1970) found a ratio of 1.2 : 1 in *Pacifastacus leniusculus* during a diving survey in Lake Tahoe, California.

F. Fecundity

An important part of the measure of reproductive potential of a population is fecundity, usually given as the number of eggs carried per female. The number of eggs carried is highly variable between species. The following illustrates the general conclusion that *Panulirus* and *Jasus* are highly fecund, and clawed lobsters carry fewer eggs than do spiny lobsters (summarized from Aiken and Waddy, 1980; Campbell and Robinson, 1983).

Genus	Range of reported fecundities
Panulirus	50,000– 900,000
Jasus	47,000–1,930,000
Homarus	2,000– 97,000
Nephrops	400– 6,000

The high value for *Jasus* is due to *Jasus verrauxi*, the largest of all spiny lobsters in which a 235 mm CL female was reported to carry nearly 2,000,000 eggs (Kensler, 1967). However, *Jasus* species tend to carry fewer eggs than do *Panulirus* of the same size. Freshwater crayfishes carry fewer eggs than do the marine lobsters, in the range of 50–200 (Momot, 1967; Abrahamsson and Goldman, 1970; Momot and Gowing, 1977a). The smaller number of eggs probably is correlated not only with smaller body size but also a greater amount of parental care by crayfish.

Egg loss during incubation can be an important factor in the determination of fecundity. Perkins (1971) reported a loss rate of 36% between egg extrusion and hatching in *Homarus americanus,* whereas for *Nephrops norvegicus,* the loss rate was 10% per month (Figueiredo and Nunes, 1965). On the other hand, in *Panulirus cygnus,* there was little egg loss during incubation (Morgan, 1972). Because egg loss does occur, it would make sense to limit the time of determination of fecundity to shortly before hatching is expected. In crayfishes, the numbers of hatched young produced are dependent upon the care of the female, and estimates of fecundity are best made close to or just after hatching (Morrissy, 1975).

Fecundity varies with environmental conditions and geographic region in most of the lobster and crayfish species studied (Morrissy, 1975; Aiken and Waddy, 1980; Campbell and Robinson, 1983).

Reproductive potential varies with density and with food resources in crayfish populations. *Orconectes virilis* showed a reduced fecundity when crowded (Momot and Gowing, 1977a,b) probably due to increased competition for food resources. A reduction in the frequency of spawning or the proportion of females spawning in the population as a response to poor food conditions was seen in *Pacifastacus leniusculus* (Goldman, 1973), *Astacus astacus* (Abrahamsson, 1973), and *Orconectes limosus* (Kossakowski, 1975). In spiny lobsters, the same generalization may be true because in *Panulirus cygnus* high population densities result in a smaller size at maturity, and small females have fewer eggs per brood and fewer broods per year (Chittleborough, 1976).

The relative fecundity of a size class of lobsters is a function of the frequency of oviposition, the number of eggs produced each time, and the proportion of the population that size class represents (Aiken and Waddy, 1980). This relative fecundity has been estimated for *Panulirus homarus* (Berry, 1971b), where it was greater for intermediate-sized females, who, despite producing fewer eggs per spawn than larger females, spawned more frequently and so contributed the most eggs. The same relationship has been found in populations of *P. argus* at Bimini (Kanciruk, 1980), Dry Tortugas (Davis, 1975), and the Florida Keys (Lyons *et al.*, 1981). *Homarus americanus* females also have the highest proportion of females extruding eggs in the largest size classes. However, intermediate size females contribute the most eggs to the population because of the greater number of ovigerous females than in larger size groups and more eggs produced per female than in the smaller size classes (Campbell and Robinson, 1983). Estimates of relative fecundity for other exploited lobster species have not been made but would seem important to fisheries management.

G. Mortality

In the analysis of population dynamics of exploited stocks, knowledge of the coefficient for instantaneous total mortality rate (Z) as well as its components of fishing and natural mortality are essential. Estimates of total mortality for exploited populations of lobsters range from about 0.5 to over 2.0. Fishing mortality (F) estimates range from 0.2 to 2.0, whereas natural mortality (M) is usually 0.1 to 0.3. The estimation of these rates usually relies on data taken from a commercial fishery, using changes in the relationship between catch and effort, the abundance of successive age classes, or the results of mark–recapture experiments. All the methods have assumptions that are met to a greater or lesser degree. Annala (1979) has compared three different types of size–frequency distribution analyses and a tag–recapture analysis. Morgan (1980a) presented a new method of size frequency analysis for use when relatively comprehensive data on the catch of various size groups and fishing effort are available.

Panulirus argus in the area around Jamaica has populations that are differentially exploited. Munro (1974) calculated Z, M, and F for stocks at the following three locations:

	Z	M	F
Port Royal (heavily fished)	1.52	0.14	1.38
South Jamaica Shelf (moderately fished)	0.50	0.23	0.27
Pedro Cays (lightly fished)	0.62	0.52	0.10

Estimates of total mortality for exploited populations of *P. argus* in Florida ranged from $Z = 1.72$ to $Z = 4.09$ (Gulf and South Atlantic Fisheries Management Councils, 1981). Natural mortality for the Florida populations probably lies in the same range as suggested by Munro for Jamaica. Lyons *et al.* (1981) suggested that the practice of using sublegal lobsters as live bait in lobster traps significantly increases mortality. Morgan (1984) gave total mortality estimates for *Panulirus cygnus* in Western Australia as 0.58 for males and 0.60 for females. There was a great variation in Z with lobster size (155 mm CL; $Z = 0.32$). The reasons for declining total mortality rate with increasing size are unknown, but they may include reduced catchability and a lower vulnerability to predators. The best estimates of instantaneous mortality rates in an exploited population of *Jasus edwardsii* in New Zealand were $Z = 1.0–1.5$, $F = 0.9–1.4$, with M assumed to be 0.1 (Annala, 1979). In *Panulirus cygnus*, M is density dependent in both juveniles (Chittleborough, 1970) and adults (Morgan, 1974a).

Mortality estimates for *Homarus americanus* have been summarized by Anthony (1980) and Campbell (1980); the coefficient of instantaneous total mortality for inshore fishing areas ranges from 1.63 to 2.80, with the average about 2.2. Fishing mortality of these stocks must thus be very high, approaching 2.0 and natural mortality is in the range of 0.05 to 0.15. Offshore stocks suffer somewhat lower mortality rates, but the trend is strongly increasing. Tag return data from Hudson Canyon showed an increase in F from 0.07 to 0.41 from 1968 to 1971 (Anthony, 1980; Fogarty *et al.*, 1982). An assessment of variation in F by year and location for offshore *H. americanus* by use of cohort analysis was performed by Fogarty *et al.* (1982). They provided estimates of provisional mortality rates ($F\Delta T$), which depend upon intermolt time interval (Δt). At smaller sizes (85–95 mm CL), Δt is about 1 year, thus $F\Delta t = F$. They found that $F\Delta t$ for the smallest size class increased from 0.16 in 1966–1971 to 0.40 in 1972–1980 in the southern New England area and from 0.03 to 0.10 during the same time period on Georges Bank. The differences in fishing mortality rates in these two regions of the offshore fishery may reflect the longer history of exploitation in southern New England or differential size segregation between areas (Fogarty *et al.*, 1982). Similar data are not available for *H. gammarus* or *Nephrops norvegicus*.

In freshwater crayfishes, mortality figures are available for *Orconectes virilis*. Momot and Gowing (1977a) followed the population dynamics of this species in three pothole lakes of Michigan and found that Z varied between 0.49 and 5.24. In general, the second and fourth years had the highest mortality. Mortality appeared to be related to density dependent factors. They concluded that fluctuations in mortality rather than in growth rate produced most of the year-to-year fluctuations in biomass accumulation.

H. Models of Exploited Populations

It is the changes in population abundance, and therefore potential yield to the fishery, that are of primary importance to the fishery biologist. Using information about the biology of the animal and the fishery, numerous attempts have been made to build models, usually mathematical or statistical ones, that will explain how long term equilibrium yields change with fishing effort. As Morgan (1980a) has pointed out, the models used for lobster populations were developed for the study of fin fish populations. This poses some problems for crustacean fisheries, although it does not invalidate their use. For example, it is relatively easy to determine the age of scale fish, but impossible to do so for Crustacea. Several models use age distribution of the population as a basic input. Because this is unavailable for lobsters, assumptions must be made about the relationship between age and size, or new models must be developed that use length rather than age composition.

Both empirical and conceptual models have been employed in the study of lobster populations. Empirical models attempt to relate trends in the yield of a fishery over a number of years with corresponding trends in environmental variables (e.g., temperature, sunspot, river discharge). Conceptual models are based on the formulation of hypotheses that explain how population size will react to changes in fishing effort. Most of the modeling studies have been directed at *Homarus americanus* and to a lesser extent, *Panulirus cygnus, P. argus,* and *Jasus edwardsii.*

1. EMPIRICAL MODELS

In many cases, the longest time series of environmental information available is that of temperature. In addition, temperature clearly affects lobster growth (see Aiken, 1980), activity (McLeese and Wilder, 1958), and feeding (Weiss, 1970). A number of investigators have shown a relationship between temperature variations and changes in lobster catch. Significant positive correlations among catch, effort, and sea surface temperature for 6 and 7 years previously have been demonstrated for *Homarus americanus* and *H. gammarus* (Dow, 1969, 1977, 1978). Conversely, *Nephrops norvegicus* catch shows a significant inverse correlation with the temperature 6 and 7 years preceeding (Dow, 1978). Flowers and Saila (1972) developed a multiple regression equation that predicted catch in the Maine lobster fishery based both on the mean annual temperature from 6, 7, and 8 years previously and on the mean temperatures from the months of January, February, and March of the previous year. Sutcliffe (1973) showed a strong correlation of the Quebec lobster catch with the April discharge of the St. Lawrence River.

More sophisticated empirical models have been summarized by Saila and Marchessault (1980). These models use statistical techniques such as autoregressive moving averages (Boudréault et al., 1977), multiple regression on principle components (DuPont and Boudréault, 1976), and a polynomial distributed lag technique (Orach-Meza and Saila, 1978). All of these techniques used a data base of a time series of catch, effort, and temperature for a restricted area such as Maine or the Magdalen Islands. When predicted landings were compared to observed landings over the period of interest, the coefficients of the models explained 80–90% of the total variance. A large proportion of the variation in lobster catch could be ascribed to changes in sea water temperature according to these models.

Only one attempt has been made to prepare an empirical model for spiny lobsters, and none has been made for freshwater crayfishes. The analysis by Saila et al. (1980) of the fishery for Jasus edwardsii in New Zealand was done with use only of data from the average catch; no other inputs such as fishing effort or environmental variables were used. The authors compared three procedures and found that an autoregressive integrated moving average model, similar to that applied by Boudréault et al. (1977) to the H. americanus fishery, provided the best fit to the actual data, and resulted in relatively unbiased forecasts.

A strong case can be built for the importance of environmental fluctuations, particularly temperature cycles, in the prediction of lobster catch. However, caution is always advisable in the use of correlations for predictive purposes. For instance, correlation of St. Lawrence River discharge and catch (Sutcliffe, 1973) may be due to temperature differences brought about by climatic conditions correlated with variations in river flows. Additionally, empirical models seem to do well for "normal" years, but years of extraordinarily high or low catches are probably very hard to predict. Nonetheless, such models allow fishing effort and supply of legal-sized lobsters to be treated as essentially unmeasurable variables whose changes are assumed to depend on other quantifiable environmental and economic variables (Orach-Meza and Saila, 1978). Additionally, the models are well suited for short-term predictions. They are thus a very valuable tool in the analysis of lobster population dynamics.

2. CONCEPTUAL MODELS

The conceptual models used in assessing fish stocks are of two basic types. Surplus yield models, based on Schaefer's (1954, 1957) work, use the concept that the production of a stock is a function of stock size. Dynamic pool models such as those of Beverton and Holt (1957) and Ricker (1958) use biological aspects of the species such as growth and mortality rates to determine the obtainable yield. The surplus yield models use only catch and effort data in estimation of population size. The analytical approach of the dynamic pool models requires the estimation of natural and fishing mortality, age at recruitment and at first capture, and some parameters of the von Bertalanffy growth equation, and the combination of all these in a deterministic model. Both have been applied to lobster fisheries.

a. *Surplus Yield Models.* Marchessault *et al.* (1976) applied both the Schaefer model and their own delayed recruitment model to the *H. americanus* fishery. Their model, a variation of Schaefer's, considers the impact of spawning stock size on recruitment to the fishery several years later, implying the existence of a stock–recruitment relationship. The delay aspect is particularly appropriate to lobster fisheries since there generally is a long period (4–7 years) between spawning and recruitment to the fishable stock. The delayed recruitment model had a considerably lower predicted maximum equilibrium yield and lower optimal level of applied effort for the Rhode Island inshore fishery than the Schaefer model (Fig. 10). They con-

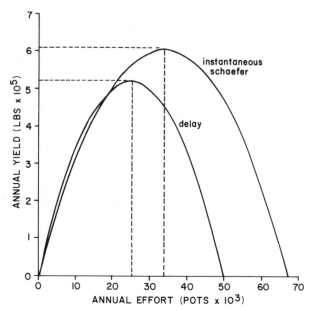

Fig. 10. Results of two surplus yield techniques, the instantaneous Schaefer model and a delayed recruitment model applied to the *Homarus americanus* fishery in Rhode Island. (From Marchessault *et al.*, 1976, in the *Journal of the Fisheries Research Board of Canada.*)

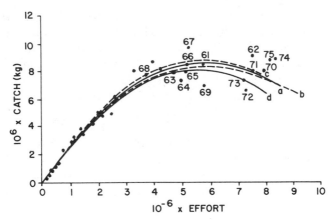

Fig. 11. Four surplus yield techniques applied to catch and effort data from the Western Australia *Panulirus cygnus* fishery. (a) Schaefer; (b) GENPROD; (c) PRODFIT; (d) delayed recruitment. (From Morgan, 1980a.)

cluded that the use of the Schaefer model for managing the Rhode Island lobster fishery may lead to overfishing and further depletion of existing stocks. Saila and Marchessault (1980) later introduced a stochastic element into the delayed recruitment model.

Four different surplus yield techniques were applied to catch and effort data for the Western Australia fishery on *Panulirus cygnus* by Morgan (1979). The four techniques used were Schaefer (1957), GENPROD (Pella and Tomlinson, 1969), PRODFIT (Fox, 1975), and Marchessault *et al.* (1976) delayed recruitment model. As Fig. 11 shows, all four provided similar solutions. Nonetheless, Morgan (1980a) pointed out that the delayed recruitment model and PRODFIT are not susceptible to bias introduced by increased variability with increased catch because they are based on the relationship between catch per unit effort and effort. These two models also appeared to give more reasonable estimates of the catchability coefficient than do the GENPROD or Schaefer models. Morgan (1979, 1980a) concluded that any of the four models provided a good description of the relationship between equilibrium catch and fishing effort up to the levels of effort so far encountered. Extrapolation beyond that level of effort is not wise. In fact, examination of Fig. 11 shows little evidence of the decline in catch predicted by the models at high values of effort. It is not clear why the delayed recruitment model was quite different from the Schaefer model when applied to *Homarus* data by Marchessault *et al.*, (1976) but not when applied to *Panulirus* by Morgan (1979).

Saila *et al.* (1979) applied Fox's (1970) modification of the Schaefer model to data from the fishery of *Jasus edwardsii* in New Zealand with similar results. It appeared that reduction of fishing effort by about 60% would give the maximum equilibrium yield of approximately 4200 metric tons (MT) per year. Similarly, application of this model to the Monroe County (Florida) *P. argus* fishery indicated an MSY of nearly 3×10^3 metric tons could be harvested using 40% of the traps in service at the time (Gulf and Atlantic Fishery Management Councils, 1981).

b. Dynamic Pool Models. The Beverton–Holt yield per recruit model has been used by several investigators, including the New England Fishery Management Council (1982) for U.S. populations from Maine to New Jersey. They subdivided the region into three areas: the Fishery Conservation Zone (3–200 mi from shore), Gulf of Maine Inshore (0–3 mi), and southern New England Inshore (0–3 mi). The values of L_∞ ranged from 195 to 270, of t_o from −0.772 to 0.500, and k ranged from 0.048 to 0.096. For inshore stocks, total mortality has been variously estimated between 1.6 and 2.8, and natural mortality generally is guessed to be in the range of 0.10–0.15. The present age at first harvest is about 4 years for offshore males and between 6 and 7 years for all others. Thomas (1973) estimated the optimal age at first harvest in Maine to be 14 years at $M = 0.10$. Optimal age at first harvest is much more sensitive to change in M than in fishing mortality (F). The analysis showed that only in the southern New England area would small increases in minimum legal size have much effect on yield per recruit. The curves clearly show that the fishery can be managed at much lower levels of exploitation than is the present practice with no loss of yield. Maximum yield per recruit is achieved at levels of F well below 0.5 (New England Fishery Management Council, 1982).

The Beverton–Holt model appears to be an adequate method of assessing yield per recruit in lobsters; however, it may be unreliable, particularly at higher fishing intensities for two main reasons (Caddy, 1977): (1) the effects of overfishing on recruitment are not considered; (2) growth is discontinuous and seasonal, and, at high fishing intensities, errors can arise if the time of application of fishing effort is not considered in relation to the annual growth period. Caddy (1977) presented a simplified approach to the yield-per-recruit analysis based on information normally collected in well-developed crustacean fisheries, namely molt frequency and size increment at molt as a function of size, length–weight relationships, and gear selection curves. His approach attempted to address the second problem mentioned above by not using the von Bertalanffy growth equation parameters, whose assumptions are not wholly appropriate for lobster growth. A similar model was developed by Ennis and Akenhead (cited by Ennis, 1980b). These efforts and others by Canadian lobster fisheries biologists provided similar results to those described for the United States fishery: yield per recruit is more sensitive to recruitment length than to changes in exploitation rate.

The cohort analysis technique to estimate fishing mortality coefficients has been applied also to *Nephrops* stocks in two major fishing areas off Iceland (Eiriksson, 1979). Maximum yield per recruit was obtained at $F = 0.35$ and $F = 0.38$, but maximum value per recruit at $F = 0.21$ and $F = 0.17$.

Yield per recruit models have been applied to three species of spiny lobsters. The *Panulirus argus* fishery in Jamaican waters was assessed by Munro (1974) using the Beverton–Holt model. At the time, it appeared that the South Jamaica Shelf was yielding about 480 g per recruit, with length at recruitment at 60 mm CL. Figure 12 shows yield per recruit curves for that recruitment size and several larger sizes. The curves indicate that changing of a size at recruitment without changing fishery effort would result in substantial increase in yield. Any change in the fishing mortality from the then current $F = 0.27$ would result in a decline in yield per recruit. In New Zealand, both the Beverton–Holt technique and an empirical model were used to estimate yield per recruit in the *Jasus edwardsii* fishery (Saila et al., 1979). They suggested that the empirical model gave more realistic results than the Beverton–Holt model. The characteristic dome shape of the Beverton–Holt curve was particularly evident in the curve at $M = 0.05$ for males, but not for females or at higher levels of natural mortality. The empirical model did not give the dome shape, but the yield per recruit was very similar to the Beverton–Holt model at all levels of F above 0.4.

Morgan (1977) applied the Ricker model to the *Panulirus cygnus* fishery in Western Australia and concluded that changes (either increase or decrease) in size at first capture would result in a decline in yield per recruit at the current level of fishing mortality. On the other hand, changes in fishing effort would make very little difference to the yield. Morgan (1980a) has provided a sensitivity analysis of the Ricker method when applied to the *Panulirus cygnus* fishery. The Ricker model was much more sensitive to errors in estimating L_∞ and M than to errors in t_0 and k. However, despite quantitative changes in the yield per recruit estimates the general trends and conclusions remained unaltered.

The yield per recruit models provide a powerful and relatively straightforward tool for fisheries analysis. However, sensitivity analyses such as those by Morgan, the difficulty of obtaining the necessary data, the inability of the Beverton–Holt method to account for changes in mortality with size or age, and the limitations of the models themselves all indicate that they must be employed with caution. An additional complication is the difficulty in mea-

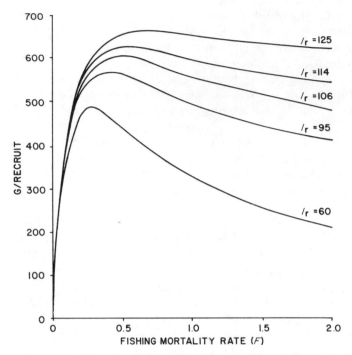

Fig. 12. Yield per recruit against fishing mortality rate for the *Panulirus argus* fishery in Jamaica calculated using the Beverton–Holt model. At the time of analysis, the length at recruitment (l_r) was 60 mm carapace length, curves for l_r at four higher values were also calculated. (From Munro, 1974.)

suring recruitment, which leads to expressing the results as yield per recruit rather than directly as yield. The evidence concerning stock–recruitment relationships is considered in the next section.

I. Stock–Larval Recruitment Relationship

In none of the clawed lobster fisheries is the relationship between parent stock, larval abundance, and subsequent catch known. Many attempts have been made to estimate larval abundance of *Homarus americanus* by use of plankton or neuston nets. Scarratt (1964) was unable to show any relationship between the first larval stage and parent stock. Scarratt (1973a) compared 3-year running means of fourth stage larval production and estimated legal stock 6 years later. He found that 36% of the variability of subsequent commercial lobster stocks was dependent upon stage IV larval production, but this is not sufficient for predictive purposes. In all the studies surveyed, it appeared that the abundance of the last larval stage was seriously underestimated relative to the size of the parent stock (e.g., Scarratt, 1964; Nichols and Lawton, 1978). This implies significant immigration of juveniles to the area, mistaken estimates of the adult stock size, or that the larval population is not being properly sampled. We lean toward the third explanation and suggest that a larval collector program similar to the one described below for *P. cygnus* might be successful. The following statement of ICES working group on *Homarus* stocks (International Council for the Exploration of the Sea, 1975) appropriately summarizes the current situation: "Any research to elucidate the stock-recruitment relationships would be most useful."

Information about recruitment in spiny lobsters is somewhat better. Phillips (1972) developed a larval collecting device that allows the measurement of relative abundance of settling stage of *Panulirus cygnus* in nursery reef areas. Data on larval recruitment, due to this device, are available from 1968 onward (Chittleborough and Phillips, 1975). Morgan et al. (1982) have supplied an index of spawning stock as catch per pot lift of ovigerous females (Fig. 13) and postulated the relationship between spawning stock

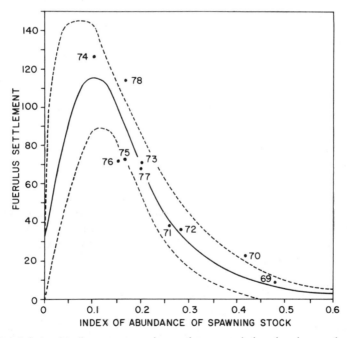

Fig. 13. Relationship between puerulus settlement and the abundance of ovigerous females for *Panulirus cygnus*. Solid line, curve fitted by the Ricker stock-recruitment equation. Dotted lines, 95% confidence intervals. (From Morgan et al., 1982.)

and puerulus settlement 9 months later using Ricker's (1958) stock–recruitment relationship. The Ricker equation fits remarkably well to the data. The nature of the biological mechanism producing this curve is not clear, but Chittleborough (1976, 1979) has suggested that female fecundity is correlated with population density.

Despite the apparent relationship between spawning stock and puerulus settlement, it may be difficult to predict recruitment to the fishery from puerulus abundance several years previously. The life cycle of this and most other shallow water palinurids is such that the juveniles live in shallow water crowded nursery reef areas. Large variations in puerulus recruitment may be obscured by density-dependent mortality in the 1–4-year old juveniles (Chittleborough and Phillips, 1975). Morgan et al. (1982) have suggested that there is an asymptotic relationship between the level of puerulus settlement and the subsequent abundance of recruits to the fishery. They feel that the level of puerulus settlement can be taken as an indicator of the likely level of the future fishery. Over a 10-year-period (1969–1979), years of very low and very high abundance were correlated with puerulus settlement 4 years earlier.

Thus, in general, stock–recruitment relationships are known only poorly in the lobster fisheries. Only in the P. cygnus fishery has there been a significant relationship postulated. No fishery has good information on recruitment to the fishable stock. The combination of a stock-dependent relationship between spawning stock and puerulus settlement and a density-dependent mortality in the juvenile phase may be the controlling factors in P. cygnus recruitment (Morgan et al., 1982). Similar relationships may hold for other shallow water dwelling palinurids, but evidence for this is lacking.

IV. THE FISHERIES

Lobsters and crayfishes make up a very valuable portion of the world's fisheries largely because of their status as a luxury food. Their relatively long life history and vulnerability to traps has allowed overexploitation to severely deplete stocks in many areas of the world. In most cases (particularly with regard to lobsters), appropriate management schemes were not implemented before overexploitation occurred. In this section, we present a brief overview of the locations and methods of fishing and some of the efforts being made to manage the fisheries.

A. Location

Homarus and Nephrops are heavily fished in the North Atlantic. The areas of greatest Nephrops fishing are (Farmer, 1975) the east and west coast of

Scotland, the North Sea, the Skagerrak and Kattegat, the Atlantic coast of France and Spain, the Irish Sea, and the Tyrrhenian and Adriatic seas of the Mediterranean. *Nephrops* is fished in depths of 20–500 m and up to 200 km from the coast. *Homarus gammarus* is fished largely in Scotland, England, France, and Norway. Ireland, Sweden, Denmark, and Spain have smaller fisheries. The *H. americanus* fishery in the United States is concentrated in the states of Maine, Massachusetts, and Rhode Island. The fishery has been divided into two parts. Inshore fishing is done in relatively shallow water (5–30 m) in the near shore areas. Offshore fishing is done at distances greater than 19 km from shore in the continental shelf and upper slope areas from Georges Bank to Virginia (Burns *et al.*, 1979). The Canadian fishery for *H. americanus* is almost entirely inshore, located largely in the Maritime provinces of Nova Scotia, Prince Edward Island, and New Brunswick. Nova Scotian landings make up almost one-half the total Canadian landings (DeWolf, 1974).

Spiny lobsters are fished in all the areas in which they are found. Spiny lobsters are generally sold for export, with only a small proportion being sold in the country of origin. The intensity of the fishery thus not only depends upon the population size and catchability of the lobsters but upon the availability of processing and transportation facilities. The largest spiny lobster fisheries are for *Panulirus argus* from Florida to Brazil, *Jasus lalandii* in southeastern Australia, *J. novahollandiae* in South Africa, *P. cygnus* in Western Australia, and *J. edwardsii* in New Zealand. The size of the fishery does not necessarily indicate its importance to the local economy. In Ecuador, for instance, the sizeable *P. gracilis* fishery makes up only a small fraction of the total value of the fishing industry (Loesch and Lopez, 1966), whereas in Tristan da Cunha, the *J. tristani* fishery is the primary source of revenue for the islands (Heydorn, 1969b; Pollock and Roscoe, 1977). In some areas, many species are caught. George (1973) lists six species of *Panulirus,* one of *Puerulus,* and one of *Palinustus* as contributing to the local fisheries of India.

Crayfishes are fished for food and for use as fish bait in most temperate regions where stocks are abundant. However, there are still many underexploited stocks, particularly in areas of western North America and in Africa, where demand has not developed. At the present time, the highest demand for crayfish as food is in the southern United States and in Europe. In the United States, the Atchafalaya Basin of Louisiana produces large quantities of *Procambarus clarkii* in years of optimal flooding (Huner and Barr, 1980). In Europe, because the outbreaks of the crayfish plague and the subsequent decline of *Astacus* spp. stocks, consumer demand has outweighed supply. As a result, many European countries import crayfish from nearby areas unaffected by plague and from North America. For instance, in 1971 France imported crayfish from Turkey, Greece, Yugoslavia, Poland, and to a lesser extent from Romania, Hungary, Switzerland, Italy, and Finland (Laurent, 1973). Sweden imports much of its crayfish from the United States.

B. Gear

The gear most used for *Homarus* capture on both sides of the Atlantic consists of traps. Generally, the traps are semicylindrical or oblong in cross section and made from wooden lath or wire mesh and weighted with metal, brick, or cement to provide stability. One or two cloth mesh funnels provide access from outside the trap. In many cases, particularly in the *H. americanus* fishery, the trap is divided into the entry compartment, often called the *kitchen*, where the bait is placed, and the second or *parlor*, where the lobster generally stays until the trap is hauled. In some areas, regulations require an escape gap of specific dimensions to allow sublegal lobsters easy egress. The traps are fished singly or in strings buoyed at each end. The boats vary markedly in size and sophistication. They usually are less than 15 m, except for the offshore *H. americanus* fishery in which boats up to 30 m are used to withstand the demands of several days at sea, rough weather, and large catches. Shortly after the discovery of the U.S. offshore lobster grounds, most of the fishing there was by otter trawl, but almost all the vessels now engaged in the directed fishery for *H. americanus* use traps.

Nephrops is fished with prawn trawls with long wings and small (30–70 mm) mesh. Mesh size is regulated in some areas, particularly where there is a directed fishery for *Nephrops*. The species may be fished for itself or be a bycatch of the whitefish or prawn fisheries. The boats used are generally side-trawlers of 10–35-m length. There is also a small creel (trap) fishery particularly where the bottom is unsuitable for trawling. The traps are single entrance, single chamber design, and set singly or in strings (Farmer, 1975).

The traps used in spiny lobster fisheries range from simple woven wicker pots to devices very similar to those used by the clawed lobster industry although the trap is not often divided into kitchen and parlor segments. In some areas, escape gaps are required. Generally, the traps are fished singly, being precisely placed by the fisherman. In small scale fisheries, traps may be used, but in many regions, lobsters are captured with cast nets, bottom set gill nets, seine nets, hand nets, or by divers with surround nets, snares, or spears (Buesa, 1965; George, 1973). Some lobsters (e.g., *Panulirus ornatus,* Chittleborough, 1974) will not enter baited traps, so alternate fishing methods are required. A small trawl fishery is developing in the Gulf of Papua during the migratory period of *P. ornatus* (Moore and McFarlane, 1977) and *P. mauritanicus* has been exploited by trawlers since 1955 off the west coast of Africa (Maigret, 1979). Bully netters using a bucket or box with a glass bottom for visibility operate with excellent success in the Bahamas during the migratory period of *P. argus*.

In all trap fisheries for lobsters, provision of bait is a significant part of the cost of fishing. Several efforts have been made to develop artificial bait, with some success (e.g., Mackie *et al.*, 1980). The use of artificial baits also allows the fisherman to present a perferred bait year around, rather than depending on the availability of bait species.

For the most part, crayfishes are fished with use of baited traps that are functionally similar to those used with marine lobsters. In the United States, conical or box-shaped wire mesh traps with funnel entrances are commonly used. Similar traps are used in Europe along with dip, bow, and trammel nets (Laurent, 1973). Electric shock techniques have also been employed to obtain crayfish. Australian crayfishermen primarily use dropnets, baited lines (Frost, 1975), scoop nets, and snares (Morrissy, 1978). The type of gear used is usually related to whether the fishing is done for sport or profit.

C. Regulations

Minimum-size regulations are enforced in all the clawed lobster fisheries. Other management rules include closed seasons, protection of berried females, escape gaps or minimum lath spacing, limited entry, and trap number limits. In Europe, the consideration of regulations is particularly complex because there is so much import and export between countries with different regulations. Size limits for *Nephrops,* for example, range from 80 to 150 mm total length and may also be expressed as tail length or weight. Minimum mesh size of trawls used in the *Nephrops* fishery is 35 to 70 mm depending upon the country (Dow, 1980). The *H. gammarus* minimum-size limit is set at between 74 and 83 mm carapace length depending upon the country. Ovigerous females are protected only in Spain, in one area in Denmark, and for a short period of the year in Portugal.

The *H. americanus* fishery of the United States and Canada is regulated largely by minimum-size limits, escape gaps in the traps, protection of ovigerous females, and in Canada, closed seasons, closed areas, and trap limitations.

The regulations placed on spiny lobster fisheries are varied, but generally include the requirement of a fishing license, a minimum size, which is measured as carapace length, total length, or tail weight, often protection of egg bearing females, closed seasons and closed areas. In Japan, closed areas are changed from year to year.

Crayfish are largely unregulated fisheries except for closed seasons, and size limits in some fisheries such as for *Pacifastacus* in Oregon and Washington, as well as *Astacus* fisheries in Europe. The *Cherax* fishery in southwestern Australia is highly regulated, with licenses, gear limits, bag limits, berried female protection, and closed seasons (Morrissy, 1978).

D. Storage, Shipment, and Sales

The tails of *Nephrops* are sold fresh, fresh frozen, or cooked. The tail meat often is removed from the shell before sale. It has been recommended that *Nephrops* either be landed alive or with the tails removed since discoloration and a bad taste caused by degeneration of the hepatopancreas develops rapidly after death. The cooked flesh can be stored 8 to 10 days at 0–2°C (Michel and Parent, 1973) or frozen for at least 2–3 months. Most of the *Nephrops* catch is marketed in the country of origin. There is some export to other European countries and a small market in the United States (Dow, 1980).

Homarus americanus and *H. gammarus* are marketed alive or, less frequently, as fresh, frozen, or canned meat. Live lobsters are shipped from the point of landing by truck to large depots where they are stored in floating crates, wooden live cars, or large tanks supplied with running sea water. Impoundments ("lobster pounds") are used, particularly in Canada and Europe for longer term storage. The pounds, small embayments separated from the sea by a concrete, wooden, or rock wall, are stocked when prices are low in May and December, and the lobsters are sold when prices are high in late summer and early spring. Losses during storage are high, often 10 to 25% or more (McLeese and Wilder, 1964). During storage and shipment, the claws are immobilized with rubber bands or pegs. Shipment is by truck or airplane, depending upon distance. Lobsters are packed in boxes with packing material and in artificial refrigerants to reduce movement, provide insulation, and keep the temperature low. Many publications are available that advise fishermen, shippers, and dealers about the best methods for handling lobsters (for example, see McLeese and Wilder, 1964; Thomas, 1965). The retail dealer grades by weight and loss of claws, with two clawed lobsters of 450–900 g bringing premium prices. In the United States, the New York market determines lobster price. In periods of short supply, prices tend to be higher there because of specialty restaurants, which must have normal supplies regardless of price. At times of oversupply, large quantities of losters can be sold in New York at relatively low prices since demand becomes very elastic at these lower prices (Dow, 1980).

Since the primary spiny lobster market is for export to the United States, Japan, and western Europe, almost all the fishermen sell to processers who remove and clean the tails before freezing, storage, and shipment. In some areas, there is a small industry developing for live shipment to consumer countries. Shipment is generally by air. Price is determined by the world market, primarily by how much the consumers in the United States are willing to pay. In the United States, crayfishes are either transported after being cooked or are processed and refrigerated before transport locally or overseas.

E. Fisheries Management

The high demand by consumers for lobster has brought about the development of very intense fisheries. Rapid transportation coupled with an increased use of freezing facilities have made the development of export fisheries in remote areas and in developing countries possible. Many of the species are now overfished because a rational management plan did not precede or accompany the development of the fishery. Rather than review all the management strategies here, we will present briefly two very different case histories.

1. HOMARUS AMERICANUS

Homarus americanus has supported an important and valuable fishery in the northeastern United States and eastern Canada for over 100 years. Since the end of World War II, the catch has fluctuated between 28,000 and 38,000 MT annually (Fig. 14). During the same period, fishing effort (measured as number of traps) has increased dramatically. Legislation in the early 1800s allowed only residents of Maine and Massachusetts to take lobsters in those states, but it was not until about 1870 that regulations protecting the resource were enacted in both the United States and Canada (DeWolf, 1974; Dow et al., 1975). These regulations prohibited the taking of egg-bearing females and soft-shelled lobsters and provided a minimum legal size. In Canada and in Maine, closed seasons were established. Starting in about 1890, the size of the catch began to decline as did the average size of the lobster caught (DeWolf, 1974). The situation became alarming enough that, in the late 1890s and early 1900s, many hatcheries were established to propagate and release larval lobsters in hope of replenishing the declining stock. No effect was seen, and hatchery efforts were abandoned by 1920. Effort increased markedly in the early 1900s with the introduction of gasoline-powered boats. Catches continued to decline until about 1930; after 1945, the somewhat increased catch can be attributed to increased effort, warmer temperatures, and the development of offshore fishing.

Present management regulations are not uniform over the range of the

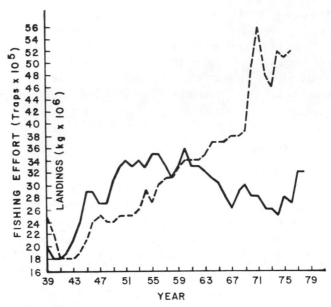

Fig. 14. Landings and fishing effort for inshore American lobsters, *Homarus americanus.* (——) Landings; (---) fishing effort. (Data for 1939–1977 from Dow, 1980; landings data for 1978–1979 from F.A.O. Fisheries Yearbooks.)

fishery. In Canada, licenses are required and the numbers of boats and traps are limited. Closed seasons, minimum-size limits, and minimum lath spacing are in force, but differ depending upon location. In all areas, egg-bearing females are protected. In the United States, each state determines the regulation concerning the fishery. Licenses are required in all states, but no limits are placed on numbers of boats or traps. There are no closed seasons. Egg-bearing females are protected, and there is a minimum size in all states. Maine does not allow the capture of large females. Escape gaps are required in most states.

In both countries, but particularly in the United States, the *Homarus* fishery is overcapitalized and very severely overexploited. A major reduction in fishing effort would be required to produce an increase in yield according to the Beverton–Holt yield per recruit curves. An increase in minimum legal size has been recommended by a state–federal management committee in the United States and if enacted should increase the yield per recruit and allow an increase in the number of lobsters reaching maturity. Suggestions of effort limitation have been made for the United States fishery, but the social and economic climate would not seem to allow such a measure at this time.

2. *PANULIRUS CYGNUS*

Panulirus cygnus, the spiny lobster of Western Australia, is a heavily fished and very valuable species. Most of the catch is exported to the United States as frozen tails. *Panulirus cygnus* has been fished since the late 1800s, and size regulations were introduced in 1897 (George et al., 1979), but the most rapid development of the fishery has taken place since the middle of the 1950s when the United States market opened up. Figure 15 shows the annual catch and effort (as number of pot lifts) during the period 1944–1980. Effort and yield in the fishery rose rapidly from 1950 until about 1960. In 1963, a limited effort policy was imposed limiting the number of licenses to 830 and the number of pots to three per each foot of boat up to a maximum of 200 per boat. Additional effort limitation measures include restrictions against replacing a boat with a larger one (which would allow more traps), geographical partitioning of the licenses into fishing zones, and closed seasons. Other regulations include minimum size, protection of egg-bearing females, and the provision of escape vents in the traps. The objectives of the management scheme are (1) optimum use of the resource, (2) reasonable economic return to the fisherman, and (3) orderly fishing (Bowen, 1980).

The restrictions on the fishery have prevented severe overcapitalization and excessive pressure on the resource. The limited entry policy has been

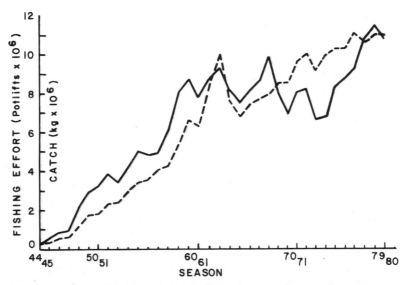

Fig. 15. Landings and fishing effort for the Western Australia lobster, *Panulirus cygnus*, from the 1944–1945 season to 1979–1980. (——) Catch; (- - -) fishing effort. (From Hancock, 1981.)

well accepted by the fishermen involved, who have experienced a period of high and relatively stable incomes that would not have occurred under an open entry management system. Additionally, enforcement is strong, and a proviso for suspension of the license of a fisherman convicted of an offense provides a great deterrent to breaking the regulations (Meany, 1979). Despite the limited entry, it became apparent that effort was increasing although not as rapidly as before 1963. The increased effort can be attributed to faster boats, more sophisticated fishing techniques, and harder work by the fishermen. This led, in 1977, to a shortening of the length of the open season by 6 weeks (Meany, 1970; Morgan, 1980b).

The management of the fishery has been successful in the eyes of both the government and the fishermen. As Morgan (1980b) pointed out, it is fortunate that limited entry was introduced before any catastrophic decline or severe overcapitalization had taken place. It is very difficult to reduce effort once a lobster fishery has become overexploited. The cost of a license in Western Australia in 1978 was approximately $1,000 per pot (Morgan, 1980b), and the cost to the government of a license buy-back scheme would be enormous. In addition to the early recognition of need for regulation, the success of Western Australia in managing the fishery is greatly facilitated by the entire resource being contained within one administrative boundary.

V. PERSPECTIVES

This chapter, in dealing with the fisheries biology of crayfishes and lobsters, has been concerned with the biological information needed for successful fisheries management. For many species, the information needed is often missing, incomplete, or not applicable to the situation at hand. Nonetheless, particularly in the last 10 years, great progress has been made towards an understanding of the ecology, behavior, and population dynamics of these crustaceans. In this section, we attempt to put some of these advances into focus and suggest where further research is needed.

Lobsters and crayfishes are usually the largest benthic decapods found in their respective communities. Their feeding habits make them important predators on other invertebrates, particularly echinoderms, annelids, molluscs, and small crustaceans. It is very likely that through their predatory interactions they play an important role in structuring the benthic community (Breen and Mann, 1976; Mann, 1977; Momot et al., 1978; Pollock, 1979; Griffiths and Seiderer, 1980). The importance of these relationships from the point of view of a fisheries biologist cannot be overstated. Intense fishing pressure, applied to only the larger animals, markedly alters the size composition and abundance of the fished population. Concurrent changes in prey and competing species are to be expected. The question has been asked (e.g., Mann, 1977) as to whether or not the changes wrought are reversible.

Momot et al. (1978) have addressed the problem of trout and crayfish management from the point of view of the trophic structure of a lake. They point out that most calculations of trophic level efficiencies assume that a lake's major food chain is through the phytoplankton, but to calculate trout production, one must take the benthic algae and detritus as a primary food source. Crayfish prey on an array of benthic organisms from invertebrates to periphyton, including a large proportion of macrophytes in their diet. Through their predatory relationships, they structure the benthic community and act as an important mechanism regulating the benthic production available to fish. Thus, an intensive fishery on crayfish may significantly alter the production of trout and perhaps other valuable species.

Competitive relationships may also be important in the management of multiple species fisheries. In New England, *Homarus americanus* appears to compete for habitat space with two species of *Cancer* crabs (Fogarty, 1976; Wang, 1982). We hypothesize that the removal of lobsters by fishing has allowed an increase in the crab populations. The crabs are competitively dominant to small lobsters but not to large ones. Thus, when large lobsters are removed, the competitive relationships are shifted, and the lobster population may become limited by the crab population rather than vice versa. The development of a fishery for these crabs may, in fact, increase the size of the lobster population.

Intraspecific competition must also be taken into account when attempting to determine possible production from a fishery. In Western Australia, for example, juvenile *P. cygnus* show density-dependent growth and mortality rates. The juvenile nursery areas act as buffers, and high levels of larval recruitment are not translated in large recruitment to the fishable stock (Chittleborough and Phillips, 1975). Only years of poor larval recruitment will be reflected in low catches several years later. Momot and Gowing (1977b) reported that, in an experimental fishery for the crayfish *Orconectes virilis* in which females were protected, there was an increase in the density of subadult females. Rather than an increase in spawning stock, this resulted in a decline because of lowered survival due to density-dependent mechanisms between age I and age II and a decreased egg production of the subsequent age II females.

A good deal of attention has been paid to models of exploited lobster populations with about equal emphasis upon surplus yield models and dynamic pool models. In the long run, the greater flexibility and predictive value of the dynamic pool models will allow them to be used with more confidence. As yet, however, growth rates are difficult to estimate and the application of the von Bertalanffy expression has been questioned. Recent attempts to develop empirical growth curves (e.g., Mauchline, 1977) will help. In some species, it may be possible to recognize cohorts (e.g., Momot and Gowing, 1977a; Jones, 1979; Nicholson, 1979; Ennis, 1980b) and to analyze growth patterns in this way. Analyses of field data and of tag–recapture experiments have their own inherent difficulties such as differential susceptibility to trapping and the flood of early recaptures in an intensive fishery (Hancock, 1977). Laboratory studies of growth rates may help, but it is difficult to be sure that the data gained in the laboratory are representative of field conditions. The differences in growth characteristics, fecundity, and size at maturity within a species from one geographic area to another make it imperative that all of these be examined closely when designing management schemes. Certainly it does not appear appropriate to apply a single set of parameters over the entire extent of a fishery as widespread as that for *H. americanus* or *Nephrops*. These kinds of considerations have not been applied to crayfish populations to any great extent. However, these fisheries are not as strongly managed, and the need for information is not as great. If needs for management of crayfish stocks increase, much the same information will be required.

The Beverton–Holt model and other similar models were developed for scale fish and use parameters of the von Bertalanffy growth equation as an input to the model. Unfortunately, it is impossible to determine the age of Crustacea, so we must assume a relationship between age and length or weight. Many authors have indicated the need for an aging technique, and this would, indeed, be very useful. However, the development of models based on length data would ease the assumptions that need to be made. Despite the questions that have been raised about the application of the von Bertalanffy growth model to crustaceans, it is important to remember that by and large it does a good job of describing the growth patterns. Because little is known about growth increments and intermolt intervals at small and large sizes, it is not wise to extrapolate beyond the range of useable data.

There is a real need to define the populations we are studying and to determine relationships between stock and recruitment. There are some indications in the *Homarus* fishery that there are differences between subpopulations. The fishery for an even more widespread species, *Panulirus argus*, spans two continents and many countries from Brazil to Florida. An ambitious study of the question of stock identity and origin of larval recruitment has taken the approach of biochemical population genetics (Menzies and Kerrigan, 1979). Preliminary results indicate that there are geographic differences in adult gene frequencies between all well separated populations, such as Jamaica and the Virgin Islands (R. Menzies, personal communication). As yet, the origins of the larval settling in the various areas are less well understood. However, it is clear that the source of the larvae has a great deal to do with management strategies. At one extreme, each population could be managed locally, at the other extreme, a Pan-Caribbean management strategy is called for. Many other widespread palinurid fisheries, for instance *Panulirus interruptus* on the west coast of North and Central America and *P. pencillatus* of the Pacific Islands, have similar problems. The larvae of clawed lobsters are not in the plankton as long as those of spiny lobsters, but the information concerning larval recruitment is equally sparse. The presence of an offshore population of *H. americanus* has raised questions about the relationship between the inshore and offshore populations and whether the intense fishing offshore will affect recruitment inshore. The relationship between parent stock and larval recruitment is not known in any of the fisheries except that for *P. cygnus*.

Knowledge of migrations are important to an understanding of the continuity between populations. Only recently has the extent of adult movements in some species become known, for instance *Panulirus argus, P. ornatus,* and *Homarus americanus.* Data summarized by Herrnkind (1980) give testimony to the wide variation in migratory behavior among species. It is evident that much more work needs to be done on movement patterns in a large number of species. Little or nothing is known in this regard for crayfish, although present indications are that populations tend to be quite localized.

Concern has been expressed, and should continue to be expressed, about the definition of a stock of unit population. However as Radway-Allen (1977) pointed out, management units need not be genetically distinct populations so long as the effects of management are large compared to the effect of interchange with adjacent subpopulations. Thus, if population statistics such as growth rates, size at maturity, and fecundity vary geographically, it makes biological sense to tailor management schemes to fit the apparent subpopulations whether or not genetic discontinuity can be shown.

The lobster fisheries offer an excellent laboratory for the study of the effects of fishing—and overfishing. Although many stocks are now heavily exploited, some, particularly spiny lobsters, are not. In addition, the slipper lobsters (Scyllaridae), a group not covered in this review, are only just beginning to be fished to any degree. Thus opportunities do exist for comparative studies on the effects of fishing on lobster populations. The offshore *Homarus americanus* fishery has been followed closely since its inception, as has the fishery for *Panulirus cygnus* in Western Australia and the trawl section of the fishery for *P. ornatus* in the Gulf of Papua. Some effects of heavy exploitation are obvious, such as a decrease in the average-size individuals in the catch, a decrease in catch per unit effort, and a decrease in population abundance of the fishable stock. This has been well documented for offshore *H. americanus* (Burns *et al.*, 1979) *Pacifastacus leniusculus leniusculus* (Miller and VanHyning, 1971), *Panulirus argus* (Beardsley *et al.*, 1975; Warner *et al.*, 1976; Davis, 1977), and *Palinurus mauritanicus* (Maigret, 1979), among other species. Other, less obvious effects may also occur. The sex ratio of a number of lobster populations varies with size, and this may be an effect of fishing and fishing regulations (Skud, 1969; Morgan, 1984). Questions have been raised (Hancock, 1977) about the genetic consequences of selective fishing techniques, for example earlier sexual maturity, reduced growth rates, reduced catchability, or other modified behavior. In effect, high fishing mortality places tremendous selective pressure on a population. To be able to follow the consequences of this would be of both scientific and practical interest. Perhaps this can be done both in natural, but fished, populations and in sanctuary areas such as in the Dry Tortugas (*P. argus*), around the French coast (*H. gammarus*), and in South Africa (*P. homarus*). The imposition of experimental fishing pressure on unfished populations, as was done by Momot and Gowing (1977b), may yield greater understanding of how the biology of these crustaceans change as a result of human activities.

REFERENCES

Abrahamsson, S. A. A. (1971). Density, growth, and reproduction in population of *Astacus astacus* and *Pacifastacus leniusculus* in an isolated pond. *Oikos* **22**, 373–380.

Abrahamsson, S. A. A. (1973). The crayfish *Astacus astacus* in Sweden and the introduction of the American crayfish *Pacifastacus leniusculus*. *Freshwater Crayfish* **1**, 27–40.

Abrahamsson, S. A. A., and Goldman, C. R. (1970). Distribution, density, and production of the crayfish *Pacifastacus leniusculus* Dana in Lake Tahoe, California–Nevada. *Oikos* **21**, 83–91.

Ache, B. W. (1977). Aspects of chemoreception in marine Crustacea. *In* "Olfaction and Taste VI" (J. LeMagnen and P. MacLeod, eds.), pp. 343–350. *Information Retrieval,* London.

Aiken, D. E. (1965). Distribution and ecology of three species of crayfish from New Hampshire. *Am. Midl. Nat* **73**, 240–244.

Aiken, D. E. (1973). Proecdysis, setal development, and molt prediction in the American lobster (*Homarus americanus*). *J. Fish. Res. Board Can.* **30**, 1334–1337.

Aiken, D. E. (1980). Molting and growth. *In* "The Biology and Management of Lobsters" (J. S. Cobb and B. F. Phillips, eds.), Vol. I, pp. 91–163.

Aiken, D. E., and Waddy, S. L. (1976). Controlling growth and reproduction in the American lobster. *Proc. World Maricult. Soc.* **7**, 415–430.

Aiken, D. E., and Waddy, S. L. (1980). Reproductive biology. *In* "The Biology and Management of Lobsters" (J. S. Cobb and B. F. Phillips, eds.), Vol. I, pp. 215–276. Academic Press, New York.

Aiken, D. E., and Waddy, S. L. (1982). Cement gland development, ovary maturation and reproductive cycles in the American lobster, *Homarus americanus. J. Crust. Biol.* **2**, 315–327.

Ameyaw-Akumfi, F., and Hazlett, B. A. (1975). Sex recognition in the crayfish *Procambarus clarkii*. *Science* **190**, 1225–1226.

Annala, J. H. (1979). Mortality estimates for the New Zealand rock lobster, *Jasus edwardsii*. *Fish. Bull.* **77**, 471–480.

Annala, J. H. (1980). New Zealand rock lobsters: biology and fishery. *Fish. Res. Div., N. Z., Occ. Publ.* 42.

Anthony, V. C. (1980). Review of lobster mortality estimates in the United States. *Can. Fish. Aquat. Sci. Tech. Rep. No. 932*, pp. 17–26.

Anthony, V. C., and Caddy, J. F. (1980). Summary. *Can. Tech. Rept. Fish. Aquat. Sci.* **932**, 183–185.

Arechiga, H., and Atkinson, R. J. A. (1975). The eye and some effects of light on locomotor activity in *Nephrops norvegicus*. *Mar. Biol. (Berlin)* **32**, 63–76.

Atema, J., and Cobb, J. S. (1980). Social behavior. *In* "The Biology and Management of Lobsters," (J. S. Cobb and B. F. Phillips, eds.), Vol. I, pp. 409–450. Academic Press, New York.

Atema, J., Jacobson, S., Karnofsky, E., Oleszeko-Szuts, S., and Stein, L. S. (1979). Pair formation in the lobster, *Homarus americanus*: Behavioral development, pheromones and mating. *Mar. Behav. Physiol.* **6**, 277–296.

Atkinson, R. J. A., and Naylor, E. (1973). Activity rhythms in some burrowing decapods. *Helgol. Wiss. Meeresunters.* **24**, 192–201.

Atkinson, R. J. A., and Naylor, E. (1976). An endogenous activity rhythm and the rhymicity of catches of *Nephrops norvegicus* (L.). *J. Exp. Mar. Biol. Ecol.* **25**, 95–108.

Barlow, J., and Ridgeway, G. J. (1971). Polymorphisms of esterase isozymes in the American lobster. (*Homarus americanus*). *J. Fish. Res. Board Can.* **28**, 15–21.

Beardsley, G. L., Costello, T. J., Davis, G. E., Jones, A. C., and Simmons, D. C. (1975). The Florida spiny lobster fishery; a White Paper. *Fl. Sci.* **38**, 144–149.

Bennett, D. B. (1980). Perspectives on European lobster management. *In* "The Biology and Management of Lobsters" (J. S. Cobb and B. F. Phillips, eds.), Vol. II, pp. 317–331. Academic Press, New York.

Bergeron, J. (1967). Contribution à la biologie du homard, *Homarus americanus* M. Edw. des Illes-de-la Madeleine. *Nat. Can.* **94**, 169–207.

Berry, P. F. (1970). Mating behavior, oviposition and fertilization in the spiny lobster *Panulirus homarus* (L.). *Oceanogr. Res. Inst. (Durban), Invest Rep.* **24**, 1–16.

Berry, P. F. (1971a). The spiny lobsters (Palinuridae) of the East Coast of South Africa. Distribution and ecological notes. *Oceanogr. Res. Inst. (Durban), Invest. Rep.* **28**, 7–75.

Berry, P. F. (1971b). The biology of the spiny lobster *Panulirus homarus* (L.) off the east coast of southern Africa. *Oceanogr. Res. Inst. (Durban), Invest Rep.* **28**, 1–76.

Berry, P. F. (1973). The biology of the spiny lobster *Palinurus delagoae* Barnard, off the coast of Natal, South Africa. *Oceanogr. Res. Inst. (Durban), Invest. Rep.* **31**, 1–27.

Berry, P. F. (1974). A revision of the *Panulirus homarus* group of spiny lobsters (Decapoda, Palinuridae). *Crustaceana* **27**, 31–42.

Beverton, R. J. H., and Holt, S. J. (1957). On dynamics of exploited fish populations. *Fish. Invest. (London), Ser. 2* **19**, 391.

Bill, R. G., and Herrnkind, W. F. (1976). Drag reduction by formation movement in spiny lobster. *Science* **193**, 1146–1148.

Bittner, G. D., and Kopanda, R. (1973). Factors influencing molting in the crayfish *Procambarus clarkii*. *J. Exp. Zool.* **186**, 7–16.

Booth, J. D. (1979). North Cape—a 'nursery area' for the packhorse rock lobster, *Jasus verreauxi* (Decapoda: Palinuridae). *N. Z. J. Mar. Freshwater. Res.* **13**, 521–528.

Bouchard, R. W. (1978). Taxonomy, distribution, and general ecology of the genera of North American crayfishes. *Fisheries* **3,** 11.

Boudréault, F. R., DuPont, J. N., and Sylvain, C. (1977). Modéles linéares de prédiction des débarquements de homard aux Iles-de-la-Madeleine (Golfe du Saint-Laurent). *J. Fish. Res. Board Can.* **34,** 379–383.

Bowen, B. K. (1980). Spiny lobster fisheries management. *In* "The Biology and Management of Lobsters" (J. S. Cobb and B. F. Phillips, eds.), Vol. II, pp. 243–264. Academic Press, New York.

Bradstock, C. A. (1950). A study of the marine spiny crayfish *Jasus lalandii* (Milne-Edwards) including accounts of autospasy. *Victoria Univ. Coll. Zool. Publ.* **7,** 1–38.

Branford, J. R. (1978). Incubation period for the lobster *Homarus gammarus* at various temperatures. *Mar. Biol. (Berlin)* **47,** 363–368.

Branford, J. R. (1979). Locomotor activity and food consumption by the lobster *Homarus gammarus. Mar. Behav. Physiol.* **6,** 13–24.

Breen, P. A., and Mann, K. H. (1976). Changing lobster abundance and the destruction of kelp beds by sea urchins. *Mar. Biol. (Berlin)* **34,** 137–142.

Briggs, P. T., and Mushacke, F. M. (1980). The American lobster and the pot fishery in the inshore waters off the south shore of Long Island, New York. *N. Y. Fish Game J.* **27,** 156–178.

Brinck, P. (1977). Developing crayfish populations. *Freshwater Crayfish* **3,** 211–228.

Brown, D. J., and Browler, K. (1977). A population study of the British freshwater crayfish *Austropotamobius pallipes* (Lereboullet). *Freshwater Crayfish* **3,** 33–49.

Brown, K. (1981). Low genetic variability and high similarities in the crayfish genera *Cambarus* and *Procambarus. Am. Midl. Nat.* **105,** 225–232.

Buesa, R. J. (1965). Biology and fishing of spiny lobster *Panulirus argus* (Latreille). *In* "Soviet–Cuban Fishing Research" (A. S. Bogdarov, ed.). Transl. from Russian by Israel Prog. Sci. Transl., Jerusalem 1969, TT69-59016, 62–77.

Burns, T. S., Clark, S. H., Anthony, V. C., and Essig, R. J. (1979). Review and assessment of the USA offshore lobster fishery. *ICES, C. M. K.* **25.**

Butler, J. A., and Pease, N. L. (1965). Spiny lobster explorations in the Pacific and Caribbean waters of the Republic of Panama. *U. S. Fish. Wildl. Serv. Spec. Sci. Rep.—Fish. No. 505,* pp. 1–26.

Caddy, J. F. (1977). Approaches to a simplified yield-per-recruit model for Crustacea, with particular reference to the American lobster, *Homarus americanus. Can. Fish. Mar. Serv., Manuscr. Rep. No. 1445.*

Caddy, J. F. (1979). The influence of variations in the seasonal temperatures regime on survival of larval stages of the American lobster (*Homarus americanus*) in the southern Gulf of St. Lawrence. *Rapp. P. V. Reun. Cons. Int. Explor. Mar.* **175,** 204–216.

Caldwell, M. J., and Bovbjerg, R. V. (1969). Natural history of the two crayfish of northwestern Iowa, *Orconectes virilis* and *Orconectes immunis. Proc. Iowa Acad. Sci.* **76,** 463–472.

Campbell, A. (1980). A review of mortality estimates of the lobster populations in the Canadian Maritimes. *Can. Fish Aquat. Sci. Tech. Rep. No. 932,* pp. 27–35.

Campbell, A. (1983). Growth of tagged American lobsters (*Homarus americanus*) in the Bay of Fundy. *Can. J. Fish. Aquat. Sci.* **40,** 1667–1675.

Campbell, A., and Robinson, D. G. (1983). Reproductive potential of three lobster (*Homarus americanus*) stocks in the Canadian Maritimes. *Can. J. Fish. Aquat. Sci.* **40,** 1958–1967.

Capelli, G. M. (1980). Seasonal variation in the food habits of the crayfish *Orconectes propinquus* (Girard) in Trout Lake, Vilas County, Wisconsin, U.S.A. (Decapoda, Astacidea, Cambaridae). *Crustaceana* **38,** 82–86.

Cassie, R. M. (1954). Some uses of probability paper for the graphical analysis of polymodal frequency distribution. *Aust. J. Mar. Freshwater Res.* **5**, 513–522.

Chapman, C. J. (1979). Some observations on populations of Norway lobster, *Nephrops norvegicus* (L.) using diving, television and photography. *Rapp. P. V. Reun. Cons. Int. Explor. Mer.* **175**, 127–133.

Chapman, C. J. (1980). Ecology of juvenile and adult *Nephrops*. *In* "The Biology and Management of Lobsters" (J. S. Cobb and B. F. Phillips, eds.), Vol. II, pp. 143–178. Academic Press, New York.

Chapman, C. J., and Howard, F. G. (1979). Field observations on the emergence rhythm of the Norway lobster *Nephrops norvegicus,* using different methods. *Mar. Biol. (Berlin)* **51**, 157–165.

Chapman, C. J., and Rice, A. L. (1971). Some direct observations on the ecology and behaviour of the Norway lobster *Nephrops norvegicus*. *Mar. Biol. (Berlin)* **10**, 321–329.

Chapman, C. J., Johnstone, A. D. F., and Rice, A. L. (1975). The behavior and ecology of Norway lobster *Nephrops norvegicus* (L.). *Proc. 9th Eur. Mar. Biol. Symp.* pp. 59–74.

Chittleborough, R. G. (1970). Studies on recruitment in the Western Australian rock lobster, *Panulirus longipes cygnus* (George: Density and natural mortality of juveniles. *Aust. J. Mar. Freshwater Res.* **21**, 131–148.

Chittleborough, R. G. (1974). Home range, homing and dominance in juvenile western rock lobsters. *Aust. J. Mar. Freshwater. Res.* **25**, 227–234.

Chittleborough, R. G. (1975). Environmental factors affecting growth and survival of juvenile western rock lobsters (*Panulirus longipes*) (Milne-Edwards). *Aust. J. Mar. Freshwater Res.* **26**, 177–196.

Chittleborough, R. G. (1976). Breeding of *Panulirus longipes cygnus* under natural and controlled conditions. *Aust. J. Mar. Freshwater Res.* **27**, 499–516.

Chittleborough, R. G. (1979). Natural regulation of the population of *Panulirus longipes cygnus* George and responses to fishing pressure. *Rapp. P. V. Reun. Cons. Int. Explor. Mer.* **175**, 217–221.

Chittleborough, R. G., and Phillips, B. F. (1975). Flucutations of year-class strength and recruitment in the western rock lobster, *Panulirus longipes*. *Aust. J. Mar. Freshwater Res.* **26**, 317–328.

Chittleborough, R. G., and Thomas, L. R. (1969). Larval ecology of western Australian marine crayfish with notes upon other palinurid larvae from the eastern Indian Ocean. *Aust. J. Mar. Freshwater Res.* **20**, 199–203.

Cobb, J. S. (1970). Effect of solitude on time between fourth and fifth larval molts in the American lobster (*Homarus americanus*). *J. Fish. Res. Board Can.* **27**, 1653–1655.

Cobb, J. S. (1971). The shelter-related behavior of the lobster *Homarus americanus*. *Ecology* **52**, 108–115.

Cobb, J. S. (1981). Behavior of the western Australian spiny lobster, *Panulirus cygnus* George, in the field and laboratory. *Aust. J. Mar. Freshwater Res.* **32**, 399–409.

Cobb, J. S., and Tamm, G. R. (1974). Social conditions increase intermolt period in juvenile lobsters *Homarus americanus*. *J. Fish. Res. Board Can.* **32**, 1941–1943.

Cobb, J. S., Tamm, G. R., and Wang, D. (1982). Behavioral mechanisms influencing molt frequency in the lobster *Homarus americanus*. *J. Exp. Mar. Biol. Ecol.* **62**, 185–200.

Cooper, R. A. (1970). Retention of marks and their effects on growth, behavior and migrations of the American lobster, *Homarus americanus*. *Trans. Am. Fish. Soc.* **99**, 409–417.

Cooper, R. A., and Herrnkind, W. F. (1971). Ecology and population dynamics of the spiny lobster, *Panulirus argus*, of St. John, Virgin Id. Research Projects conducted during Tektite 2 (1970–1971). (U.S. Dept. Interior, Washington) VI, 34–57.

Cooper, R. A., and Uzmann, J. R. (1971). Migrations and growth of deep-sea lobsters, *Homarus americanus. Science* **171**, 288–290.

Cooper, R. A., and Uzmann, J. R. (1977). Ecology of juvenile and adult clawed lobsters *Homarus americanus, Homarus gammarus,* and *Nephrops norvegicus. In* "Workshop on Lobster and Rock Lobster Ecology and Physiology" (B. F. Phillips and J. S. Cobb, eds.), pp. 187–208.

Cooper, R. A., and Uzmann, J. R. (1980). Ecology of juvenile and adult *Homarus. In* "The Biology and Management of Lobsters" (J. S. Cobb and B. F. Phillips, eds.), Vol. 11, pp. 97–142. Academic Press, New York.

Cooper, R. A., Clifford, R. A., and Newell, C. D. (1975). Seasonal abundance of the American lobster, *Homarus americanus,* in the Boothbay Region of Maine. *Trans. Am. Fish. Soc.* **104**, 669–674.

Covich, A. P. (1977). How do crayfish respond to plants and Mollusca as alternate food resources? *Freshwater Crayfish* **3**, 165–179.

Crawford, D. R., and DeSmidt, W. J. J. (1922). The spiny lobster, *Panulirus argus* of southern Florida, its natural history and utilization. *Bull. U.S. Bur. Fish.* **38**, 281–310.

Crawshaw, L. I. (1974). Temperature selection and activity in the crayfish, *Orconectes immunis. J. Comp. Physiol.* **95A**, 315–322.

Dall, W., and Barclay, M. C. (1977). Induction of viable ecdysis in the western rock lobster by 20-hydroxyecdysone. *Gen. Comp. Endocrinol.* **31**, 323–334.

Davis, G. E. (1974). Notes on status of spiny lobsters, *Panulirus argus* at Dry Tortugas, FL. *Fl. State Sea Grant Rep. SUSF-SG-74-201.* 64 pp.

Davis, G. E. (1975). Minimum size of mature spiny lobsters, *Panulirus argus,* at Dry Tortugas, Florida. *Trans. Am. Fish. Soc.* **104**, 675–676.

Davis, G. E. (1977). Effects of recreational harvest on a spiny lobster, *Panulirus argus,* population. *Bull. Mar. Sci.* **27**, 223–236.

Davis, G. E. (1979). Management recommendations for juvenile Spiny lobsters, *Panulirus argus,* in Biscayne National Monument, Florida. *South. Fl. Res. Ctr. Rep. M-530.* 32 pp.

DeWolf, G. A. (1974). The lobster fishery of the Maritime Provinces: Economic effects of regulations. *Bull. Fish. Res. Board Can.* **187**, 1–59.

Dow, R. L. (1969). Cycles of geographic trends in seawater temperature and abundance of American lobster. *Science* **164**, 1060–1063.

Dow, R. L. (1974). American lobsters tagged by Maine commercial fishermen, 1957–1959. *Fish. Bull.* **72**, 622–623.

Dow, R. L. (1977). Relationship of sea surface temperature to American and European lobster landings. *J. Cons. Int. Explor. Mer.* **37**, 186–191.

Dow, R. L. (1978). Effects of sea surface temperature cycle on landings of American, European and Norway lobsters. *J. Cons. Int. Explor. Mer.* **38**, 271–272.

Dow, R. L. (1980). The clawed lobster fisheries. *In* "The Biology and Management of Lobsters" (J. S. Cobb and B. F. Phillips, eds.), Vol. II, pp. 265–316. Academic Press, New York.

Dow, R. L., Bell, F. W., and Harriman, D. M. (1975). Bioeconomic relationships for the Maine lobster fishery with consideration of alternative management schemes. *NOAA Tech. Rep. NMFS SSRF No. 683,* 1–44.

Drach, P. (1939). Mue et cycle d'intermue chez les Crustaces Decapodes. *Ann. Inst. Oceanogr.* **19**, 103–392.

Drach, P., and Tchernigovtzeff, C. (1967). Sur la méthode de détermination des stades d'inter-mue et son application générale aux crustacés. *Vie Milieu* **18A**, 595–610.

DuPont, J. N., and Boudréault, E. R. (1976). Prédiction des débarquements de homard aux Iles-de-la-Madeleine: II. Regression multiple sur les composantes principales. *Rech. Div. Pêch. Marit., Min. Ind. Commer. Cah. Inf.* **73**, 67–101.

Dybern, B. I. (1973). Lobster burrows in Swedish waters. *Helgol. Wiss. Meeresunters.* **24**, 401–414.

Eiriksson, H. (1979). A study of the Icelandic *Nephrops* fishery with emphasis on stock assessments. *Rapp. P. V. Reun. Cons. Int. Explor. Mer.* **175**, 270–279.

Elner, R. W., and Jamieson, G. S. (1979). Predation of sea scallops, *Placopecten magellanicus* by the rock crab, *Cancer irroratus* and the American lobster, *Homarus americanus*. *J. Fish. Res. Board Can.* **36**, 537–543.

Emmel, V. E. (1906). The relation of regeneration to the molting process of the lobster. *R. I. Comm. Inland Fish. Annu. Rep.* **38**, 98–114.

Engle, J. M. (1979). Ecology and growth of juvenile California spiny lobster *Panulirus interruptus* (Randall). Ph.D. Thesis, Univ. of Southern California, Los Angeles.

Ennis, G. P. (1973a). Endogenous rhythmicity associated with larval hatching in the lobster *Homarus gammarus*. *J. Mar. Biol. Assoc. U. K.* **53**, 531–538.

Ennis, G. P. (1973b). Behavioral responses to changes in hydrostatic pressure and light during larval development of the lobster, *Homarus gammarus*. *J. Fish. Res. Board Can.* **30**, 1349–1360.

Ennis, G. P. (1973c). Food, feeding and condition of lobsters, *Homarus americanus*, throughout the seasonal cycle in Bonavista Bay, Newfoundland. *J. Fish. Res. Board Can.* **30**, 1905–1909.

Ennis, G. P. (1975). Observations on hatching and larval release in the lobster *Homarus americanus*. *J. Fish. Res. Board Can.* **32**, 2210–2213.

Ennis, G. P. (1980a). Size–maturity relationships and related observations in Newfoundland populations of the lobster (*Homarus americanus*) *Can. J. Fish. Aquat. Sci.* **37**, 945–956.

Ennis, G. P. (1980b). Recent and current research on growth of lobsters in the wild. *Can. Fish. Aquat. Sci. Tech. Rep. No. 932*, pp. 10–15.

Evans, P. D., and Mann, K. H. (1977). Selection of prey by American lobsters (*Homarus americanus*) when offered a choice between sea urchins and crabs. *J. Fish. Res. Board Can.* **34**, 2203–2207.

Fair, J. J. (1980). U.S. surveys of lobster larvae, *Can. Fish. Aquat. Sci. Tech. Rep. No. 932*, pp. 153–156.

Farmer, A. S. D. (1974a). The development of the external sexual characteristics of *Nephrops norvegicus* (L.) (Decapoda: Nephropidae). *J. Nat. Hist.* **8**, 241–255.

Farmer, A. S. D. (1974b). Reproduction in *Nephrops norvegicus* (Decapoda: Nephropidae). *J. Zool.* **174**, 161–183.

Farmer, A. S. D. (1975). Synopsis of biological data on the Norway lobster *Nephrops norvegicus* (Linnaeus, 1758). *FAO Fish. Synop.* **112**, 1–97.

Fielder, D. R. (1964). The spiny lobster, *Jasus lalandii* (H. Milne-Edwards), in South Australia. I. Growth of captive animals. *Aust. J. Mar. Freshwat. Res.* **15**, 77–92.

Fielder, D. R. (1965). A dominance order for shelter in the spiny lobster, *Jasus lalandii* (H. Milne-Edwards). *Behaviour* **24**, 236–245.

Figueiredo, M. J., and Nunes, M. C. (1965). The fecundity of the Norway lobster, *Nephrops norvegicus* (L.) in Portuguese waters. *I.C.E.S., Shellfish Comm. Doc.* **34**, 1–5.

Figueiredo, M. J., and Thomas, H. J. (1967). On the biology of the Norway lobster, *Nephrops norvegicus* (L.). *J. Cons. Int. Explor. Mer.* **31**, 89–101.

Fitzpatrick, J. F., and Hobbs, H. H., III. (1968). The Mississippi as a barrier to crawfish dispersal. *Am. Zool.* **8**, 807.

Flint, R. W. (1975). Growth in a population of the crayfish *Pacifastacus leniusculus* from a subalpine lacustrine environment. *J. Fish. Res. Board Can.* **32**, 2433–2440.

Flint, R. W. (1977). Seasonal activity, migration, and distribution of the crayfish, *Pacifastacus leniusculus*, in Lake TAhoe. *Am. Midl. Nat.* **97,** 280–292.

Flint, R. W., and Goldman, C. R. (1975). The effects of a benthic grazer on the primary productivity of the littoral zone of Lake Tahoe. *Limnol. Oceanogr.* **20,** 935–944.

Flint, R. W., and Goldman, C. R. (1977). Crayfish growth in Lake Tahoe: Effects of habitat variation. *J. Fish. Res. Board Can.* **34,** 155–159.

Flowers, J. M., and Saila, S. B. (1972). An analysis of temperature effects on the inshore lobster fishing. *J. Fish. Res. Board Can.* **29,** 1221–1225.

Fogarty, M. J. (1976). Competition and resource partitioning in two species of *Cancer* (Crustacea, Brachyura). M.S. Thesis, Univ. of Rhode Island, Kingston.

Fogarty, M. J., Borden, D. V. D., and Russell, H. J. (1980). Movements of tagged American lobster, *Homarus americanus* off Rhode Island. *Fish. Bull.* **78,** 771–780.

Fogarty, M. J., Cooper, R. A., Uzmann, J. R., and Burns, T. (1982). Assessment of the U.S.A. offshore American lobster (*Homarus americanus*) fishery. *Int. Council. Expl. Sea. C.M., 1982/K:14*, pp. 21.

Fox, W. W. (1970). An exponential yield model for optimizing exploited fish populations. *Trans. Am. Fish. Soc.* **99,** 80–88.

Fox, W. W. (1975). Fitting the generalized stock production model by least squares and equilibrium approximation. *Fish. Bull.* **73,** 23–36.

Fraser, J. H. (1965). Larvae of *Nephrops norvegicus* in the Scottish area, 1935–1964. *ICES, C.M. 1965, 10.*

Frost, J. V. (1975). Australian crayfish. *Freshwater Crayfish* **2,** 87–96.

George, M. J. (1973). The lobster fishery resources of India. *Spec. Publ., Cent. Mar. Fish. Res. Inst., Cochin-II, India*, pp. 570–580.

George, R. W. (1958). The status of the "white" crayfish in Western Australia. *Aust. J. Mar. Freshwater Res.* **8,** 476–490.

George, R. W. (1974). Coral reefs and rock lobster ecology in the Indo-West Pacific region. *Proc. 2nd Coral Reef. Symp.,* **1,** 321–325.

George, R. W., and Main, A. R. (1967). The evolution of spiny lobsters (Palinuridae): A study of evolution in the marine environment. *Evolution* **21,** 803–821.

George, R. W., and Morgan, G. (1979). Linear growth stages in the rock lobster *Panulirus versicolor* as a method for determining size at first physical maturity. *Rapp. P. V. Reun. Cons. Int. Explor. Mar.* **175,** 182–185.

George, R. W., Morgan, G. R., and Phillips, B. F. (1979). The western rock lobster, *Panulirus cygnus. R. Soc. West. Aust. Proc.* **62,** 45–51.

Gibson, F. A. (1967). Irish investigation of the lobster *Homarus gammarus. Ir. Fish. Invest., Ser. B* **1,** 13–45.

Goldman, C. R. (1973). Ecology and physiology of the California crayfish *Pacifastacus leniusculus* (Dana) in relation to its suitability for introduction to European waters. *Freshwater Crayfish* **1,** 105–120.

Grey, K. A. (1979). Estimates of the size of first maturity of the western rock lobster, *Panulirus cygnus,* using secondary sexual characteristics. *Aust. J. Mar. Freshwater Res.* **30,** 785–791.

Griffiths, C. L., and Seiderer, J. L. (1980). Rock lobsters and mussels—limitations and preferences in a predator-prey interaction. *J. Exp. Mar. Biol. Ecol.* **44,** 95–109.

Gulf and South Atlantic Fisheries Management Councils (1981). Draft environmental impact statement for the spiny lobster fishery of the Gulf of Mexico and South Atlantic. Mimeo. *Natl. Mar. Fish. Serv., St. Petersburg, Florida.*

Gunderson, K. R. (1969). Preliminary results of field tagging experiments on lobster, *Homarus vulgaris,* in Norwegian waters. *ICES, C.M. K.* **38.**

Halgren, B. A. (1976). Management of the lobster, *Homarus americanus,* resource of the continental shelf, canyons and slope of the northern portion of management area IV. Prog. Rep. U.S. Dep. Comm. NDAA Contract 03-4-043-359, U.S. Dep. Fish Game.

Hallback, H., and Warrenm A. (1972). Food ecology of the lobster, *Homarus vulgaris* in Swedish waters. Some preliminary results. *ICES, C.M. K.* **29.**

Hancock, D. A. (1977). Population ecology and growth (Rapporteur's report). *In* "Workshop on Lobster and Rock Lobster Ecology and Physiology" (B. F. Phillips and J. S. Cobb, eds.). CSIRO (Aust.) Div. Fish. Oceanog. Circ. **7,** 279–286.

Hancock, D. A. (1981). Research for Management of the rock lobster fishery of Western Australia. *Proc. Gulf Carib. Fish. Inst.* **33,** 207–229.

Havinga, B. (1938). Krebse und Weichtiere. Handbuch der Seefisch. *Nordeuropas* **3,** 1–47.

Hayes, W. A. (1975). Behavioral components of social interactions in the crayfish *Procambarus gracilis* (Bundy) (Decapoda, Cambaridae). *Proc. Okla. Acad. Sci.* **55,** 1–5.

Hazlett, B. A., Rittschof, D., and Rubenstein, D. (1974). Behavioral biology of the crayfish *Orconectes virilis.* I. Home Range. *Am. Midl. Nat.* **92,** 301–319.

Henry, K. A. (1951). Spring Creek crayfish migrations. *Oreg. Fish. Comm. Res. Briefs* **3,** 48–55.

Herrnkind, W.F. (1969). Queueing behavior of spiny lobsters. *Science* **164,** 1425–1427.

Herrnkind, W. F. (1980). Spiny lobsters: Patterns of movement. *In* "The Biology and Management of Lobsters" (J. S. Cobb and B. F. Phillips, eds.), Vol. I, pp. 349–407. Academic Press, New York.

Herrnkind, W. F., and Cummings, W. C. (1964). Single file migrations of the spiny lobster, *Panulirus argus* (Latreille). *Bull. Mar. Sci. Gulf Carib.* **14,** 123–125.

Herrnkind, W. F., VanderWalker, J., and Barr, L. (1975). Population dynamics, ecology and behavior of spiny lobster, *Panulirus argus,* of St. John, U.S. Virgin Islands: Habitation and pattern of movements. Results of the Tektite Program, Vol. 2. *Sci. Bull. Nat. Hist. Mus. Los Angeles* **20,** 31–45.

Hewett, C. J. (1974). Growth and moulting in the common lobster (*Homarus vulgaris* Milne-Edwards). *J. Mar. Biol. Assoc. U. K.* **54,** 379–391.

Heydorn, A. E. F. (1965). The rock lobster of the South African west coast *Jasus lalandii* (H. Milne-Edwards). I. Notes on the reproductive biology and the determination of minimum size limits for commercial catches. *S. Afr. Div. Sea Fish. Invest. Rep.* **53,** 1–32.

Heydorn, A. E. F. (1969a). Notes on the biology of *Panulirus homarus* and on length weight relationships of *Jasus lalandii. S. Afr. Div. Sea Fish. Invest. Rep.* **69,** 1–22.

Heydorn, A. E. F. (1969b). The South African rock lobster *Jasus tristani* at Vema sea mount, Gough Island and Tristan da Cunha. *S. Afr. Div. Sea Fish Invest. Rep.* **73,** 1–20.

Hillis, J. P. (1971). Effects of light on *Nephrops* catches. *ICES Shellfish Benthos Cttee. K3,* 7 pp.

Hillis, J. P. (1972). Juvenile *Nephrops* caught in the Irish Sea. *Nature (London)* **238,** 280–281.

Hillis, J. P. (1974). A diving study of Dublin Bay prawns *Nephrops norvegicus* (L.) and their burrows off the east coast of Ireland. *Ir. Fish. Invest., Ser. B* **12,** 1–8.

Hillis, J. P. (1979). Growth studies on the prawn, *Nephrops norvegicus. Rapp. P. V. Reun Cons. Int. Explor. Mer.* **175,** 170–175.

Hindley, J. P. R. (1977). A review of some aspects of the behavior of juvenile and adult palinurids. *In* "Workshop on Lobster and Rock Lobster Ecology and Physiology" (B. F. Phillips and J. S. Cobb, eds.), pp. 133–142.

Hirtle, R. W., and Mann, K. H. (1978). Distance chemoreception and vision in the selection of prey by American lobster, *Homarus americanus. J. Fish. Res. Board Can.* **35,** 1006–1008.

Hobbs, H. H., III. (1974). Observations on the cave-dwelling crayfishes of Indiana. *Freshwater Crayfish* **2,** 405–413.

Hobbs, H. H., III (1977). Studies of the cave crayfish, *Orconectes immunis immunis* Cope *(Decapoda: Cambaridae). II. Growth. Cave Res. Found. 19th Annu. Rep.,* pp. 31–32.

Hobbs, H. H., III, Thorp, J. M., and Anderson, G. E. (1976). The freshwater decapod crustaceans (Palaemonidae: Cambaridae) of the Savannah River Plant, S.C. *Publ. Savanna River Plant, Aiken, South Carolina.*

Holthuis, L. B. (1974). Biological results of the University of Miami deep-sea expeditions: 106. The lobsters of the superfamily Nephropidea of the Atlantic Ocean (Crustacea: Decapoda). *Bull. Mar. Sci.* **24,** 723–884.

Hopkins, C. L. (1967). Growth rate in a population of the freshwater crayfish *Paranephrops. planifrons* White. *N. Z. J. Mar. Freshwater Res.* **1,** 464–474.

Hughes, J. T., Sullivan, J. J., and Shleser, R. (1972). Enhancement of lobster growth. *Science* **177,** 1110–1111.

Hughes, J. T., and Matthiessen, G. C. (1962). Observations on the biology of the American lobster, *Homarus americanus. Limnol. Oceanogr.* **7,** 414–421.

Huner, J. V. (1977). Introduction of the Louisiana red swamp crayfish, *Procambarus clarkii* (Girard): An update. *Freshwater Crayfish* **3,** 193–202.

Huner, J. V., and Barr, J. E. (1980). Red Swamp crayfish. Biology and exploitation. *L. State Univ. Sea Grant Publ. No. LSU-T-80-001.*

Huner, J. V., and Romaire, R. P. (1980). Size at maturity as a means of comparing populations of *Procambarus clarkii* (Girard) (Crustacea: Decapoda) from different habitats. Manuscript.

International Council for the Exploration of the Seas (1975). Report of the working group on *Homarus* stocks. *ICES C.M. K.* **38,** 1–19.

Johnson, M. W. (1974). On the dispersal of lobster larvae into the east Pacific barrier (Decapoda, Palinuridae). *Fish. Bull.* **72,** 639–647.

Jones, R. (1979). An analysis of a *Nephrops* stock using length composition data. *Rapp. P. V. Reun. Cons. Int. Explor. Mer.* **175,** 259–269.

Jorgensen, O. M. (1925). The early stages of *Nephrops norvegicus,* from the Northumberland plankton, together with a note on the post-larval development of *Homarus vulgaris. J. Mar. Biol. Assoc. U. K.* **13,** 870–879.

Kanciruk, P. A. (1980). Ecology of juvenile and adult Palinuridae (spiny lobsters). *In* "The Biology and Management of Lobsters" (J. S. Cobb and B. F. Phillips, eds.), Vol. II, pp. 59–96. Academic Press, New York.

Kanciruk, P., and Herrnkind, W. F. (1976). Autumnal reproduction in *Panulirus argus* at Bimini, Bahamas. *Bull Mar. Sci.* **26,** 417–432.

Kanciruk, P., and Herrnkind, W. (1978). Mass migration of spiny lobster *Panulirus argus* (Crustacea: Palinuridae): Behavior and environmental correlation. *Bull. Mar. Sci.* **28,** 601–623.

Kensler, C. B. (1967). Size at first maturity in females of the spiny lobster *Jasus verreauxi* (Crustacea: Decapoda: Palinuridae). *N. Z. Mar. Freshwater Res.* **1,** 327–340.

Konikoff, M. (1977). Study of the life history and ecology of the red swamp crawfish *Procambarus clarkii,* in the lower Atchafalaya Basin floodway. *Final Rep. U. S. Fish. Wildl. Serv. Dep. Biol. Univ. Southwest. La.,* 81 pp.

Kossakowski, J. (1973). The freshwater crayfish in Poland. *Freshwater Crayfish* **1,** 18–28.

Kossakowski, J. (1975). Crayfish *Orconectes limosus* in Poland. *Freshwater Crayfish* **2**, 31–47.

Krouse, J. S. (1973). Maturity, sex ratio and size composition of a natural population of American lobster, *Homarus maericanus*, along the Maine coast. *Fish. Bull.* **71**, 165–173.

Krouse, J. S. (1977). Completion report of lobster tagging project. Comm. Fish. Res. Dev. Act. Contract 3-228-R. Maine Dep. Mar. Res., Augusta, Maine.

Krouse, J. S. (1980). Summary of lobster, *Homarus americanus*, tagging studies in American waters (1898–1978). *Can. Fish. Aquat. Sci. Tech. Rep. No. 932*, pp. 135–140.

Krouse, J. S. (1981). Movement, growth and mortality of American lobsters, *Homarus americanus*, tagged along the coast of Maine. *NOAA Tech. Rep. NMFS SSRF—747*, pp. 1–12.

Kubo, J., and Ishiwata, N. (1964). On the relationship between activity of Japanese spiny lobster and underwater light intensity. *Bull. Jpn. Soc. Sci. Fish.* **30**, 884–888.

LaCaze, C. G. (1976). Crawfish farming. *L. Wildl. Fish. Comm., Baton Rouge, Bull. No. 7*, 27 pp.

Lang, F., Govind, C. K. Costello, W. J., and Greene, S. I. (1977). Developmental neuroethology: Changes in escape and defensive behavior during growth of the lobster. *Science* **197**, 682–685.

Laurent, P. J. (1973). *Astacus* and *Cambarus* in France. *Freshwater Crayfish* **1**, 70–78.

Lindberg, R. B. (1955). Growth, population dynamics and field behavior in the spiny lobster, *Panulirus interruptus*. *Univ. Calif. Publ. Zool.* **59**, 157–248.

Little, E. E. (1975). Chemical communication in maternal behaviour of crayfish. *Nature (London)* **255**, 400–401.

Little, E. E. (1976). Ontogeny of maternal behavior and brood pheromone in crayfish. *J. Comp. Physiol.* **112A**, 133–142.

Loesch, H., and Lopez, C. (1966). Observaciones sobre la langosta de la costa continental del Ecuador. *Inst. Nac. Pesc. Ecuador Bol. Cient. Tec.* **1**, 1–30.

Loew, E. R. (1976). Light and photoreceptor degeneration in the Norway lobster, *Nephrops norvegicus* (L.). *Proc. R. Soc. London, Ser. B* **193**, 31–44.

Lowery, R. S., and Mendes, A. J. (1977). The biology of *Procambarus clarkii* in Lake Naivasha, Kenya; with a note on its distribution. *Freshwater Crayfish,* **3**, 203–210.

Lund, W. A., and Stewart, L. L. (1970). Abundance and distribution of larval lobsters, *Homarus americanus*, off the coast of southern New England. *Proc. Natl. Shellfish. Assoc.* **60**, 40–49.

Lyons, W. G., Barber, D. G., Foster, S. M., Kennedy, F. S., Jr., and Milano, G. R. (1981). The spiny lobster, *Panulirus argus*, in the middle and upper Florida Keys: Population structure, seasonal dynamics, and reproduction. *Fla. Dept. Nat. Res. Mar. Res. Publ. No. 38*, pp. 1–38.

Mackie, A. M., Grant, P. T., Shelton, R. G. J., Hepper, B. T., and Walne, P. R. (1980). The relative effectiveness of natural and artificial baits for the lobster, *Homarus gammarus*, laboratory and field trials. *J. Cons. Int. Explor. Mer.* **39**, 123–129.

McKoy, J. L., and Esterman, D. B. (1981). Growth of rock lobsters (*Jasus edwardsii*) in the Gisborne region, New Zealand. *N. Z. J. Mar. Freshwater Res.* **15**, 121–136.

McLeese, D. W., and Wilder, D. G. (1958). The activity and catchability of the lobster (*Homarus americanus*) in relation to temperature. *J. Fish. Res. Board Can.* **15**, 1345–1354.

McLeese, D. W., and Wilder, D. G. (1964). Lobster storage and shipment. *Bull. Fish. Res. Board Can.* **147**, 1–69.

Magnuson, J. J., Capelli, G. M., Lorman, J. G., and Stein, R. A. (1975). Consideration of crayfish or macrophyte control. *In* "The Proceedings of a Symposium on Water Quality

Management Through Biological Control" (P. L. Brezonik and J. L. Fox, eds.), Rep. ENV. 07-75-1. Univ. of Florida, Gainesville.

Maigret, J. (1979). Etat du stock de langostes roses (*Palinurus mauritanicus* Gruvel, 1911) au large des cotes de Muaritanie. *Invest Pesq.* **43,** 83–94.

Mann, K. H. (1977). Destruction of kelp beds by sea urchins: A cyclical phenomenon or irreversible degradation? *Helgol. Wiss. Meeressunters.* **30,** 455–467.

Marchessault, G.D., Saila, S. B., and Palm, W. J. (1976). Delayed recruitment models and their application to the American lobster (*Homarus americanus*) fishery. *J. Fish. Res. Board Can.* **33,** 1779–1787.

Mason, J. C. (1970). Maternal–offspring behavior of the crayfish, *Pacifastacus trowbridgi* (Stimpson). *Am. Midl. Nat.* **84,** 463–473.

Mason, J. C. (1975). Crayfish production in a small woodland stream. *Freshwater Crayfish* **2,** 449–470.

Matthews, J. P. (1962). The rock lobster of southwest Africa (*Jasus lalandii*) (Milne-Edwards). Size frequency, reproduction, distribution and availability. *Invest. Rep. Mar. Res. Lab., S.W. Afr.* **7,** 1–61.

Mauchline, J. (1977). Growth of shrimp, crabs and lobsters—an assessment. *J. Cons. Int. Explor. Mer.* **37,** 162–169.

Mead, A. D., and Williams, L. W. (1903). Habits and growth of the lobster, and experiments in lobster culture. *R. J. Comm. Inland Fish. Annu. Rep.* **33,** 57–83.

Meany, T. F. (1979). Limited entry in the western Australian rock lobster and prawn fisheries: An economic evaluation. *J. Fish. Res. Board Can.* **36,** 789–798.

Menzies, R. A., and Kerrigan, J. M. (1979). Implications of spiny lobster recruitment patterns of the Caribbean—a biochemical genetic approach. *Proc. Gulf Carib. Fish. Inst.* **31,** 164–178.

Merkle, E. E. (1969). Home range of crayfish *Orconectes juvenalis*. *Am. Midl. Nat.* **81,** 228–235.

Michel, C., and Parent, B. (1973). Essais de traitement des langoustines par des agents conservateurs. *Sci. Peche.* **229,** 9–17.

Miller, G. C., and VanHyning, J. M. (1971). The commercial fishery for freshwater crayfish, *Pacifastacus leniusculus* (Astacidae), in Oregon, 1893–1956. *Res. Rep. Fish. Comm. Oreg.* **2,** 77–89.

Mitchell, J. R. (1971). Food preferences, feeding mechanism and related behavior in phyllosoma larvae of the California spiny lobster, *Panulirus interruptus* (Randall). M.S. Thesis, San Diego State Univ., San Diego, California.

Mobberly, W. D., Jr., and Pfrimmer, R. J. (1967). Distribution of crayfish in a roadside ditch. *Am. Midl. Nat.* **78,** 82–88.

Moller, T. H., and Naylor, E. (1980). Environmental influence on locomotor activity in *Nephrops norvegicus* (Crustacea: Decapoda). *J. Mar. Biol. Assoc. U. K.* **60,** 103–113.

Momot, W. T. (1966). Upstream movement of crayfish in an intermittent Oklahoma stream. *Am. Midl. Nat.* **75,** 150–159.

Momot, W. T. (1967). Population dynamics and productivity of the crayfish *Orconectes virilis* in a small lake. *Am. Midl. Nat.* **78,** 55–81.

Momot, W. T., and Gowing, H. (1977a). Production and population dynamics of the crayfish *Orconectes virilis* in three Michigan lakes. *J. Fish. Res. Board Can.* **34,** 2041–2055.

Momot,W. T., and Gowing, H. (1977b). Results of an experimental fishery on the crayfish *Orconectes virilis*. *J. Fish. Res. Board Can.* **34,** 2041–2055.

Momot, W. T., and Jones, P. D. (1977). The relationship between biomass, growth rate, and annual production in the crayfish, *Orconectes virilis*. *Freshwater Crayfish* **32,** 3–31.

Momot, W. T., Gowing, H., and Jones, P. D. (1978). The dynamics of crayfish and their role in ecosystems. *Am. Midl. Nat.* **99**, 10–35.

Moore, R., and MacFarlane, J. W. (1977). Migration patterns and life history of *Panulirus ornatus* in Papua New Guinean waters. *In* "Workshop on Lobster and Rock Lobster Ecology and Physiology" (B. F. Phillips and J. S. Cobb, eds.). CSIRO (Aust.) Fish. Oceanog. Circ. **7**, 30.

Morgan, G. R. (1972). Fecundity in the western rock lobster *Panulirus longipes cygnus* (George) (Crustacea: Decapoda: Palinuridae). *Aust. J. Mar. Freshwater Res.* **23**, 133–141.

Morgan, G. R. (1974a). Aspects of the population dynamics of the western rock lobster, *Panulirus cygnus* George. I. Estimation of population density. *Aust. J. Mar. Freshwater Res.* **25**, 235–248.

Morgan, G. R. (1974b). Aspects of the population dynamics of the western rock lobster, *Panulirus cygnus* George. II. Seasonal changes in the catchability coefficient. *Aust. J. Mar. Freshwater Res.* **25**, 249–259.

Morgan, G. R. (1977). Aspects of the population dynamics of the western rock lobster and their role in management. Ph.D. Thesis, Univ. of Western Australia.

Morgan, G. R. (1978). Locomotor activity in the western rock lobster *Panulirus longipes cygnus. Aust. J. Mar. Freshwater Res.* **29**, 169–174.

Morgan, G. R. (1979). An assessment of the stocks of the western rock lobster *Panulirus cygnus* using surplus yield models. *Aust. J. Mar. Freshwater Res.* **30**, 355–365.

Morgan, G. R. (1980a). Population dynamics of spiny lobsters. *In* "The Biology and Management of Lobsters" (J. S. Cobb and B. F. Phillips, eds.), Vol. II, pp. 189–217. Academic Press, New York.

Morgan, G. R. (1980b). Increases in fishing effort in a limited-entry fishery—the western rock lobster fishery 1963–1976. *J. Cons. Int. Explor. Mer.* **39**, 82–87.

Morgan, G. R. (1980c). Population dynamics and management of the western rock lobster fishery. *Mar. Policy, Jan. 1980,* pp. 52–60.

Morgan, G. R. (1984). Estimates of mortality rate of the western rock lobster *Panulirus cygnus* using length–frequency data. *J. Cons. Int. Explor. Mer.* (in press).

Morgan, G. R., Phillips, B. F., and Joll, L. M. (1982). Stock and recruitment relationships in *Panulirus cygnus,* the commercial rock (spiny) lobster of Western Australia. *Fish. Bull.* **80**, 475–486.

Morrissey, T. D. (1971). Movements of tagged lobsters, *Homarus americanus,* liberated off Cape Cod, Massachusetts. *Trans. Am. Fish. Soc.* **100**, 117–120.

Morrissy, N. M. (1974). Ecology of marron, *Cherax tenuimanus* (Smith), introduced into some farm dams near Boscabel in the great southern area of the wheat belt region in Western Australia. *Fish. Res. Bull. (West. Aust. Mar. Res. Lab.), No. 12,* pp. 1–55.

Morrissy, N. M. (1975). Spawning variation and its relationship to growth rate and density in the marron, *Cherax tenuimanus* (Smith). *Fish. Res. Bull. (West. Aust. Mar. Res. Lab.), No. 16,* pp. 1–32.

Morrissy, N. M. (1976). Aquaculture of marron, *Cherax tenuimanus* (Smith). Part 2. Breeding and early rearing. *Fish. Res. Bull. (West. Aust. Mar. Res. Lab.), No. 17,* pp. 1–32.

Morrissy, N. M. (1978). The amateur marron fishery in southwestern Australia. *Fish. Res. Bull. West. Aust.* **21**, 1–44.

Morrissy, N. M. (1980). Production of marron in Western Australia wheatbelt farm dams. *Fish. Res. Bull. (West. Aust. Mar. Res. Lab.), No. 24,* pp. 1–79.

Morrissy, N. M., and Caputi, N. (1981). Use of catchability equations for population estimation of marron, *Cherax tenuimanus* (Smith) (Decapoda: Parastacidae). *Aust. J. Mar. Freshwater Res.* **32**, 213–225.

Munro, J. L. (1974). The biology, ecology, exploitation and management of Caribbean reef fishes. Scientific Report of ODA/UNI fisheries ecology research project, 1962–1973. Part VI. The biology, ecology and bionomics of Caribbean reef fishes—crustaceans (spiny lobsters and crabs). *Univ. West Indies Zool. Dep. Res. Rep.* **3,** 1–57.

Naylor, E., and Atkinson, R. J. A. (1976). Rhythmic behavior of *Nephrops* and some other marine crustaceans. *In* "Perspectives in Experimental Biology" (P. Spencer Davis, ed.), Vol. I, pp. 135–143. Pergamon, Oxford.

New England Fishery Management Council (1982). Draft American lobster fishery management plan. Mimeo, pp. 1–129. Saugus, Massachusetts.

Newman, G. G., and Pollock, D. E. (1974a). Growth of the rock lobster *Jasus lalandii* and its relationship to benthos. *Mar.Biol. (Berlin)* **24,** 339–346.

Newman, G. G., and Pollock, D. E. (1974b). A mass stranding of rock lobsters *Jasus lalandii* (H. Milne-Edwards, 1837) at Eland Bay, South Africa (Decapoda: Palinuridea). *Crustaceana* **26,** 1–5.

Nichols, J. H., and Lawton, P. (1978). The occurrence of the larval stages of the lobster *Homarus gammarus.* (Linnaeus, 1758) off the northwest coast of England in 1976. *J. Cons. Int. Explor. Mer.* **38,** 234–243.

Nicholson, M. D. (1979). The use of length–frequency distributions for age determination of *Nephrops norvegicus. Rapp. P. V. Reun. Cons. Int. Explor. Mer.* **175,** 176–181.

Niemi, A. (1977). Population studies on the crayfish *Astacus astacus* L. in the river Pyhajoki, Finland. *Freshwater Crayfish* **3,** 81–99.

Orach-Meza, F. L., and Saila, S. B. (1978). Application of a polynomial distributed lag model to the Maine lobster fishery. *Trans. Am. Fish. Soc.* **197,** 402–411.

Peacock, N. A. (1974). A study of the spiny lobster fishery of Antigua and Barbuda. *Proc. Gulf Carrib. Fish. Inst.* **26,** 117–130.

Pella, J. J., and Tomlinson, P. K. (1969). A generalized stock production model. *Bull. Int. Am. Trop. Tuna Comm.* **13,** 421–496.

Penn, G. H. (1943). A study of the life history of the Louisiana red crayfish, *Procambarus clarkii. Ecology* **24,** 1–18.

Perkins, H. C. (1971). Egg loss during incubation from offshore northern lobsters (Decapoda: Homaridae). *Fish. Bull.* **69,** 451–453.

Perkins, H. C. (1972). Developmental rates at various temperatures of embryos of the northern lobster (*Homarus americanus*) (Milne-Edwards). *Fish. Bull.* **70,** 95–99.

Phillips, B. F. (1972). A semi-quantitative collector of the puerulus larvae of the western rock lobster *Panulirus longipes cygnus* George (Decapoda: Palinuridea). *Crustaceana* **22,** 147–154.

Phillips, B. F. (1981). The circulation of the southeastern Indian Ocean and the planktonic life of the western rock lobster. *Oceanogr. Mar. Biol. Annu. Rev.* **19,** 11–39.

Phillips, B. F., and Sastry, A. N. (1980). Larval ecology. *In* "The Biology and Management of Lobsters" (J. S. Cobb and B. F. Phillips, eds.), Vol. II, pp. 11–57. Academic Press, New York.

Phillips, B. F., Campbell, N. A., and Rea, W. A. (1977). Laboratory growth of early juveniles of western rock lobster, *Panulirus longipes cygnus. Mar. Biol. (Berlin)* **39,** 31–39.

Phillips, B. F., Rimmer, D. W., and Reid, D. D. (1978). Ecological investigation of the late stage phyllosoma and puerulus larvae of the western rock lobster *Panulirus longipes cygnus. Mar. Biol. (Berlin)* **45,** 347–357.

Phillips, B. F., Brown, P. A., Rimmer, D. W., and Reid, D. D. (1979). Distribution and dispersal of the phyllosoma larva of the western rock lobster, *Panulirus cygnus,* in the southeastern Indian Ocean. *Aust. J. Mar. Freshwater Res.* **30,** 773–783.

Phillips, B. F., Cobb, J. S. and George, R. W. (1980). General Biology. In "The Biology and Management of Lobsters" (J. S. Cobb and B. F. Phillips, eds.), Vol. I, pp. 1–82. Academic Press, New York.

Pippitt, M. R. (1977). Mating behavior of the crayfish Orconectes nais (Faxon, 1885) (Decapoda: Astacoidea). Crustaceana 33, 265–271.

Pollock, D. E. (1979). Predator–prey relationships between the rock lobster Jasus lalandii and the mussel Aulacomya ater at Robben Island on the Cape West Coast of Africa. Mar. Biol. (Berlin) 52, 347–356.

Pollock, D. E., and Roscoe, M. J. (1977). The growth at moulting of crayfish Jasus tristani at Tristan da Cunha, South Atlantic. J. Cons. Int. Explor. Mer. 37, 144–146.

Pratten, D. J. (1980). Growth in the crayfish Austropotamobius pallipes (Crustacea: Astacidae). Freshwater Biol. 10, 401.

Radway-Allen, K. (1977). Population density and recruitment. In "Workshop on Lobster and Rock Lobster Ecology and Physiology" (B. F. Phillips and J. S. Cobb, eds.), pp. 287–292.

Rice, A. L. and Chapman, C. J. (1971). Observations on the burrows and burrowing behavior of two mud-dwelling decapod crustaceans, Nephrops norvegicus and Goneplax rhomboides. Mar. Biol. (Berlin) 10, 330–342.

Ricker, W. E. (1958). Handbook of computations for biological statistics of fish populations. Bull. Fish. Res. Board Can. 119, 1–300.

Rimmer, D. W., and Phillips, B. F. (1979). Diurnal migration and vertical distribution of phyllosoma larvae of the western rock lobster, Panulirus cygnus George. Mar. Biol. (Berlin) 54, 109–124.

Roberts, T. W. (1944). Light, eyestalk, chemical and certain other factors as regulators of the community activity for the crayfish Cambarus virilis Hagen. Ecol. Monogr. 14, 359–392.

Rogers, B. A., Cobb, J. S., and Marshall, N. M. (1968). Size comparison of inshore and offshore larvae of the lobster, Homarus americanus, off southern New England. Proc. Natl. Shellfish. Assoc. 58, 78–81.

Russell, H. J. (1980). The determination of growth rates for American lobsters. Can. Fish. Aquat. Sci. Tech. Rep. No. 932, pp. 1–8.

Russell, H. J., Borden, D. V. D., and Fogarty, M. J. (1978). Management studies of lobster resources. U.S. Dept. Comm. NOAA Contract 03-4-043-360. R.I. Div. Fish. Wild., Providence, Rhode Island.

Saila, S. B., Annala, J. H., McKoy, J. L., and Booth, J. D. (1979). Application of yield models to the New Zealand rock lobster fishery. N. Z. J. Mar. Freshwater Res. 13, 1–11.

Saila, S. B., and Flowers, J. M. (1965). A simulation study of sex ratios and regulation effects with the American lobster, Homarus americanus. Proc. Gulf Caribb. Fish. Inst. 18, 66–78.

Saila, S. B., and Flowers, J. M. (1968). Movements and behavior of berried female lobsters displaced from offshore areas to Narragansett Bay, Rhode Island. J. Cons. Int. Explor. Mer. 31, 342–351.

Saila, S. B., and Flowers, J. M. (1969). Geographic morphometric variation in the American lobster. Syst. Zool. 18, 330–338.

Saila, S. B., and Marchessault, G. D. (1980). Population dynamics of clawed lobsters. In "The Biology and Management of Lobsters" (J. S. Cobb and B. F. Phillips, eds.), Vol. II, pp. 219–241. Academic Press, New York.

Saila, S. B., Wigbout, M., and Lermit, R. J. (1980). Comparison of some time series models for the analysis of fisheries data. J. Cons. Int. Explor. Mer. 39, 44–52.

Santucci, R. (1926). Lo sviluppo e l'ecologia post-embrionali dello "Scampo" (*Nephrops norvegicus* (L)) net Tirreno e nei Mari Nordici. *Mem. R. Com. Talazzogr. Ital.* **125**, 1–36.

Scarratt, D. J. (1964). Abundance and distribution of lobster larvae (*Homarus americanus*) in Northumberland Strait. *J. Fish. Res. Board Can.* **21**, 661–679.

Scarratt, D. J. (1968a). An artificial reef for lobsters, *Homarus americanus*. *J. Fish. Res. Board Can.* **28**, 1733–1738.

Scarratt, D. J. (1968b). Distribution of lobster larvae (*Homarus americanus*) off Pictou, Nova Scotia. *J. Fish. Res. Board Can.* **25**, 427–430.

Scarratt, D. J. (1972). The effects on lobsters, *Homarus americanus*, of raking Irish moss, *Chondrus crispus*. *ICES, C.M. 192/K:36.*

Scarratt, D. J. (1973a). Abundance, survival and vertical and diurnal distribution of lobster larvae in Northumberland Strait, 1962–63, and their relationships with commercial stocks. *J. Fish. Res. Board Can.* **30**, 1818–1824.

Scarratt, D. J. (1973b). Lobster populations on a man-made rocky reef. *ICES, C.M. K. 47.*

Scarratt, D. J., and Elson, P. F. (1965). Preliminary trials of a tag for salmon and lobsters. *J. Fish. Res. Board Can.* **22**, 421–423.

Schaefer, M. B. (1954). Some aspects of the dynamics of populations important to the management of the commercial fisheries. *Bull. Int. Am. Trop. Tuna Comm.* **1**, 27–56.

Schaefer, M. B. (1957). A study of the dynamics of the fishery for yellow fin tuna in the eastern tropical Pacific Ocean. *Bull. Int. Am. Trop. Tuna Comm.* **2**, 245–285.

Simpson, A. C. (1961). A contribution to the bionomics of the lobster (*Homarus vulgaris* Edw.) on the coast of North Wales. *Fish. Invest. (London) Ser. 2,* **23**, 1–28.

Simpson, A. C. (1965). Variations in the catches of *Nephrops norvegicus* at different times of day and night. *Rapp. P. V. Reun. Cons. Int. Explor. Mer.* **156**, 186–189.

Sims, H. W., and Ingle, R. M. (1966). Caribbean recruitment of Florida's spiny lobster populations. *Q. J. Fla. Acad. Sci.* **29**, 207–242.

Skud, B. E. (1969). The effect of fishing on size composition and sex ratio of offshore lobster stocks. *Fisk. Dir. Skr. Ser. Havunders.* **15**, 259–309.

Skud, B. E., and Perkins, H. C. (1969). Size composition, sex ratio and size at maturity of offshore northern lobster. *U. S. Fish. Widl. Serv., Spec. Sci. Rep. Fish. No. 598,* pp. 1–10.

Smale, M. J. (1978). Migration, growth and feeding in the Natal rock lobster *Panulirus homarus* (Linnaeus). *Oceanogr. Res. Inst. (Durban), Invest. Rep.* **47**, 56.

Smith, D. G. (1981). Evidence for hybridization between two crayfish species (Decapoda: Cambaridae:*Orconectes*) with a comment on the phenomenon in Cambarid crayfish. *Am. Midl. Nat.* **105**, 405–407.

Smith, E. M. (1976). Food burial behavior of the American lobster (*Homarus americanus*). M.S. Thesis, Univ. of Connecticut, Storrs.

Smith, E. M. (1977). Some aspects of catch/effort, biology and the economics of the Long Island Sound lobster fishery during 1976. *NOAA, NMFS, Comm. Fish. Res. Dev. Act. Final Rep. Conn. Proj. No. 3-253-R-1,* pp. 1–97.

Smith, P. J., McKoy, J. L., and Machin, P. J. (1980). Genetic variation in the rock lobsters *Jasus edwardsii* and *J. novaehollandiae*. *N. Z. J. Mar Freshwater Res.* **4**, 55–63.

Squires, H. J., Tucker, G. E., and Ennis, G. P. (1971). Lobsters (*Homarus americanus*) in Bay of Islands, Newfoundland, 1963–1965. *Fish. Res. Board Can. Manuscr. Rep. Ser. (Biol.)* p. 1151.

Stasko, A. (1980). Lobster larval surveys in Canada. *Can. Fish. Aquat. Sci. Tech. Rep. No. 932,* pp. 157–165.

Stein, R. A. (1976). Sexual dimorphism in crayfish chelae: Functional significance related to reproductive activities. *J. Zool.* **54**, 220–227.

Stein, R. A. (1977a). Selective predation, optimal foraging, and the predator–prey interaction between fish and crayfish. *Ecology* **58**, 1237–1253.

Stein, R. A. (1977b). External morphological changes associated with sexual maturity in the crayfish (*Orconectes propinquus*). *Am. Midl. Nat.* **97**.

Stein, R. A., and Magnuson, J. J. (1976). Behavioral response of crayfish to a fish predator. *Ecology* **57**, 751–761.

Stevenson, J. R. (1972). Changing activities of the crustacean epidermis during the molting cycle. *Am. Zool.* **12**, 373–380.

Stewart, L. L. (1972). The seasonal movements, population dynamics, and ecology of the lobster, *Homarus americanus*, off Ram Island, Conn. Ph.D. Thesis, Univ. of Connecticut, Storrs.

Street, R. J. (1969). The New Zealand crayfish *Jasus edwardsii* (Hutton, 1875). *N. Z. Mar. Dep. Fish. Tech. Rep.* **30**, 1–53.

Street, R. J. (1971). Rock lobster migration off Otago. *N. Z. Commer. Fish., June,* pp. 16–17.

Sutcliffe, W. H. (1952). Some observations on the breeding and migration of the Bermuda spiny lobster, *Panulirus argus. Proc. Gulf Carib. Fish. Inst.* **4**, 64–69.

Sutcliffe, W. H. (1956). Effect of light intensity on the activity of the Bermuda spiny lobster, *Panulirus argus. Ecology* **37**, 200–201.

Sutcliffe, W. H. (1973). Correlations between seasonal river discharge and local landings of American lobster (*Homarus americanus*) and Atlantic halibut (*Hippoglossus hippoglossus*) in the Gulf of St. Lawrence. *J. Fish Res. Board Can.* **30**, 856–859.

Sweat, D. E. (1968). Growth and tagging studies on *Panulirus argus* (Latreille) in the Florida keys. *Fla. Board Cons. Mar. Res. Lab. Tech. Ser.* **57**, 1–30.

Symonds, D. J. (1972). Infestation of *Nephrops norvegicus* (L.) off the northeast coast of England. *Fish. Invest. London, Ser. 2* **27**, 1–35.

Tcherkashina, N. Ya. (1975). Distribution and biology of crayfish of genus *Astacus* (Crustacea:-Decapoda:Astacidea) in the Turkman waters of the Caspian Sea. *Freshwater Crayfish* **2**, 553–556.

Tcherkashina, N. Ya. (1977). Survival, growth, and feeding dynamics of juvenile crayfish (*Astacus leptodactylus cubanicus*) in ponds and the River Don. *Freshwater Crayfish* **3**, 95–100.

Templeman, W. (1935). Local differences in the body proportions of the lobster, *Homarus americanus. J. Biol. Board Can.* **1**, 213–226.

Templeman, W. (1936a). Local differences in the life history of the lobster (*Homarus americanus*) on the coast of the maritime provinces of Canada. *J. Biol. Board Can.* **2**, 41–88.

Templeman, W. (1936b). Further contributions to mating in the American lobster. *J. Biol. Board Can.* **2**, 223–226.

Templeman, W.(1937). Habits and distributions of larval lobster (*Homarus americanus*). *J. Biol. Board Can.* **3**, 343–347.

Templeman, W., and Tibbo, S. N. (1945). Lobster investigations in Newfoundland 1938–1941. *Newfoundland Dep. Nat. Resour. Res. Bull. (Fish)* **16**, 1–98.

Thomas, H. J. (1954). Observations on the recaptures of tagged lobsters in Scotland. *ICES, C.M.* 7.

Thomas, H. J. (1965). Handling lobsters and crabs. *Rapp. P.V. Cons. Int. Explor. Mer.* **156**, 35–40.

Thomas, H. J., and Davidson, C. (1962). The food of the Norway lobster *Nephrops norvegicus* (L). *Mar. Res. Scotland 1962(3)* pp. 1–15.

Thomas, J. C. (1973). An analysis of the commercial lobster (*Homarus americanus*) fishery along the coast of Maine, August 1966 through December 1970. *U.S. Nat. Mar. Fish. Serv. SSRF-667*, pp. 1–57.

Thomas, M. L. H. (1968). Overwintering of American lobsters, (*Homarus americanus*) in burrowing in Biddeford River, Prince Edward Island. *J. Fish. Res. Board Can.* **25,** 2525–2527.

Tomlinson, P.K., and Abramson, N. J. (1961). Fitting a von Bertalanffy growth curve by least squares. *Calif. Dep. Fish Game Fish. Bull.* **116,** 1–69.

Tracy, M. L., Nelson, K., Hedgecock, D., Shleser, R. A., and Pressick, M. L. (1975). Biochemical genetics of lobsters: Genetic variation and structure of American lobster (*Homarus americanus*) populations. *J. Fish. Res. Board Can.* **32,** 2091–2101.

Unestam, T. (1973). Significance of disease on freshwater crayfish. *Int. Symp. Freshwater Crayfish.* **1,** 135–150.

Uzmann, J. R. (1970). Use of parasites in identifying lobster stocks. *Proc. 2nd Int. Conf. Parasitol.* **56,** 12–20.

Uzmann, J. R., Cooper, R. A., and Pecci, K. J. (1977). Migration and dispersion of tagged American lobsters, *Homarus americanus,* on the southern New England continental shelf. *NOAA Tech. Rep. NMFS, SSRF No. 705.*

Van Engel, W. A., (1980). Maturity and fecundity in the American lobster, *Homarus americanus—*A review. *Can. Fish. Aquat. Sci. Tech. Rep. No. 932,* pp. 51–58.

Van Herp, F., and Bellon-Humbert, C. (1978). Setal development and molt prediction in the larvae and adults of the crayfish *Astacus leptodactylus* (Nordmann, 1842). *Aquaculture* **14,** 289–301.

von Bertalanffy, L. (1938). A quantitative theory of organic growth. *Hum. Biol.* **10,** 181–213.

Vranckx, R., and Durliat, M. (1978). Comparisons of the gradient of setal development of uropods and scaphognathites in *Astacus leptodactylus. Biol. Bull. (Woods Hole, Mass.)* **155,** 627–639.

Wang, D. (1982). The behavioral ecology of competition among three decapod species, the American lobster *H. americanus,* the Jonah crab. *C. borealis* and the rock crab, *C. irroratus,* in rocky habitats. Ph.D. Diss., Univ. of Rhode Island, Kingston.

Warme, J., Cooper, R. A., and Slater, R. (1978). Bioerosion in submarine canyons. *In* "Submarine Canyon Fan and Trench Sedimentation" (D. J. Stanley and G. Kelling, eds.), pp. 65–70. Dowden, Hutchenson and Ross, Stroudberg, Pennsylvania.

Warner, R. E., Combs, C. L., and Gregory, D. R. (1976). Biological studies of the spiny lobster, *Panulirus argus* (Decapoda:Palinuridae) in South Florida. *Proc. Gulf Carib. Fish. Inst.* **29,** 166–183.

Watson, P. S. (1974). Investigations on the lobster *Homarus gammarus,* in Northern Ireland, a progress report, 1972–73. *ICES, C.M. K. 20.*

Weiss, H. M. (1970). The diet and feeding behavior of the lobster, *Homarus americanus,* in Long Island Sound. Ph.D. Thesis, Univ. of Connecticut, Storrs.

Westman, K. (1975). On crayfish research in Finland. *Freshwater Crayfish* **2,** 65–75.

Wilder, D. G.(1963). Movements, growth and survival of marked and tagged lobsters liberated in Egmont Bay, Prince Edward Island. *J. Fish. Res. Board Can.* **20,** 305–318.

Williams, D. D., Williams, N. E., and Myers, H. B. N. (1974). Observations on the life history and burrow construction of the crayfish Cambarus fodiens (Cottle) in a temporary stream in southern Ontario. *Can. J. Zool.* **52,** 365–370.

Williamson, D. I. (1956). The plankton of the Irish Sea, 1951 and 1952. *Bull. Mar. Ecol.* **4,** 87–144.

Wolff, T. (1978). Maximum size of lobster (*Homarus*) (Decapoda: Nephropidae). *Crustaceana* **34,** 1–14.

Yoza, K., Nomura, K., and Miyamoto, H. (1977). Observations upon the behavior of the lobster and the top shell caught by the bottom-set gillnet. *Bull. Jpn. Soc. Sci. Fish.* **43,** 1269–1272.

Zeitlin-Hale, L., and Sastry, A. N. (1978). Effects of environmental manipulation on the locomotor activity and agonistic behavior of cultured juvenile American lobsters, *Homarus americanus. Mar. Biol. (Berlin)* **47,** 369–379.

Culture of Crustaceans: General Principles

ANTHONY J. PROVENZANO, JR.

I. PURPOSES OF CULTIVATION

Cultivation of crustaceans for research dates from the early nineteenth century and generally falls into one of four categories: systematics, biology of larvae, provision of experimental material, or provision of food for other research organisms.

Laboratory rearing to understand systematic relationships dates from the early decades of the last century. By keeping planktonic forms alive from a molt to megalopa stage and by hatching eggs from a crab, Thompson (1828) was able to establish the relationship of the genus *Zoea* as larvae of brachyuran crabs. He also established the crustacean affinities of barnacles by observing their newly hatched larvae. The differences between adults of some taxa are frequently the result of modifications for a peculiar life style. Thus, various morphological differentiations often obscure phyletic relationships; but examination of earlier life stages, which tend to be conservative in many ways, may reveal similar larvae from apparently diverse adults.

The major works of Gurney (1939, 1942) and others contributed to the classification of the order Decapoda. A very large number of publications in the intervening decades has been devoted to description of larval stages of various taxa, although, in most groups, the larvae are still unknown for a majority of species. Examples of the use of information on larvae for revision of classification abound in literature of recent decades, and the continuing explosion of new life history descriptions promises to provide significant new data for workers in this fertile field.

A recent review of the use of such data in systematics was made by Felder et al. (1985) who gave examples from many of the decapod families. The same work presented the first comprehensive review of the characteristics of postlarval stages and showed the usefulness of this information in the study of phylogeny, which demonstration was made possible by the accumulation of many individual specific studies. Detailed summary of the use of larval characters in study of classification of brachyuran crabs was presented by Rice (1980) and expanded upon by Gore (1985). For some groups of crustaceans, there is not sufficient information to use larvae in classification, as in the Stomatopoda (Provenzano and Manning, 1978; Morgan and Provenzano, 1979); but as the number and percentage of known life histories increases, the usefulness of the data in systematics also increases.

Laboratory rearing for purpose of identification of planktonic forms is now standard procedure. Nearly any planktonic crustacean larva can, with suitable technique, be captured in good condition, returned to the laboratory, and reared to metamorphosis. In this way, larvae of even rare forms or burrowing species (the adults of which are difficult to collect) can be described (Goy and Provenzano, 1978). Alternatively, egg-bearing adults may be captured and the eggs hatched in the laboratory. Roberts (1975) reviewed methods for the rearing of decapod larvae on a research scale. Kinne (1977) also has given a review of culture methods and general examples of taxa that have been reared for research, rather than for commercial purposes.

Once methodologies are sufficiently developed to permit routine rearing, laboratory culture may provide experimental material for special purposes. It is extremely difficult to collect large numbers of very young juvenile or postlarval stages of most species from natural habitats, especially for studies in which similar ages or physiological history of the specimens, or both, is critical to the study. Hazlett and Provenzano (1965), using cultured specimens, were able to study the ontogeny of shell inhabiting behavior in hermit crabs. Forward (1977) was able to study shadow responses in brachyuran larvae. Physiological requirements of larvae, including nutritional requirements, and their responses to changes in salinity and temperature ranges, to cyclic variations, to pollutants, and to hormones are routinely investigated by use of laboratory-hatched larvae (Bookhout and Costlow, 1970, 1975; Bookhout and Monroe, 1977; Buchanan et al., 1970; Bigford, 1978; Epifanio, 1971; McConaugha, 1979; McConaugha and Costlow, 1981; Provenzano and Goy, 1976; Sulkin and Epifanio, 1975; Sulkin et al., 1980).

Kinne (1977) has reviewed the rearing of various crustaceans and their use as food or bioassay organisms. Other than the decapods, the principal groups receiving attention for these purposes have been branchiopods, especially Artemia, and the freshwater cladocerans, benthic and planktonic copepods, and barnacles. Details of methods and apparatus for rearing of each major group are given by Kinne. Larvae of these groups normally feed on phytoplankton, though adults may not. In addition to their use in basic research, including food chain studies, copepods have been used in assay work and have been grown in large quantities as food for other organisms, principally marine fishes (Fujita, 1973; Kitajima, 1973; Hanaoka, 1973; Iwasaki, 1973; Turk et al., 1982).

Efficiency in large-scale cultures may be markedly improved by new techniques, such as substitution of rice bran for phytoplankton, as has been done for Daphnia (de Pauw et al., 1981), Artemia (Sorgeloos et al., 1980), Acartia (Turk et al., 1982), and Penaeus (Ishida, 1967). Other groups are less well-studied. The peracarids, which do not have free larvae, are generally not difficult to maintain and propagate. Mysids have been used as assay organisms (Nimmo et al., 1977), but isopods and amphipods are less commonly exploited. Euphausiids are mainly plankton feeders and are easily subject to damage upon capture. Although they are of commercial importance in the wild, cultivation has been done only for research purposes.

Crustaceans are cultured commercially as food for man and as food for other organisms, such as fishes, as bait for fishing, for restocking of natural habitats, or for the ornamental or pet trade. Culture of crustaceans for human food probably predates any other culture purpose. The high value, rapid growth rate, and wide availability of suitable species of shrimps and prawns are among the factors contributing to the popularity of modern shrimp farming efforts. Growth of the shrimp fishing industry over the last few decades has contributed to both the increasing demand and to the decreasing supply relative to that demand. As fishable populations are exploited to their maximum in one major ground after another, the potential supply from the fishing industry appears to be unable to meet the demand. As is true for many other types of fish and other seafood, producers have begun turning to artificial culture to help meet the demand.

General principles of aquaculture and an overview of general practices for various groups of marine and freshwater species were summarized by Bardach et al. (1972) and more recently by Wickins (1982). Development of aquaculture techniques, as well as the industry itself, has accelerated rapidly in the decade since the Bardach data were compiled, but many of the basic facts and principles as presented by those authors are still valid. All crustaceans reared for food are members of the order Decapoda, the most important of these being the larger shrimps, lobsters, and crabs. Although marine crustacean fisheries provide some 1.3 million MT of harvested product (representing about 2% of total fishery products by weight) and a very high percentage by value (Pillay, 1979), crustacean aquaculture is relatively insignificant on a global basis (Pillay, 1979). Total worldwide aquaculture of fish and shell fish doubled in the last one-half of the 1970s to about 10% of total fishery landings. Prospects are good for another doubling by 1985 (Glude, 1978). In the last few years, shrimp and prawn farming has exploded. In Ecuador, where the Incas were farming shrimp 400 years ago, shrimp production from farming jumped from 13% of the total shrimp production in 1976 to 75% by 1981 (Rosenberry, 1983, p. 5), so we may expect crustacean culture to increase markedly during the coming decade.

Rearing of shrimps and crayfishes for bait purposes is at present restricted to a few small projects (Avault, 1973; Huner, 1979). The species and methods selected usually permit marketing of a smaller, younger size than in food products. Moreover, it is generally true that species reared for a nonfood purpose will bring a higher price, per marketed unit, than a comparable item sold for the mass food market. Hence, the potential production or value per unit time may be greater.

Large-scale rearing of crustaceans for stocking of natural habitats has been practiced for shrimps, crabs, lobsters, and crayfishes, and in most instances, they have been used for the purpose of replenishing or supplementing overfished populations. Except for crayfish, evidence of the effectiveness of this procedure is scant, but economic analysis of such stocking on an artificial tideland showed a profit equal to nearly one-half the gross revenues (Kurata, 1981). Stocking of large, fenced, natural areas with shrimp postlarvae or juveniles is also practiced with good result (Kittaka, 1981).

Deliberate introduction and establishment of crustacean species in new areas also have been successful. The large freshwater caridean, *Macrobrachium lar*, was introduced to the Hawaiian Islands by the liberation of adults, which then produced larvae capable of surviving marine salinities and colonizing other islands in the chain (Goodwin et al., 1977, p. 198). North American crayfishes have been introduced successfully in Japan, Hawaii, Kenya, Sudan, Uganda, and Spain, as well as in numerous European localities (Huner, 1977).

Crustaceans are reared as food for other organisms in the ornamental fish industry and also in many commercial and research aquaculture projects.

The Branchiopoda is the group of greatest utility for this purpose, with *Artemia* being the most important for larvae of freshwater and saltwater fishes and for other crustaceans. Deliberate fertilization of culture ponds to promote zooplankton is a form of indirect, mixed culture. Mass scale rearing of cladocerans in sewage or waste lagoons for the multiple purpose of water improvement and fish food production is widely practiced (Norman et al., 1979).

In the ornamental or pet trade, marine carideans from tropical habitats and a few species of crabs are marketed directly as aquarium specimens in small numbers, but at moderately high prices. Perhaps the most exploited species in recent years has been terrestrial hermit crabs of the genus *Coenobita*. The popular appeal of hermit crabs and the biological characteristics of the land hermit in particular generally make it well suited for the pet trade. The scale of the industry in the United States has caused alarm among environmental groups in Florida, where overcollecting may deplete the natural populations. Few of the crustacean species marketed for the pet or ornamental trade are currently being reared commercially, but there are several factors in favor of culture in the near future. Restrictions on collecting or exporting or both have been instituted by several governments of the former collecting areas. As with many other animals (and plants) used to supply domestic consumer needs, the wild populations frequently cannot withstand the pressure of intensive collection or fishing, and ultimately, if the supply is to meet demand, commercial scale culture will have to be a major source, as it is already for the freshwater tropical fish industry.

II. CULTURE SYSTEMS AND THEIR MANAGEMENT

Kinne (1977, p. 580) listed the principal factors for breakthrough in aquatic animal cultivation as (1) knowledge of environmental and nutritive requirements of the organisms studied, (2) timing of changes in environment and nutrition concurrent with changes in growth or developmental stage or physiological state, or all of these factors (3) management of water quality (4) design of culture system (5) assessment and control of system's carrying capacity, (6) avoidance or counteraction of disease, and (7) reconstruction of essential ecosystem characteristics regarding the flow of energy and matter.

Kinne (1976) has given an excellent review of the major types of culture systems used for marine organisms, and nearly all of this is applicable to freshwater, as well as to saltwater, crustacean culture. Physical systems range in size from depression slides or watch glasses to ponds and lagoons. General categories of systems, as defined by the nature of the water supply or movement, are open, semiopen, and closed.

In open systems, the water supply is essentially continuous, and little or no effort is expended to maintain water quality, because the incoming water must be of adequate quality initially. The culture medium is discharged, usually without treatment, after a minimal residence time in the system. Advantages of open systems are the potentially rapid flow rates, that can flush away metabolites from the culture organisms, and limited expense in maintenance of water quality. Disadvantages include the possible introduction of contaminants in water supply of uncertain and usually highly variable quality and the increasing limitations on discharge of environmentally undesirable effluents. For commercial scale culture, the advantages may outweigh the disadvantages of open systems, except in special circumstances. For crustacean culture, there is no free food supply in the incoming water as for oyster culture.

Semiopen or closed systems are those in which there is some modification of the water quality during capture time and usually some amount of recycling before discharge. As a reuse of water approaches 100%, the classification approaches that of a closed system. Closed systems are those in which a very high proportion of the culture medium is retained and recycled, incoming water consisting only of that necessary to make up for evaporation or very limited discharge losses, or both.

The range of enclosure types used in commercial cultivation is of necessity different from that used in research cultivation. The smallest systems or containers are for larval production, and even those may range from a few 10s of liters to thousands of liters. Semiintensive systems include ponds in which monoculture is practiced, and these range in area from much less than a hectare to 10s of hectares. At present, closed systems for crustacean culture are mainly experimental or pilot scale, and intensive systems actually putting out a marketed product are few. A number of firms are in the process of planning or constructing large-scale closed systems for shrimp and prawn production (Rosenberry, 1982). The subject of use of recirculated water systems has been thoroughly reviewed by Muir (1982) for aquaculture generally and by Wickins (1982) for crustaceans in particular.

Natural waters may be drawn from surface supplies or from wells. Surface waters are frequently contaminated by agricultural or industrial chemicals inimical to crustaceans. For example, many chemical insecticides used extensively in agriculture for control of insect pests are also capable of killing crustaceans, even in very dilute concentrations. Biological contaminants may include spores, eggs, or fry of crustacean disease organisms, or predators. In some types of shrimp farming in Asia, for example, culture ponds are filled with tidal water flowing through nets of various meshes to exclude large predators, but the eggs and fry of predatory fishes enter the ponds through the finest practical netting and quickly grow in size and number so as to affect seriously the survival of the cultured prawns. Water drawn from wells, on the other hand, is usually free of biological contaminants and hence more desirable for culture if obtainable in adequate volume. Well water may occasionally be low in oxygen content; but, if it is otherwise suitable, restoration of oxygen by simple mechanical means is feasible. Water for larval rearing must be of higher quality than that used for pond culture. In general, high quality of water means freedom from nitrogenous wastes, as well as other contaminants. If well water is not available, surface water may be filtered and otherwise treated to render it usable; but this is practical only for hatchery purposes in most instances, as the volume required for pond culture precludes extensive treatment of incoming water.

Within moments after entering a culture system, particularly one in which animals are being grown or maintained, captive water will begin to undergo various physical and chemical changes that will render it unsuitable for culture use, if these changes are not controlled. Among the physical changes are an increase in turbidity caused by particles of food, excrement, and other detritus produced in the culture vessels. In open ponds or other systems exposed to light, turbidity from phytoplankton production is almost certain, unless the system is open and has a relatively fast turnover rate. Other physical changes include possible variations in temperature as incoming water heats up or cools off, and variations in salinity caused by rainfall or evaporation.

Chemical changes are of two basic types, those caused by basic metabolism and those induced by ectocrine production. Metabolic processes will result in ammonia excretion, which may cause reduced growth or survival unless the ammonia is converted through the nitrite cycle to nitrate or is taken up by phytoplankton, or both. Though numerous studies of effects of ammonia on other aquatic organisms have been published, only recently has this factor been investigated for crustaceans (Delistraty et al., 1977; Armstrong et al., 1978; Wickins, 1976). Nitrite may depress larval growth in Macrobrachium rosenbergii at levels of only 1.8 mg/liter (Armstrong et al., 1976) and may kill adults in 3–4 weeks at levels of 15.4 mg/liter (Wickens, 1976); but its accumulation is normally prevented by bacterial conversion to nitrate, which is apparently harmless to crustaceans, even in very high concentration (Hinsman, 1977; A. J. Provenzano, unpublished).

Production of carbon dioxide and the consequent lowering of pH tends to decrease the toxicity of ammonia, because lower pH shifts the equilibrium towards the less toxic ionized ammonium NH_{4+} (Armstrong et al., 1978). However, the carbon dioxide itself is detrimental, and aeration, in addition to supplying oxygen, tends to remove carbon dioxide.

Reduced amounts of oxygen as a result of high biomass or organic decomposition in ponds, or both, is another commonly encountered problem and is perhaps the most frequent cause of large-scale mortality in crustacean culture. Feeding by *Penaeus japonicus* may be depressed at oxygen levels below 4 ppm (Liao, 1969).

Ectocrines may be of several types, but the most important for commercial crustacean culture may be those that tend to stimulate reproductive behavior or signal molting processes. Growth-inhibiting ectocrines, which are well known for fin fishes, are unproven for commercially cultivated crustaceans, though reduced growth under crowded conditions is well-documented for several species.

Maintenance of water quality is essential for both open and closed systems. Water supply for open systems may be filtered, aerated, heated, cooled, or left untreated, depending upon the quality of the incoming water and the requirements for culture. Unfiltered water may be used for rearing organisms that are capable of filtering phytoplankton, as, for example, bivalve mollusks; but for crustacean culture, unfiltered and untreated water usually means introduction of harmful biological contaminants. Methods for filtration of gross, minute, and intermediate particles were well described by Kinne (1976), Wheaton (1978), and Muir (1982).

The essential character of closed systems calls for water treatment as a condition for successful culture. Principal factors to be controlled are oxygen level, temperature, salinity where applicable, concentration of nitrogenous wastes and dissolved organics, and occasionally numbers of pathogenic organisms. Kinne (1976) and Muir (1982) have reviewed methods of biological, mechanical, and physiochemical water treatment, as applied in aquatic animal culture.

Other than gross mechanical filtration, the most common treatment is oxygenation. Compressed air introduced by piping, mechanical aeration, pumping and spraying water, "tumbling," and physical agitation of surface water by paddling machines are all practical methods. Under conditions of exceptionally high biomass in ponds, early morning oxygen depletion may be relieved temporarily by the addition of potassium permangante (Hughes, 1971) or by mechanical aeration, but the underlying cause of such depletion is too much organic load in a heavily stocked pond. The condition can be corrected only by reducing the biomass of the cultured crop or by flushing the pond with clean water to reduce the total organic matter in the water or by both methods.

Mechanical treatment is applied for temperature control and for removal of particulates. Though turbidity per se seldom is a problem for cultured crustaceans, failure to control particulate matter may result in local accumulation and decomposition of food and waste materials, resulting in deterioration of water quality. Sedimentation and mechanical filtration are the most frequently used methods for turbidity control.

Biological water treatment is essentially concerned with the removal of ammonia and nitrite by bacteria of the genera *Nitrobacter* and *Nitrosa-*

monas. This removal of poisonous nitrogenous wastes, or their conversion to relatively harmless nitrate, is the key factor in control of water quality for aquaculture systems. Algae have been used for improving water quality for larval rearing (Manzi *et al.,* 1977). It is known that some algae take up ammonia directly and, hence, help to detoxify the media, but there remains the possibility that beneficial effects may result from ectocrine excretion (e.g., production of antibiotics) or supplemental nutrition even though of minor caloric significance. Nevertheless, according to hatchery workers, there is no clear-cut advantage of green water (phytoplankton) techniques over clear water methods, so long as the quality of management is sufficient (New, 1982).

The physicochemical treatment of water serves to remove excess dissolved organic matter from the system. Activated carbon contractors, as well as foam separation methods (Kinne, 1976), serve this function. Aeration is an essential treatment of all closed systems, as provision of oxygen and removal of excess CO_2 must be accomplished.

Chemical content of culture water may affect culture of crustaceans. For example, water hardness, especially calcium ion levels, even in some natural waters may adversely affect crayfish and *Macrobrachium.* For marine forms, the maintenance of optimum salinity is accomplished by partial water changes or by addition of freshwater to compensate for evaporation or by both methods.

Disease control in small systems may be facilitated by UV treatment of the circulating water and by gross filtration to remove particulate matter before the biological filtration. Chemical prophylaxis is possible for many bacterial, fungal, and protozoeal diseases of crustaceans, but addition of chemicals to the culture system may destroy the beneficial flora of the biological filter as well (Levine and Meade, 1976). Accordingly, the crop should be removed for treatment, a sterile filter system should be made ready for replacement, or temporary use of an alternative technique for removal of ammonia may be practical (Johnson and Sieburth, 1974).

III. CRITERIA FOR AQUACULTURE SPECIES: CHARACTERISTICS OF CRUSTACEANS

General requirements for species selection for aquaculture have been presented by Bardach *et al.* (1972). Aside from obvious qualities, such as size and acceptability in the marketplace, desirable biological features include reproductive habits amenable to control or manipulation, hardy larval stages, acceptable feeding habits or requirements, and adaptability to crowding.

As with fishes and some other cultured organisms, the crustaceans include (1) species with numerous small eggs and a long and delicate larval development and (2) other species with a few large eggs, from which the young hatch either in an advanced larval stage or as a postlarva, this situation being analogous to that of live-bearing species of fishes. The mode of reproduction is extremely important, for fecundity and hatchery technology are affected thereby. Examples of species with abbreviated development include the freshwater crayfishes and some brachyuran crabs that hatch as postlarvae. At the other extreme are the penaeid shrimps, with hundreds of thousands of small eggs that yield nauplii, and the palinurid lobsters, which hatch as relatively long-lived phyllosoma larvae.

Reproductive behavior, age to maturity, fecundity, size of the egg, and self-sufficiency of the larva all affect the feasibility of breeding in captivity. Although complete control over the life cycle is not an absolute requirement of aquaculture, it is a practical necessity for large-scale, long-term, economically successful propagation. The restrictions on production imposed by lack of control over seed supply ordinarily will inhibit growth of the industry. Obtaining seed stock from the wild is politically, ecologically, and economically undesireable, and the irregularity of natural recruitment is a factor that cannot be tolerated by a capital-intensive industry.

Although mode of reproduction is very important in the selection of species for culture, it is never the only consideration. For example, there are species of fishes that have delicate marine larvae, which are difficult or impossible to rear on a large scale with present technology; yet commercial culture of some species (mullet, eels, and milkfish) is conducted on a vast scale. The source of fry is wild stocks. Similarly, some crustaceans are cultured by capture of wild seed stocks. These include penaeid shrimps in Asia and South America and carideans in India. Thus, given an adequate supply of wild fingerlings or fry, large scale culture is possible; but, without complete control over the life cycle, genetic improvement of stock is impossible and production may fluctuate from year to year as a function of the erratic availability of wild seed stock. In general, when life cycle control of a culture species is complete, its large-scale culture flourishes. When the seed supply is limited to wild stock, culture may be practiced, but the industry is seriously handicapped.

One of the principal attractive features of the giant freshwater prawn *Macrobrachium rosenbergii* is that it can be bred in captivity with a minimum of technique and trouble. In the last decade, a beginning has been made in developing genetic information about various stocks of this widely distributed caridean. It has been successfully hybridized with another species, suggesting possibilities for enhancement of gene pools for selection (Sankolli *et al.*, 1982). On the other hand, penaeid shrimp culture, the technique for which was developed in the late 1940s, has been dependent until very recently entirely upon the capture of wild, mated, and ripe females transported to the laboratory or hatchery for spawning. These females had mated in the field with males of unknown quality, and hence, selection and genetic improvement of stocks has been impossible. Breeding of these shrimps in captivity has been achieved, and penaeids can be routinely mated and spawned in captivity on a large scale (see Chapter 5 of this volume).

Although most crabs can be mated under controlled conditions and their larvae reared, for species such as the spiny lobsters, artificial propagation appears to be some years away. Few species with phyllosoma larvae have been reared in the laboratory through all larval stages, though there are exceptions. The palinurids seem to have a long planktonic life, and hence the most probable source of culturable young is likely to be juveniles or postlarvae captured at particularly suitable sites. Legal and political problems often accompany the taking of wild seed for artificial culture, but government sponsored or cooperative projects, or both, could avoid this problem.

Time to sexual maturity and frequency of spawning are additional reproductive factors that vary considerably from one taxon to another. Extremes are represented by *Homarus,* which may take in excess of 2–5 years to reach sexual maturity and then may require many months, even more than 1 year, from mating to laying and hatching of eggs. *Macrobrachium,* on the other hand, may mature in 4–6 months, will hatch eggs approximately 3 weeks after mating, and can produce multiple broods in a single season.

Fecundity may vary greatly even within genera. Small eggs normally mean a delicate early stage larva and usually a protracted larval life. These eggs require more sophisticated and lengthy hatching technology than for species that have relatively large, lecithotrophic eggs from which a relatively advanced, hardy larva may hatch, thereby requiring less hatchery expertise. Again, extreme examples are the spiny lobsters, which have defied hatchery production, and astacid crayfish, which hatch directly as nonpelagic juveniles.

Feeding habits are of great significance in species selection. In all forms of animal husbandry, herbivores are preferred over carnivores. Feeding close to the base of the food pyramid permits larger crops at less cost, whereas by their ecological nature, predators are energy expensive to produce and frequently require protection from one another. Among crustaceans, as in many other aquatic and marine groups, filter feeders and detritivores (such as *Artemia* and *Penaeus*) are less difficult to feed, house, and produce than are carnivores, such as lobsters and most crabs.

Molting and cannibalism are two intimately associated phenomena in crustacean culture. Though molting is a normal process, it increases the probability of cannibalism. Cannibalism is common among most crabs, lobsters, and many shrimps, but not among filter feeding crustaceans, such as branchiopods. It is a primary obstacle to culture of the homarid lobsters and most crabs, though some poorly studied species may be exceptions (Brownell *et al.*, 1977). Attempts to reduce cannibalism include using special habitats or shelters, developing complete diets, providing open spaces for refuge of molting animals, and increasing turbidity to reduce interindividual aggression. Certain molt stages may make an individual more susceptible to being attacked.

Adaptability to crowding may involve many aspects of crustacean biology other than feeding habits. Social behavior, intraspecific aggression, and resistance to disease, and other factors vary greatly among crustaceans, of even relatively closely related taxa. For example, the large prawn *Macrobrachium carcinus* tends to be solitary, more aggressive, and cannibalistic, and hence, it is less suited to crowding than is its equally large congener *M. rosenbergii*. Additional examples may be found among penaeid shrimps.

Density-dependent behaviors are closely linked to the subject of cannibalism among crustaceans. In other aquatic animals, such as fin fishes and filter-feeding mollusks, space requirements per individual can be very limited, so long as there is adequate oxygenation and food supply, and such species can be cultured with individuals in continuous physical contact. With crayfishes, as density increases, aggressive behavior may increase to a point, then decrease again as the interindividual behavior breaks down (Bovbjerg and Stephen, 1975). Mock *et al.* (1973) have shown that providing abundant food and a continuous current for orientation will result in high survival in *Penaeus*. In ponds, total production of *Macrobrachium rosenbergii* may increase as the stocking rate of juveniles reaches 20/m^2, but the increase is at the expense of lower survival, lower growth, and smaller mean size (Willis and Berrigan, 1977). In intensive culture of postlarvae and juveniles, densities higher than 1000/m^2 may yield survival high enough for practical stocking or nursery production prior to pond stocking (Sandifer and Smith, 1977).

IV. SEED SUPPLIES

The problem of a source of juveniles for crustacean culture is solved in various ways according to the biology of the taxon, the site for the culture, and the state of hatchery technology for the type of organism involved. Freshwater crayfishes offer one example of the simplest kind of reproduction. Adults stocked in ponds breed without assistance from man; the females carry masses of large eggs, which hatch as virtually independent juveniles. Thus, no hatchery technology or procedure is involved. It is, nevertheless, possible to breed selected individuals simply by isolating them together, so controlled matings are technically feasible.

Another type of nonhatchery seedling supply is wild stock, as in the centuries-old culture of penaeid prawns in Asian ponds. Here the postlarvae and juveniles of the desired species as well as an uncontrolled assortment of competitors and predators are admitted on flood tides through screened influent weirs. The juvenile prawns then are retained in the pond by placing a fine screen at the opening, which allows some water to leave the pond but which retains the young shrimps and fishes. The mixed crop is then simply retained, or minimal management is applied, and the prawns are harvested on outgoing tides some weeks or months later. By empirical means, the times of year and phases of the moon associated with peak abundances of larvae are known, and, hence, the ponds are allowed to flood only at these selected times. Some species of prawns, such as *M. rosenbergii,* migrate up rivers and are concentrated at the foot of dams, where they may be captured in large enough numbers for stocking in ponds.

These relatively unsophisticated methods of obtaining seedlings have numerous disadvantages, and large scale industrial operations cannot be established safely with such methods. Thus, it is necessary to select for culture those species amenable to controlled reproduction. A few crustaceans, such as the homarid lobsters, that have a relatively brief and uncomplicated larval development, can be reared easily, for the larvae are large and robust and will take prepared foods. Other species such as the penaeid prawns have more complicated larval series and require a greater variety of mostly natural feeds, such as phyto- and zooplankton, that may be supplemented with prepared foods for the later larval stages. Hatchery technology for crustacean culture, as opposed to laboratory scale experimentation, has developed only within the last 2 decades, and for the first 10 years was restricted essentially to homarid lobsters and Japanese penaeid shrimps. Over the last decade, the rapid development of aquaculture in general and crustacean rearing technology in particular has made large-scale production of crustacean seedlings of a wide variety of species feasible.

Obtaining larvae from captive breeders during the normal breeding season is seldom a problem, but most hatchery operators prefer to control reproduction at all times during the year. Technology for out of season spawning of crustaceans is not as well developed as for molluscs and fin fishes, but some common principles have been applied successfully to a variety of species. Manipulation of photoperiod and temperature in combination with adequate diet is often sufficient to induce spawning out of season. Little (1968) showed that long daylight, coupled with high temperatures, was sufficient to induce winter spawning in *Palaemonetes pugio,* normally a summer spawner. The same general principles have been applied to other palaemonid shrimps, such as *Macrobrachium,* as well as to a variety of other decapods. For example, Sulkin *et al.* (1976) induced winter spawning of the blue crab, *Callinectes sapidus,* by elevating temperature, feeding the crabs, and allowing the photoperiod to simulate winter conditions.

With the exception of branchiopod culture, which is conducted primarily either in large salterns or in large ponds for fish food production, the technology for producing postlarval crustaceans is generally similar for a broad range of species. Nearly all decapods are carnivorous in the larval stages, the most well-known exceptions being the early postnaupliar stages of penaeids. Additional species (e.g., noncommercial caridean shrimps, and some crabs) have been reported to develop wholly or partly on phytoplankton. Larvae of *Cancer magister* have been reared to the fifth stage on a diet consisting of two species of diatoms (Hartman and Letterman, 1978). Nevertheless, an animal protein diet seems to be essential for larvae of most decapods of current commercial interest. Fish eggs, shredded fish flesh, and other fresh foods have been used to supplement the diet for later larval stages of some species.

Among the most pressing problems in hatchery operation or in the production of decapod juveniles in hatcheries is the need for a suitable live or moving food particle. At present, *Artemia* nauplii are used extensively in commercial and research operations. Other live foods, such as rotifers, are usable in part for some species (Sulkin and Epifanio, 1975). Artificially compounded flake diets have been developed more recently and show considerable promise, particularly when used as a supplement with live foods. Freeze-dried ground oyster meat, when supplemented with live *Artemia,* has given 78% survival to metamorphosis of *M. rosenbergii* (Murai and Andrews, 1978).

Unfortunately, little is known about the qualitative food requirements of decapod larvae, and not much is known about quantitative requirements, although some publications give clues for a few decapods (Sulkin and Epifanio, 1975; Sulkin and Norman, 1976; Provenzano and Goy, 1976), and empirical methods have been developed for others.

Various approaches to maintenance of water quality for hatchery operations have been used successfully. These range from seminatural "green water" techniques to filtered, closed, or semiclosed systems. Development of more efficient hatchery techniques is probably not a major constraint for penaeids and carideans, but may be necessary for species not yet being reared commercially. For example, the phytoplankton requirement of penaeids no longer includes elaborate timing procedures of simultaneous phytoplankton stock cultures. The phytoplankton can be grown in advance, concentrated, and frozen in predetermined pack sizes for later use (Brown, 1972; Mock, 1974).

V. GROWOUT

Extensive culture involves relatively large areas that are stocked at relatively low population densities. Extensive culture may involve a single species of commercial interest (monoculture) or several species in the same space and time (polyculture). The advantages of monoculture include simplified managment of culture conditions for a single species, including especially harvesting procedures. This method often, but not always, involves artificial feeding, and, as the population density of the stock is increased, extensive monoculture approaches intensive monoculture. Because it is the interests of the culturist to obtain the maximum return for his time and capital, he is usually trying to crowd the largest possible crop into the available facilities. As the population density is increased, however, the problems of water quality and disease are intensified, so the culturist is operating in a balance between too much stock and too little. Disadvantages of monoculture, as in agriculture, include reliance on a single crop that may be subject to failure under the conditions of any particular year, and the generally less-than-maximum efficiency of utilization of all the food energy naturally available within the extensive system under conditions of monoculture. Traditional crayfish farming in Louisiana is an example of extensive crustacean monoculture.

Extensive polyculture, on the other hand, offers the possibility of stocking compatible species that are capable of using different food resources within the system. Frequently, polyculture systems require little or no supplemental or artificial feeding other than fertilizers. Additional advantages include the production of crops with different markets and possibly different responses to environmental fluctuations, and hence a more stable income for the culturist. Numerous crustaceans are grown in polyculture with other organisms, for example, penaeids with milkfish or mullet and even with pompano (Tatum and Trimble, 1978). Polyculture systems have been proposed for *Homarus* in combination with shell fish, fin fishes, and other species, and the large brachyuran *Scylla serrata* is polycultured with fish in Asia (Bardach *et al.,* 1972, p. 668).

The degree to which the system is managed in extensive culture is largely a function of the intensity of stocking, whether monoculture or polyculture. Thus, a lightly stocked pond may require little or no manipulation between time of stock and time of harvest in an extensive polyculture operation, whereas a relatively intensively stocked pond may require daily monitoring of water quality and a feeding schedule of at least once per day.

Problem areas in extensive polyculture relate to preventing the introduction of predators or competitors, whereas in monoculture water quality control, particularly maintenance of adequate aeration, becomes critical as stocking densities approach the maximum carrying capacity of the system. In addition, proper estimates of biomass of the crop are difficult but essential for calculation of feeding rates, especially since feeding is a major portion of operating costs (Huang et al., 1976). If population estimates based on initial stocking rate, growth, and estimated mortality are low, insufficient food will be provided, and the crop will not reach market size in the projected time. If biomass or population estimates are too high, then excess feed will be put into the pond, resulting in a direct waste of funds, and unconsumed food will cause deterioration of water quality.

Intensive culture refers to procedures and systems in which the crop, almost always a monoculture, is heavily stocked and is usually completely or almost completely dependent upon provisional feeding, the addition of food directly to the system. The major advantage of intensive culture is that it has the highest production per unit space and the maximum control, in theory, over the animal and the environment. Water quality control is always essential, and it is most critical when intensive culture approaches or becomes completely closed. Heating becomes feasible, and harvesting becomes a simple and low-cost part of the operation. A major problem with all intensive culture systems is that, should disease strike, it can spread very rapidly thoughout the crop. The intensive monitoring of the system and the high degree of control over the operation may make it possible to observe and treat such disease much more quickly and effectively than might be possible in a more natural environment. Very intensive monoculture of some fish species has been achieved, and yields per unit area on the order of many thousands of kg/ha are possible because of the feasibility of using the vertical dimension within the system for fin fishes. Unfortunately, crustaceans, because of their tendency towards cannibalism and the apparent requirement of most of them for isolation during the molting process, cannot be stacked in the same way as can fish. Nevertheless, much progress has been made in the development of intensive, closed systems for crustacean culture, especially for lobsters, prawns, and shrimps (for review, see Van Olst et al., 1977; Mahler et al., 1974). For penaeids and freshwater prawns, at least, the technology may be available already for economic intensive culture. In 1982, at least three North American companies claimed to have achieved this capability (Rosenberry, 1982). These developments are almost entirely the result of work done in the last decade. Engineering solutions can be made for water quality control and for feeding systems, as well as for harvesting, and for the provision of individual rearing compartments for such species as Homarus; but the fundamental problems still to be solved are the high capital costs and the social behavior of the animals to be cultured.

REFERENCES

Armstrong, D. A., Stephenson, M. F., and Knight, A. W. (1976). Acute toxicity of nitrite to larvae of the giant Malaysian prawn, *Macrobrachium rosenbergii*. *Aquaculture* **9**, 39–46.

Armstrong, D., Chippendale, D., Knight, A. W., and Colt, J. E. (1978). Interaction of ionized and unionized ammonia on short term survival and growth of prawn larvae, *Macrobrachium rosenbergii*. *Biol. Bull. (Woods Hole, Mass.)* **154**, 15–31.

Avault, J. (1973). Crayfish farming in the United States. *In* "Freshwater Crayfish: Papers from the First International Symposium on Freshwater Crayfish, Austria, 1972" (S. Abrahamson, ed.), pp. 240–250. Student Litteratur, Lund.

Bardach, J., Ryther, J., and McLarney, W. O. (1972). "Aquaculture: The Farming and Husbandry of Freshwater and Marine Organisms." Wiley (Interscience), New York.

Bigford, T. E. (1978). Effect of several diets on survival, development, time growth of laboratory-reared spider crab, *Libinia emarginata* larvae. *Fish. Bull.* **76**, 59–64.

Bookhout, C. G., and Costlow, J. D. (1970). Nutritional effects of *Artemia* from different locations on larval development of crabs. *Helgol. Wiss. Meeresunters.* **20**, 435–442.

Bookhout, C. G., and Costlow, J. D. (1975). Effects of Mirex on the larval development of blue crabs. *Water, Air, Soil Pollut.* **4**, 113–126.

Bookhout, C. G., and Monroe, R. J. (1977). Effects of Malathion on the development of crabs. *In* "Physiological Responses of Marine Biota to Pollutants" (F. J. Vernberg, A. Calabrese, F. P. Thurberg, and W. B. Vernberg, eds.), pp. 439–459. Academic Press, New York.

Bovbjerg, R. V., and Stephen, S. L. (1975). Behavioral changes with increased density in the crayfish, *Orconectes virilis*. *In* "Freshwater Crayfish: Papers from the Second International Symposium, Baton Rouge, Louisiana, 1974" (J. Avault, ed.), pp. 429–442. Louisiana State Univ., Baton Rouge.

Brown, A. (1972). Experimental techniques for diatoms used as food for larval *Penaeus aztecus*. *Proc. Natl. Shellfish. Assoc.* **62**, 21–25.

Brownnell, W. N., Provenzano, A. J., and Martinez, M. (1977). Culture of the West Indian sponge crab (*Mithrax spinosissimus*) at Los Roques, Venezuela. *Proc. World Maricult. Soc.* **8**, 157–167.

Buchanan, D. V., Milleman, R. E., and Stewart, N. E. (1970). Effects of the insecticide Sevin on various stages of the dungeness crab *Cancer magister*. *J. Fish. Res. Board Can.* **27**, 93–104.

Delistraty, D. A., Carlberg, J. M., and Van Olst, J. C., and Ford, R. F. (1977). Ammonia toxicity in cultured larvae of the American lobster, *Homarus americanus*. *Proc. World Maricult. Soc.* **8**, 647–672.

de Pauw, N., Lauresy, P., and Morales, J. (1981). Mass cultivation of *Daphnia magna* Straus on rice bran. *Aquaculture* **25**, 141–152.

Epifanio, C. E. (1971). Effects of dieldrin in seawater on the development of two species of crab larvae, *Leptodius floridanus* and *Panopeus herbstii*. *Mar. Biol. (Berlin)* **11**, 356–362.

Felder, D. L., Martin, J. W., and Goy, J. W. (1985). Patterns in early postlarval developments of decapods. *Symp. Crustacean Growth; Crustacean Issues* **2**, 163–225.

Forward, R. B., Jr. (1977). Occurrence of a shadow response among brachyuran larvae. *Mar. Biol. (Berlin)* **39**, 331–341.

Fujita, S. (1973). Importance of zooplankton mass culture in producing marine fish seed for fish farming. *Bull. Plankton Soc. Jpn.* **20**, 49–53. (in Jpn., Engl. Abstr.).

Glude, J. B. (1978). The contribution of fisheries and aquaculture to world and U.S. food supplies. In "Drugs and Food from the Sea—Myth or Reality?" P. N. Kaul and C. J. Sindermann, eds.), pp. 235–247. Univ. of Oklahoma, Norman.

Goodwin, H. L., Hanson, J. A., Trimble, W. C., and Sandifer, P. A. (1977). Freshwater prawn farming (genus Macrobrachium) in the Western Hemisphere. A state-of-the-art review and status assessment. In "Shrimp and Prawn Farming in the Western Hemisphere." (J. A. Hanson and H. L. Goodwin, eds.), pp. 193–200. Bowden, Hutchinson and Ross, Stroudsburg, Pennsylvania.

Gore, R. H., (1985). Molting and growth in decapod larvae. Symp. Crustacean Growth, Crustacean Issues 2, 1–65.

Goy, J. W., and Provenzano, A. J. (1978). Larval development of the rare borrowing mud shrimp Naushonia crangonoides Kingsley (Decapoda: Thalassinidea; Laomediidae). Biol. Bull. (Woods Hole, Mass.) 154, 241–261.

Gurney, R. (1939). "Bibliography of the Larvae of Decapod Crustacea." Ray Society, London.

Gurney, R. (1942). "Larvae of Decapod Crustacea." Ray Society, London.

Hanaoka, H. (1973). Cultivation of three species of pelagic microcrustacean plankton. Bull. Plankton Soc. Jpn. 20, 19–29. (in Jpn., Engl. Abstr.).

Hartman, M. C., and Letterman, G. R. (1978). An evaluation of three species of diatoms as food for Cancer magister larvae. Proc. World Maricult. Soc. 9, 271–276.

Hazlett, B. A., and Provenzano, A. J. (1965). Development of behavior in laboratory reared hermit crabs. Bull. Mar. Sci. 15, 616–633.

Hinsman, C. (1977). Effect of nitrite and nitrate on the larval development of the grass shrimp Palaemonetes pugio Holthuis. Unpublished Master's Thesis, Department of Oceanography, Old Dominion Univ., Norfolk, Virginia.

Huang, W. Y., Wang, J. K., and Fujimura, T. (1976). A model for estimating prawn populations in ponds. Aquaculture 8, 57–70.

Hughes, J. S. (1971). Pond management: Key to fish farming success. Am. Fish Farmer World Aquacult. News 2, 10–15.

Huner, J. V. (1977). Introductions of the Louisiana Red Swamp crayfish, Procambarus clarkii (Girard): An update. In "Freshwater Crayfish: Papers from the Third International Symposium on Freshwater Crayfish, Kuopio, Finland, 1976" (O. V. Lindquist, ed.), pp. 193–202. Univ. of Kuopio, Finland.

Huner, J. V. (1979). Exploitation of freshwater crayfishes in North America. Fisheries 3, 2–5 and 16–19.

Ishida, M., (1967). On the culture of Penaeus japonicus zoeal stage on rice bran diet. Fukuoka Prefecture Buzen Fish. Exp. Stn. Res. Ser. Rep., 1961 pp. 88–97 (in Jpn.).

Iwasaki, H. (1973). Problems in the cultivation and mass culture of marine copepods. Bull. Plankton Soc. Jpn. 20, 72–73.

Johnson, P. W., and Sieburth, M. N. (1974). Ammonia removed by selective ion exchange, a back-up system for microbiological filters in closed-system aquaculture. Aquaculture 4, 61–68.

Kinne, O. (1976). Cultivation of marine organisms: Water quality, management and technology. "Marine Ecology" Vol. 3, Part 1, pp. 19–300. Wiley, New York.

Kinne, O. (1977). Cultivation of animals. "Marine Ecology" Vol. 3, Part 2, pp. 579–1293. Wiley, New York.

Kitajima, C. (1973). Experimental trials on mass cultures of copepods. Bull. Plankton Soc. Jpn. 20, 54–60.

Kittaka, J. (1981). Large scale production of shrimp for releasing in Japan and in the United States and the results of the releasing program at Panama City, Florida. Kuwait Bull. Mar. Sci., 1981 pp. 149–163.

Kurata, H. (1981). Shrimp fry releasing techniques in Japan with special reference to the artificial tideland. Kuwait Bull. Mar. Sci., 1981 pp. 117–147.

Levine, G., and Meade, T. L. (1976). The effects of disease treatment on nitrification in closed system aquaculture. *Proc. World Maricult. Soc.* **7,** 483–493.

Liao, I. C. (1969). Study on the feeding of Karuma prawn *Penaeus japonicus* Bate. *Tungkang Mar. Lab., Collect. Repr.* **1** (1969–1971), 17–24.

Little, G. (1968). Induced winter breeding and larval development in the shrimp, *Palaemonetes pugio* Holthuis (Caridea: Palaemonidae). Studies on decapod larval development. *Crustaceana,* Suppl. **2,** 19–26.

McConaugha, J. R. (1979). The effect of 20-hydroxyecdysone on survival and development of first and third stage *Cancer anthonii* larvae. *Gen. Comp. Endocrinol.* **37,** 421–427.

McConaugha, J. R., and Costlow, J. D. (1981). Ecdysone regulation of larval crustacean molting. *Comp. Biochem. Physiol.* **68A,** 91–93.

Mahler, L. E., Groh, J. E., and Hodges, C. W. (1974). Controlled environment aquaculture. *Proc. World Maricult. Soc.* **5,** 379–384.

Manzi, J. J., Maddox, M. B., and Sandifer, P. A. (1977). Algal supplement enhancement of *Macrobrachium rosenbergii* (De Man) larviculture. *Proc. World Maricult. Soc.* **8,** 207–223.

Mock, C. (1974). Larval culture of penaeid shrimp at the Galveston Biological Laboratory. *NOAA Tech. Rep., NMSF Circ.* No. *388,* pp. 33–40.

Mock, C., Neal, R. A., and Salser, B. A. (1974). A closed raceway for the culture of shrimp. *Proc. World Maricult. Soc.* **4,** 247–259.

Morgan, S. G., and Provenzano, A. J. (1979). Development of pelagic larvae and postlarva of *Squilla empusa* Say (Crustacea:Stomatopoda), with an assessment of larval characters within the Squillidae. *Fish. Bull.,* **77,** 61–90.

Muir, J. F. (1982). Recirculated water systems in aquaculture. *In* "Recent Advances in Aquaculture" (F. J. Muir and R. J. Roberts, eds.), pp. 87–177. Westview Press, Boulder, Colorado.

Murai, T., and Andrews, J. W. (1978). Comparison of feeds for larval stages of the giant prawn (*Macrobrachium rosenbergii*). *Proc. World Maricult. Soc.* **9,** 189–194.

New, M., ed. (1982). "Giant Prawn Farming." Developments in aquaculture and fisheries science, **10.** Elsevier, Amsterdam.

Nimmo, D. W. R., Bahner, L. H., Rigby, R. A., Sheppard, J. M., and Wilson, A. J. (1977). *Mysidopsis bahia:* An estuarine species suitable for life cycle toxicity tests to determine the effects of a pollutant. *In* "Aquatic Toxicology and Hazard Evaluation" (F. L. Mayer and J. L. Hamelink, eds.), pp. 109–116. ASTM STP 634, Am. Soc. for Testing and Materials, Philadelphia, Pennsylvania.

Norman, K. E., Blakely, J. B., and Chew, K. K. (1979). The occurrence and utilization of the cladoceran *Moina macrocopa* (Straus) in a kraft pulp mill treatment lagoon. *Proc. World Maricult. Soc.* **10,** 116–121.

Pillay, T. V. R. (1979). The status of aquaculture. *In* "Advances In Aquaculture, Papers from the 1976 FAO Technical Conference of Aquaculture, Kyoto, Japan." *FAO Fish. Rep. R.* **36,** 13.

Provenzano, A. J., and Goy, J. W. (1976). Evaluation of a sulfate lake strain of *Artemia* as a food for larvae of the grass shrimp, Palaemonetes pugio. *Aquaculture* **9,** 343–350.

Provenzano, A. J., and Manning, R. B. (1978). Studies on development of stomatopod Crustacea. II. The later larval and early postlarval stages of *Gonodactylus oerstedii* Hansen reared in the laboratory. *Bull. Mar. Sci.* **28,** 297–315.

Rice, A. L., (1980). Crab zoeal morphology and its bearing on the classification of the Brachyura. *Trans. Zool. Soc. Lang.* **35,** 271–424.

Roberts, M. H., Jr. (1975). Culture techniques for decapod crustacean larvae. *In* "Culture of Marine Invertebrate Animals" (W. L. Smith and M. H. Chanley, eds.), pp. 209–227. Plenum, New York.

Rosenberry, R. (1982). *Aquacult. Dig.* **7,** 1–24.

Rosenberry, R., ed. (1983). Special report: Shrimp farming in Ecuador. *Aquacult. Dig.* **8,** 4–7.

Sandifer, P. A., and Smith, T. I. J. (1977). Intensive rearing of postlarval Malaysian prawns, *Macrobrachium rosenbergii,* in a closed cycle nursery system. *Proc. World Maricult. Soc.* **8,** 225–235.

Sankolli, N., Shakuntalt Shenoy, Jalikal, D. R., and Dmelkar, G. B. (1982). Crossbreeding of the giant freshwater prawns, *Macrobrachium rosenbergii* (De Man) and *M. malcolmsonii* (H. Milne-Edwards). *In* "Giant Prawn Farming" (M. New, ed.), pp. 91–98. Elsevier, Amsterdam.

Sorgeloos, P., Baeza-Mesa, M., Bossyut, E., Bruggeman, E., Dobbelier, J., Versichale, D., Lavina, E., and Bernardino, A. (1980). The culture of *Artemia* on rice bran: The conversion of a waste-product into highly nutritive animal protein. *Aquaculture* **21,** 393–396.

Sulkin, S. D., and Epifanio, C. E. (1975). Comparison of rotifers and other diets for rearing early larvae of the blue crab, *Callinectes sapidus* Rathbun. *Estuarine Coastal Mar Sci.* **3,** 109–113.

Sulkin, S. D., and Norman, K. (1976). A comparison of two diets in the laboratory culture of the zoeal stages of the brachyuran crabs *Rhithropanopeus harrissii* and *Neopanope* sp. *Helgol. Wiss. Meeresunters.* **28,** 183–190.

Sulkin, S. D., Branscomb, E. S., and Miller, R. E. (1976). Induced winter spawning and culture of larvae of the blue crab, *Callinectes sapidus* Rathbun. *Aquaculture* **8,** 103–113.

Sulkin, S. D., Van Heukelem, W. F., Kelly, P., and Van Heukelem, L. (1980). The behavioral basis of larval recruitment in the rab, *Callinectes sapidus* Rathbun. A laboratory investigation of ontogenetic changes in geotoxic and barokinesis. *Biol. Bull. (Woods Hole, Mass.)* **159,** 402–417.

Tatum, W. M., and Trimble, W. C. (1978). Monoculture and polyculture pond studies with pompano (*Trachinotus carolinus*) and penaeid shrimp (*Penaeus aztecus, P. duorarum* and *P. setiferus*). *Proc. World Maricult. Soc.* **9,** 433–446.

Thompson, J. V. (1928). "Zoological Researches and Illustrations; or Natural History of Nondescript or Imperfectly Known Animals in a Series of Memoirs: Illustrated by Numerous Figures. Memoir I. On the Metamorphosis of the Crustacea, and on Zoea, Exposing Their Singular Structure and Demonstrating That They Are Not, as Has Been Supposed, a Peculiar Genus, but the Larva of Crustacea!!—with Two Plates," pp. 1–11, pls. I, II. King and Ridings, Cork.

Turk, P. E., Krejci, M. E., Won Tack Yang, (1982). A laboratory method for the culture of *Acartia tonsa* (Crustacea:Copepoda) using rice bran. *J. Aquaric. Aquat. Sci.* **3,** 25–27.

Van Olst, J. C., Carlberg, J. M., and Ford, R. F. (1977). A description of intensive culture systems for the American lobster (*Homarus americanus*) and other cannibalistic crustaceans. *Proc. World Maricult. Soc.* **8,** 271–292.

Wheaton, F. W. (1978). "Aquaculture Engineering." Wiley, New York.

Wickens, J. F. (1976). The tolerance of warm water prawns to recirculated water. *Aquaculture* **9,** 19–37.

Wickins, J. F. (1982). Opportunities for farming crustaceans in western temperate regions. *In* Recent Advances in Aquaculture, (J. F. Muir and R. J. Roberts, eds.), pp. 87–178. West View Press, Boulder, Colorado.

Willis, S. P., and Berryman, M. E. (1977). Effects of stocking size and density on growth and survival of *Macrobrachium rosenbergii* (De Man) in Ponds. *Proc. World Maricult. Soc.* **8,** 251–264.

5

Commercial Culture of Decapod Crustaceans

ANTHONY J. PROVENZANO, JR.

I. HISTORICAL DEVELOPMENT AND GENERAL INTRODUCTION

Aquaculture, the growing of aquatic organisms for food and other purposes, has a long history in some parts of the world. Attempts at carp culture

269

THE BIOLOGY OF CRUSTACEA, VOL. 10
Copyright © 1985 by Academic Press, Inc.

in China date from 2000 B.C., and oyster culture was practiced by the Romans. The origins of crustacean culture probably date back 4 or 5 centuries in the Philippines and Peru, where extensive polyculture has been practiced. In these countries, broad, shallow ponds were flooded at selected times of the month and year, and abundant postlarval penaeid shrimps were carried into the ponds, along with their numerous enemies and competitors. Screened gates then retained the shrimps and fishes which were allowed to grow almost without management. Yields were low, but so were costs after the initial pond construction.

As long as stockable seedlings were available only from the natural environment, the amount stocked and time of stocking could not be closely controlled. The factor that changed the nature of shrimp culture and that marked the beginning of the modern era of crustacean aquaculture was the work of Dr. M. Fujinaga of Japan, who reared the larvae of *Penaeus japonicus* from females spawned in the laboratory. A second major large-scale development in the technology of rearing crustaceans, as well as many fin fishes, was the discovery that the brine shrimp *Artemia* is a convenient and suitable food for a wide variety of species. Because of the storability of the cysts, which permits the production of large numbers of live, fat, small food particles on demand, this organism now occupies a key position in the culture of fin fishes, as well as most decapod crustaceans. Gurney was among the first to recognize the potential of this food, but the first remarkable application was the rearing to metamorphosis of the brachyuran blue crab, *Callinectes sapidus,* by Costlow and Bookhout (1959). From that date, hundreds of decapod species of many taxa, cold water and tropical, have been reared to metamorphosis with the aid of *Artemia*. Although other natural foods and a variety of artificially compounded diets are now known to be satisfactory for some species, *Artemia* is still a basic laboratory food for larvae of many fishes and crustaceans.

About the time of the introduction of *Artemia* for scientific, as well as commercial, crustacean culture, the life cycle of an important species of caridean shrimp came under control (Ling and Merican, 1961). During the decade of the 1960s, during which the availability of *Artemia* and early publications were promoting a flood of research efforts on the rearing of decapods for descriptive and other scientific purposes, the development of salt water aquarium hardware and water treatment technology was setting the stage for advancement in hatchery techniques that could be applied to a variety of species.

The culture of crustaceans is done primarily for one of the four following basic purposes: (1) for scientific studies on the organisms themselves, for example, their morphological or physiological changes with time and their ecological requirements; (2) for studies in which the crustaceans are being used as assay organisms; (3) for use of the crustaceans as food for other organisms, such as the culture of branchiopods for fish food; (4) or for use of the crustaceans as food for human beings.

Most crustaceans cultured, including virtually all those grown for use as human food, are members of the order Decapoda, which contains nearly all the larger crustaceans. This diverse group includes so many species as to offer unlimited possibilities for biological, ecological, and gustatory investigations. The flesh of crustaceans is almost universally prized, though there are locally important taboos, and a few people may have allergic reactions to protein components. Toxicity of a few crab species has been reported.

II. BIOLOGY OF DECAPODS AS RELATED TO CULTURE

The order Decapoda is the most diverse of the crustacean orders, and thus the variation in characteristics that makes species more or less adaptable to commercial culture is extreme. There are some general requirements for any cultured species, but a species deficient in one of the desired features may be so well-suited in other ways that the fault may be tolerated. Among the biological considerations for species selection are general size, reproductive habits, trophic level, social or behavioral characteristics, growth rates, and other habits or special requirements, including adaptability to unnatural environments.

A. Size

Virtually all species that are cultured commercially or that are being considered seriously for culture at the present time attain relatively large size. Usually, this means that the ratio of meat to waste is higher, processing costs are less, and prices are more favorable than for similar, but smaller, species or market sizes. A general trend associated with large size is rapid growth in the juvenile phase, which is very important from the practical viewpoint. For some species, the value of the crop may be close to zero for a considerable portion of the culture period, may be low to moderate as the population reaches the minimum commercially acceptable size, and may rise rapidly thereafter. This will be discussed in more detail in the section on growth characteristics.

B. Reproductive Habits

Parthenogenesis, which is common in some other crustacean groups, is believed to be nonexistent among the Decapoda. Protandrous hermaphroditism has been reported in the anomuran *Emerita* (Subramoniam, 1981) and in the caridean families Pandalidae, Hippolytidae, Crangonidae, Atyidae, and the monogeneric Campylonotidae, (Butler, 1964; Couturier-Bhaud, 1974; Yaldwyn, 1966). Evidence for hermaphroditism among penaeids is weak (Perez Farfante, 1978), but the axiid *Calocaris macandreae* (Thalassinidea) is apparently a functional protandrous hermaphrodite throughout its life (Yaldwyn, 1960).

With these few known exceptions, virtually all species of decapods are dioecious. Some show marked sexual dimorphism, which may be important commercially. For example, the males of some species, such as the xanthid stone crab, *Menippe mercenaria,* have much larger claws or a generally larger size, making them more desirable for the market. The deep-water red crab, *Geryon quinquedens,* of the western Atlantic continental shelf and slope is currently marketable only at a size that excludes nearly all females from the present catch (Wigley *et al.,* 1975).

A common factor in reduced size or growth of females of many species, relative to males, is the energy demand of gametogenesis. For some species, growth of the sexes has been shown to be equal until the onset of gonadal development, at which time the males may continue to grow relatively rapidly, but the females slow their growth as energy is diverted to gonadal development. In many brachyuran crabs, males are often larger than females, perhaps for the same reason, i.e., the energy requirements of ovarian development. Penaeids appear to be the exception in which females grow larger, if not faster, than males.

Control over the reproductive cycle is nearly imperative for successful commercial cultivation. Nevertheless, some large scale culture enterprises have succeeded without this control. For example, in Ecuador, the extraordinary development of penaeid culture that is based on low-density techniques using wild seed has resulted in production of at least 8000 metric tons (MT) per annum in 1981, with increases expected as management improves and hatcheries come into use (Sandifer, in press). As many as 60,000 ha were expected to be in production by the end of 1982 (Rosenberry, 1982), nearly all stocked from wild seed. However, the stock is essentially unselected and uncontrolled with respect to genetic background.

Though large-scale culture can occasionally be effected without control over reproduction, the benefits of genetic improvement, freedom from environmental fluctuations, and freedom from erratic availability of wild seed stock demand captive reproduction in the long run. General behavioral patterns for reproduction of most large groups of decapods are known, but variation even within families may be significant in a practical way. For example, in a review of brachyuran reproductive behavior, Hartnoll (1969) showed that although some species may have a terminal ecdysis, others, apparently closely related, may not, and so they may continue to grow after sexual maturity (see Chapter 3, Volume 2, and Chapter 2, Volume 9, of this treatise).

Patterns of reproductive behavior within other groups, such as the caridean shrimps, are sufficiently uniform as to permit some broad generalizations. For example, we may expect the female to be receptive to mating only at or after nuptial molts, which may or may not be interspersed with nonbreeding molts. Maturation and generation times depend upon a number of physiological and ecological factors. The frequency of brood production in the field may have little to do with capabilities in captive conditions. For example, some species will have only one or two broods in nature because of variations in seasonal temperature and light regimes, but under appropriate environmental manipulations, they may be induced to breed year round. Little (1968) induced winter spawning of the palaemonid grass shrimp *Palaemonetes pugio* by lengthening the photoperiod and raising temperature. This technique has been known among culturists to be effective for other species as well, and it is generally recognized as useful for many decapods. Hence, the biology of field specimens and populations is only a clue to the potential of the species and does not define what may be accomplished in culture.

Fecundity among decapods varies extensively. Within families and genera, there may be a wide range of potential. For example, within the palaemonid genus *Macrobrachium,* there are species with less than 100 eggs per brood, and others with 10s of thousands, perhaps more than 100,000 in larger females (Holthuis, 1952). Fecundity may be related to both egg size and to type of larval development. Those species with a few large eggs often have a relatively short larval development and hatch at a very advanced age, in some instances as postlarvae. For such species, propagation in captivity may be relatively simple (Dobkin, 1969; Williamson, 1968). Unfortunately, many such species of decapods with abbreviated development are of small size or are otherwise unsuitable for commercial culture.

One of the problems with cold water species in general and the homarid lobsters in particular is the long maturation time (time to first maturity and incubation time under field conditions). This may affect the fishery far more than it need affect culture. For example, in the field, *Homarus* may require 5 to 8 years to reach reproducing size; the female may not lay her eggs until 9 months after copulation, and the eggs may take as much as another 9 months to develop to hatching. Under artificially warm conditions, these excessive periods may be drastically reduced (Hughes et al., 1972), time to maturity being little more than 2 years.

At the other extreme, another widely cultured form, the caridean prawn *Macrobrachium rosenbergii,* may reach sexual maturity in as little as 4 months after metamorphosis and may be induced to breed as many as 6 times in a single year, with incubation time less than 3 weeks for each brood.

C. Trophic Levels

The suitability of a species for commercial culture is determined in large part by the trophic position of the species, that is, its food habits. In order of general desirability, cultured species may be filter feeders, detritivores, herbivores, omnivores, or predators. Those feeding at low trophic levels require less expensive food energy, and they usually present fewer problems in culture than species at the predator levels. The latter not only require expensive animal protein but also may exhibit other undesirable features, such as cannibalism, related to their general food habits.

The variety of filter feeders that support fisheries is extensive [for example, the anomuran red crab *Pleuroncodes* of the American West Coast (Kato, 1974; Lyubimova et al., 1973; Eddie, 1977), and increasingly the Antarctic krill *Euphausia superba*]. Some species of decapods are filter feeders, or at least they combine filter feeding with detritivorous habits. Among these are hermit crabs, porcellanid and galatheid crabs, and a few species of shrimps (Fryer, 1977; Hunte, 1977). However, few if any crustacean filter feeders other than branchiopods are among those species presently being considered for commercial culture.

Because of the superior characteristics of *Artemia* as a food for fishes as well as for crustaceans, and its hardiness, rapid growth, and the high prices of this food, commercial harvesting of wild populations of brine shrimp has been supplemented very recently by cultured production, both in indoor artificial tanks (Jahnig, 1977) and in outdoor pools or ponds. Helfrich (1973) suggested large-scale *Artemia* production in tropical salt lagoons, and experiments have been made in shore-based tanks on large scale production of *Artemia* fed with algae cultured by the use of nutrient-rich deep ocean water (Tobias et al., 1979). Comprehensive reviews of ecology, culture methods, and uses in aquaculture have been provided by Persoone et al. (1980). Although brine shrimps have been demonstrated frequently to be superior to compounded diets and other fresh food materials and to be capable of supporting high growth rates of commercially desirable species, such as lobsters (Shleser and Gallagher, 1974; Rosemark, 1978) and shrimps (Sick, 1976), the responsible factor(s) remains unidentified (Sandifer and Williams, 1980).

Detritivores and omnivores have much in common, and the distinction between them is perhaps artificial. Numerous commercially important species fall into this group, among them many of the penaeid shrimps and some crabs. Detritivores actually may be predators to some extent upon the very young stages of benthic organisms found in sediments, and it may be in order to obtain such small organisms that some crustaceans consume the sediments.

Predators among crustaceans may specialize in their prey, but often they do not, having instead rather broad feeding abilities and taking whatever happens to be abundant at the time. Thus, the Atlantic blue crab, *Callinectes sapidus,* may feed on fresh or decaying fish, oysters and other invertebrates, algae, and vascular plants (Darnell, 1958; Van Engel, 1958). It can hardly find an environment without suitable food, yet it is usually considered very much a predator, at least at the large sizes, and is highly aggressive towards other crabs, including members of its own species.

D. Social Characteristics

Aside from commensal species, there are essentially three important social patterns among the commercially important crustaceans. Solitary species are those in which individuals are seldom in company with others of the same species or, at least, maintain relatively great individual distances, for example, *Homarus.* Social species are those that may be found in relatively high densities and that seem to tolerate one another very well under most circumstances (*Penaeus*); schooling species are those that often or usually are found in very high densities (euphausiids, red crabs, some penaeids). The intraspecific behaviors of the species are of great concern in artificial culture where, by the very nature of the process, densities of organisms are much higher than in the natural habitat, and, hence, intraspecific aggression or the lack of it may determine the feasibility of maintaining dense populations over a major portion of the cycle. For example, there may be inhibition of growth of individuals through some social effect as well as reduction of populational biomass via selective mortality of faster growing individuals.

Other factors of concern to the culturist include the general style of life of the organism, special habitat requirements if any, and adaptability to unnatural conditions, whether crowding or variations in water quality. Unlike fin fish, most commercially interesting crustaceans (the branchiopods excepted) are benthic, rather than pelagic, in habit. This means that bottom space is a critical or limiting factor, or both, in the culture facility. Means of compensating for this include the provision of vertical or horizontal layers of netting or other artificial substrates (Smith and Sandifer, 1975; Goyert and Avault, 1979; Van Olst *et al.,* 1980; Sandifer *et al.,* 1982).

E. Growth Characteristics

Because of the molting process and the particular requirements and vulnerability of newly molted crustaceans, this group presents special problems to the culturist. For example, fin fishes may often be crowded together vertically, as well as horizontally, in ponds or tanks, so long as adequate water flow or water quality is maintained. Benthic crustaceans, on the other hand, cannot be kept so close together without physical barriers to protect the newly molted individual. Many systems have been designed in attempts to make intensive culture of crustaceans in high densities economically feasible, but these are presently of limited application in commercial production (Van Olst et al., 1977).

Moreover, at very high densities, growth may be affected in ways perhaps not related to available food or water quality. McSweeny (1977) has suggested that social interactions at very high population densities inhibit growth. Although very high biomass yield from Macrobrachium tanks can be achieved, the size of the largest individuals reared in those conditions is never close to that in less dense populations, even when food is apparently sufficient. On the other hand, Goyert and Avault (1979) report higher growth rate of crayfishes at $40/m^2$ than at $10/m^2$ and suggest that the difference may relate to change in behavior from antagonistic to passive at the higher density.

F. Nutrition

Various aspects of nutrition in crustaceans in general have been reviewed elsewhere in this series (see especially Chapter 4, Volume 5, and Chapter 3, Volume 8). Despite the efforts of recent workers to determine the nutritional requirements of various decapods in culture, the current state of knowledge is very inadequate. Nutritional aspects of crustacean aquaculture have been addressed by Provasoli (1976) and by Conklin (1980).

Although most so-called nutritional studies have been dietary studies, there have been advances in understanding basic requirements. Zein-Eldin and Meyers (1973) and New (1976a,b) have reviewed the general problems in shrimp nutrition including dietary work, and New (1980) has presented an extensive bibliography. For penaeids, many of the requirements are known (Kanazawa et al., 1970, 1971; Kitabayashi et al., 1971; Deshimaru and Shigeno, 1972; Kittaka, 1976; Forster, 1976; Wickins, 1982). Andrews et al. (1972) studied the influence of dietary proteins and energy levels on growth and survival of penaeid shrimps. Colvin and Brand (1977) have demonstrated the change in protein requirements during growth of penaeids in artificial systems.

Effects of fatty acids on growth (Fenucci et al., 1981) and maturation (Middleditch et al., 1980) have also been summarized. Most of the useful work on caridean dietary requirements has been done using Macrobrachium and Palaemon (Forster and Gabbott, 1971; Forster and Beard, 1973).

Biddle (1977) has reviewed the nutrition of freshwater prawns. When fed only formula feeds in the laboratory, *Macrobrachium* grows less rapidly than when pond-reared. This suggests the need for at least small amounts of an unidentified growth factor, perhaps normally provided by live food. Younger prawns eat more per unit weight and convert or assimilate more efficiently than do larger specimens, but dietary energy requirements are still unknown. In addition to environmental factors, such as temperature, energy requirements are affected by stage of development, the form in which the diet is presented, and the specific energy sources in the diet. Stephenson and Simmons (1976) found caloric requirements of *Macrobachium* larvae to be seven times that of juveniles, whereas Iwai (1976) showed that juvenile *M. rosenbergii* increase their efficiency of energy assimilation up to 6 months of age. Sick and Beaty (1974) found that late stage larvae of *M. rosenbergii* showed higher assimilation efficiencies on dry frozen diet than on food in flake or gel form.

Unlike fish, crustaceans generally seem unable to tolerate high levels of dietary fat. Growth can be inhibited if prawns are given more than 15% corn oil or cod liver oil (Forster and Beard, 1973), and higher omega-6 fatty acids may be responsible (Sick and Andrews, 1973). Apparently, both lobsters and prawns benefit from omega-3 acids (Castell and Covey, 1976; Joseph and Williams, 1975; Sandifer and Joseph, 1976). Total dietary lipids probably should not exceed 10% of the diet (dry weight), and neutral fat (free oil) supplements should not exceed 5 to 7%. The ratio of fat to carbohydrates is also important. Clifford and Brick (1978) showed that a fat : carbohydrate ratio of 1 : 3 or 1 : 4 results in more efficient utilization of dietary protein than does a ratio of 1 : 1 or 1 : 2, which results in excessive mobilization of protein for catabolic purposes.

The importance of minerals, vitamins, cholesterol, and fiber in crustacean diets was also reviewed by Biddle (1977). Dietary requirements for ascorbic acid has been demonstrated for several species of penaeids (Deshimaru and Kuroki, 1976; Guary et al., 1976; Lightner et al., 1977; Margarelli and Colvin, 1978). Sandifer (in press) summarized recent findings in requirements of shrimp and prawns.

Requirements of lobsters and development of artificial diets for lobsters have received much attention recently (Conklin, 1976; Conklin et al., 1977). Shleser and Gallagher (1974) and Castell and Budson (1974) studied effects of dietary protein levels. The role of lipids was examined by Castell and Covey (1976), and the requirements of juvenile lobsters for cholesterol were investigated by Castell et al. (1975). An extensive review of recent progress in nutritional requirements of lobsters and other decapods was given by Conklin (1980). A complete diet for lobsters seems to have been achieved (Conklin, 1980; D'Abramo et al., 1981).

Little is known about the nutritional needs of brachyurans in culture, partly because so few species are subject to large-scale culture attempts. Adelung and Ponat (1977) have attempted to develop an optimal diet for the green crab, *Carcinus maenas*, the needs of which may be expected to parallel those of other Portunidae. Lasser and Allen (1976) reported essential amino acid requirements for the dungeness crab, *Cancer magister*.

The nutritional problems of larvae are less well understood than are those of adults. Sick (1976; see also Sick and Beaty 1975) studied protein and amino acid requirements for *Macrobrachium* larvae. Nearly all decapod species reared in commercial quantities can and do grow well on *Artemia* nauplii, at least early in their development. For later stages, various compounded flake diets have been developed, which can be used to incorporate vitamins, attractants, specific proteins, and other nutrients and that have the advantage of long shelf life and convenient storage (Meyers and Brand, 1975; Sick, 1976). Inert diets for larval penaeids include soy cake (Hirata *et al.*, 1975), yeast (Villegas *et al.*, 1980; Watanabe, 1980), and others (Sandifer, in press). Freeze-dried foods have been used successfully for *Palaemonetes* (Sandifer and Williams, 1980).

Availability of natural foods in ponds may be increased by organic or inorganic fertilization, which results in higher plankton and benthos productivity, and this is the least expensive method of increasing food supply for cultured decapods. For relatively small scale culture operations, waste agricultural products, grains, vegetables, and the like can be provided to species such as *Macrobrachium*, and trash fish or chopped mollusks have been used for penaeids. This approach is often unsatisfactory because of the large amount of residual debris, handling costs, and difficulties of applying the large amounts of such food required, as well as inadequate supplies and composition of most such commodities. Hence, the increasing reliance on dry, artifically compounded diets, which reduce the volume of feed handled, can be applied mechanically to large areas, and can incorporate virtually any nutritional component required. The first artificial diet for large-scale caridean culture was chicken feed, used for *Macrobrachium* (Fujimura, 1966, 1972).

Once the optimal chemical composition of diets or the nutritional requirements of the species are determined, a selection of abundant, inexpensive commodities may be made to optimize the composition of the artificially compounded diet. Factors other than nutritional composition of the diet may affect the cost of feed. Thus, it may be necessary to substitute a less expensive and less desirable component for a more expensive one for commercial scale operations.

In extensive culture under seminatural conditions, supplemented feeds need not contain all essential nutrients. As the intensity of culture increases,

the diet provided must approach meeting the complete nutritional requirements of the species. Complete diets that are adequate to meet all the
nutritional needs of commercially cultured decapods in totally artificial systems have been developed for lobsters, penaeids, and *Macrobrachium*.

In addition to dietary composition, addressed by a variety of workers
according to species, other aspects of nutrition in aquaculture include the
physical form of the ration, feeding rates, and conversion efficiencies in the
culture systems.

The physical form of the artificial feed is important. Unlike fin fishes,
which can swallow large particles rapidly, crustaceans tend to nibble their
food. Thus, the particle must remain intact long enough to be found and
consumed, but must retain its attractiveness until eaten. Compressed pellets,
adequate for fish, are not suited for decapods, as they break up within minutes and the resulting detritus is largely unavailable to many species. Many
artificial foods lose significant nutrients, including vitamins and free amino
acids, through rapid leaching (Goldblath *et al.*, 1980). Use of various binders is necessary to stabilize the form of the pelleted or extruded feed long
enough to permit crustaceans to find and consume it. Such binders must
hold feed components together without excessive loss of nutrients, yet must
permit diffusion of attractants (Meyers and Zein-Eldin, 1973). Extruded diets
have been developed, which keep their form for a relatively long time.

Efficiency of artificial or supplemental diets is expressed either as percentage of feed used in production of biomass or as conversion ratio of feed to
biomass produced. Thus, a feed with an efficiency of 33% would have a
conversion value of 3 : 1, three units (dry or wet weight) of feed to each one
unit of biomass produced. Conversions as high as 2 : 1 have been achieved
for some artificial diets in tanks, and even higher conversions are possible in
ponds, where large but unquantified amounts of natural foods are also
available to the crop.

G. Diseases

Particularly those affecting fisheries and commercial culture, diseases of
crustaceans are treated elsewhere in this series (see Volume 6) and have also
been reviewed, especially as relating to lobsters, by Stewart (1980). Among
the most important factors contributing to disease in artificial culture is the
stress imposed by abnormally high population densities and the potential for
rapid spread of disease, once it does appear. Although some diseases of
freshwater crustaceans have been known for many years, only recently has
there been emphasis on diseases of marine crustaceans and knowledge of
these is still very incomplete. Diagnosis is imperative for control measures,
which may consist of chemical prophylaxis, breaking of critical life cycle

patterns for the pathogen, or other means, such as environmental manipulations. Preventive measures universally acknowledged as being important in any disease control program include especially the maintenance of adequate water quality, since stress induced by poor water quality will facilitate the onset of disease and by itself cause problems. Moderation of other stresses, such as extremes of temperature and low oxygen, is imperative. Nutrition is frequently incomplete or inadequate, especially in intensive systems, and can induce the appearance of several nutritional or deficiency diseases and also lower resistance of stock to infection.

Environmental contaminants may be a serious source of losses among cultured crustaceans. Among the most important of these are various insecticides (Couch, 1978). Couch and Nimmo (1974) have suggested interactions between chlorinated hydrocarbon contamination and the prevalence of viral disease in shrimps.

At present, there are recognized diseases for most of the major species of crustaceans being cultured, and disease may be a serious cause of mortality. Sindermann (1977), in reviewing the status of disease diagnosis and control for North American marine aquaculture, devoted one-half of his book to crustaceans. *Penaeus japonicus,* according to Korringa (1976), is relatively free from serious disease, yet in the United States there are at least 14 recognizable diseases among cultured penaeid shrimps (Sindermann, 1977, p. 10). In a recent review of infectious and noninfectious diseases of commercial penaeids, Couch (1978) has shown that viruses, bacteria, fungi, protozoans, helminths, and nematodes may cause disease. He included a summary of abiotic disease factors, such as chemical pollutants, heavy metals, and environmental stresses. A smaller number of disease agents is known for *Macrobrachium,* perhaps in part because until recently little work on this group has been done. At least 10 diseases of potential significance to the culture of blue crabs have been identified, and a like number is known for lobsters. Diseases of crayfishes and freshwater shrimps, as well as penaeids, have recently been summarized by Johnson (1975, 1977). None of these is normally so serious as to totally prevent successful culture, but for penaeids at least, disease has been ranked second only to nutritional and reproductive requirements in limiting aquacultural success (Couch, 1978).

III. CULTURE OF SHRIMPS AND PRAWNS

Species of shrimps differ greatly in their maximum size and growth rate, according to taxa and habitats, but a high percentage of marine, brackish, and freshwater habitats contain species of potential interest. The areas of major interest remain in the tropics and warm temperature zones, where the most rapidly growing species and the most favorable growing conditions exist. This has not prevented the exploration of the suitability of cold water and of slower growing species and the development of artificial environments, to permit extended growth in otherwise unsuitable areas.

The simplest and oldest methods of shrimp culture are those in which young shrimps are captured and confined in coastal ponds for some months before harvesting. Often, no supplemental food is provided, the species stocked are mixed with predators and competitors, and production per unit area per year is low. Such culture dates from at least 5 centuries ago in southeast Asia and is still practiced in many places today, where access to capital and developed technology is limited. While production is low, costs also are low, and the profit can be relatively high compared with that from other activities. Development of monoculture shrimp farming has been practical only in recent decades, as sources of stockable fry of given species have become available through development of hatchery technology.

The beginning of modern intensive shrimp farming dates from the work of a Japanese scientist, who succeeded in spawning penaeid shrimps in captivity and in rearing the larvae (Fujinaga, 1969; Hudinaga, 1942). The first efforts occurred in 1934. Twenty-five years later, in 1959, the first pilot hatchery and farm was established in Japan (Bardach et al., 1972). In the last 20 years, an explosive development of shrimp rearing activity has occurred, and today commercial shrimp farms are found in Asia, North and South America, Africa, Australia, and Europe.

The culture of shrimps in freshwater ponds became possible on a large scale when the essential requirements of brackish water for the larvae of the giant Malaysian prawn, Macrobrachium rosenbergii, were discovered (Ling and Merican, 1961; Ling, 1969). Prior to that time, this and other species could be reared only on a limited basis, in areas where large numbers of postlarvae or juveniles were accessible (George et al., 1968; Ibrahim, 1962). Today, M. rosenbergii males, which can reach 400 g, are being reared or being proposed for rearing in a large and increasing number of tropical and temperate countries. In areas where natural habitats are too cold for survival or growth year round, the prawn can be grown in heated effluents or in geothermal waters (Yee, 1971, 1972; Cohen et al., 1976; Johnson, 1979).

Despite the large number of species of shrimps known to science and the continuing interest in exploring possibilities of some of the lesser known forms, the majority of efforts at shrimp culture are being devoted to a very small number of penaeids (approximately a dozen species) and an even smaller number of carideans (M. rosenbergii and related species, plus a handful of other, mainly marine, species). Traditional methodologies for shrimp culture, including criteria for pond siting and methods for construction and harvesting, were reviewed by Bardach *et al.* (1972). Rapidly developing new techniques have been reviewed by Wickens (1982).

A major distinction between penaeids and caridean shrimps, as far as aquaculture is concerned, relates to reproduction and early development. Williamson (1968) has reviewed the significance of type of development as a factor in shrimp farming. All penaeids spawn their eggs freely into the water at some time after mating, and the eggs hatch at the primitive nauplius stage, whereas all carideans attach the eggs to the pleopods of the female immediately after spawning, where they are carried until hatching at a stage always more advanced than nauplius. This difference has profound implications for breeding programs and hatchery procedures for artificial culture. For example, until very recently, controlled mating of penaeid prawns in captivity had not been achieved. Rather, ripe females, already mated were and, in some instances still are, collected on the fishing grounds and placed into aerated containers for transport back to the hatchery. Spawning usually occurs within 24–48 hours and may occur even aboard the collecting vessel. The larvae may then be placed into rearing tanks, according to procedures well described by a number of authors (Hanson and Goodwin, 1977; Kittaka, 1977; Heinen, 1974; Bardach *et al.*, 1972).

In contrast, carideans have been mated routinely under laboratory conditions, and advance selection of particular females to mate with particular males is not difficult. A wide variety of carideans representing at least five families has been so mated in captivity. Some species have a transparent or translucent carapace, which permits viewing of the developing gonad in the female and, hence, facilitates the selection of ripe specimens. Others, such as *Macrobrachium carcinus,* because of an opaque carapace, are more difficult to stage with respect to gonad maturation, and attempted controlled matings frequently result in loss of the unripe females when they are placed with rather aggressive and cannibalistic males. Normally, a female molting in preparation for mating will be protected by an adult male, whereas at other times either sex may be attacked at the time of molt.

A. PENAEID CULTURE

Most penaeid culture is currently conducted in ponds of widely varying degrees of control and management. A few enterprises have used large natural areas, in which predators may be controlled to some degree and in which stocking from hatcheries is practiced (Anonymous, 1972). General methodology, as practiced in Japan and by many tropical American enterprises, has been summarized by Bardach *et al.* (1972), Korringa (1976), Kittaka (1977), and others. Although most production is still accomplished in large outdoor ponds, there has been considerable progress made in development of intensive culture systems for penaeids. Mock et al. (1974) and Parker et al. (1974) have developed designs for intensive culture systems. Mahler et al. (1974) and Salser et al. (1978) have reported on a totally closed, high-intensity production scale project that appears to have great promise. Such intensive systems require additional study before large-scale use can be anticipated, because problems concerning disease at high densities and adequate nutrition are not yet well understood.

Procedures for spawning penaeids under controlled conditions are now available. Recently, maturation and spawning of several penaeid species has been obtained through unilateral eyestalk ablation, a major breakthrough (Aquacop, 1975, 1976; Arnstein and Beard, 1975; Primavera, 1978; Santiago, 1977). Environmental manipulation also has been successful (Laubier-Bonichon and Laubier, 1979). Using recirculating systems, Beard and Wickens (1980) showed that Penaeus monodon may spawn multiple broods within one intermolt from a single impregnation but that a new impregnation is needed after a molt. Methods for artificial insemination are now available for penaeids (Persyn, 1977) and carideans (Sandifer and Lynn, 1980), and in vitro fertilization has also been achieved for penaeids (Clark et al., 1973) and carideans (A. V. Berg and P. A. Sandifer, in preparation).

Methodologies of rearing larval penaeids have been described in detail by Mock (1972) and by others. Hanson and Goodwin (1977) give an extensive bibliography. For all species, the general procedures are similar. The rapidly molting naupliar stages do not feed, but the early protozoeal stages feed on phytoplankton, with zooplankton becoming important in later stages. Rotifers, Artemia, minced flesh of fish or clams, and other natural foods, as well as some prepared diets that can be kept in suspension, are used until the postlarval stage, which usually occurs after 9–14 days. Several modifications of the larval rearing methods have been developed, ranging from the use of wild mixed phytoplankton and zooplankton cultures that are supplemented in later stages by Artemia, minced fish, and mollusk flesh, to more intensive and more highly controlled methods, such as those described by Mock and Murphy (1970), in which selected species of algae are grown separately and fed as needed, either fresh or frozen.

After metamorphosis, the juveniles may be maintained in ponds or nursery areas for a time, then transferred to growing ponds, but very young postlarvae may be placed directly into growout ponds, especially if these have been recently filled or cleaned and contain a minimum number of predators.

Growth rates and feeding habits do differ markedly among species of penaeids. The more primitive "white" shrimps of the genus Penaeus, subgenus *Litopenaeus,* which is represented by at least five American species, have an open thelycum and tend to lose their spermatophores easily, but generally grow very rapidly, as do certain of the closed-thelycum type, such as *P. monodon* of the Pacific. Other species, such as the Gulf of Mexico brown shrimp, *P. aztecus,* and pink shrimp, *P. duorarum,* grow more slowly.

Species of main concern vary according to geographical area. *Penaeus japonicus, P. monodon,* and a large number of other species of *Penaeus,* as well as *Metapenaeus ensis,* are popular candidates in Asia. *Penaeus stylirostris* and *P. vannamei* are most favored along the western coasts of Central America. *Penaeus setiferus* and *P. schmitti* are the two most popular endemics of the eastern American coasts. Similar species have been selected for other areas, such as Australia and southeastern South America (Boschi and Scelzo, 1974).

Despite the availability in most areas of species of penaeids suitable for pond culture, a few species have been introduced to new areas because of their superior growth characteristics or other considerations that make them apparently superior candidates for culture. For example, *P. japonicus* has been cultured in France and elsewhere outside its normal range, and *P. stylirostris* and *P. vannamei* of the eastern Pacific have been tried in western Atlantic waters. Several species from the Gulf of Mexico have been cultured experimentally in Tahiti.

Any introduction of exotic species may be controversial because of fears of ecological upset or introduction of diseases, and the literature is replete with examples of such disasters. Possible danger to native species from an escaped cultured exotic must be weighed, case by case, against the benefits to society of wider utilization of particularly desirable species for domestication. There is no excuse, however, for accidental introduction of diseases or parasites, because it is now possible to avoid moving potentially contaminated stocks by using only hatchery-produced postlarvae that can be grown in a parasite- and disease-free environment.

Major research priorities for penaeid farming (Hanson and Goodwin, 1977, pp. 112–116) include the need for better control over the reproductive process, improved commercial feeds, and better diagnosis and control of disease, all of which have received much attention and have advanced significantly over the past few years.

B. Caridean Culture

Though many caridean species have been reared from egg to metamorphosis and even through the entire life cycle, serious mass culture efforts have been restricted to a very small number of species. Commercial caridean culture today is essentially that of a single species of palaemonid shrimp, *Macrobrachium rosenbergii,* which occurs widely throughout the Indo-Pacific faunal region. The adults live in freshwater ponds and streams, but the larvae develop only in brackish water. Other species of the genus have some potential and have been cultured experimentally, but the large size, fast growth, and many other favorable features of this particular shrimp have resulted in its introduction to many countries and an explosive development of its culture during the last decade. Its culture does not compete with that of penaeids, inasmuch as it is grown primarily in freshwater, though it is capable of growing well in salinities up to 7 ppt. Moreover, the product is distinguishable from penaeids in appearance, texture, and flavor.

The review of Bardach *et al.* (1972) was prepared at the very beginning of a successful pilot culture of this species. In the intervening years, the amount of new information and the extension of culture efforts to new areas and to larger scales has been remarkable. A useful compilation of data was presented by Goodwin *et al.* (in Hanson and Goodwin, 1977). Ling (1977) and New (1982) have presented additional information concerning the culture of this prawn.

Modern freshwater prawn culture, still a young though rapidly growing activity, owes its beginning to Ling, who discovered the necessity of salt water for the survival of the larvae of this freshwater shrimp. In frustration after many failures, he added soy sauce to the culture water and found that the larvae lived longer than in any previous attempts. During the following decade, development of mass culture methods permitted the establishment of a pilot hatchery and pond growout operation in Hawaii (Fujimura, 1966, 1972; Fujimura and Okamoto, 1970), which served as a model for those in many other locations.

Approximately one-fourth of *Macrobrachium* species is found in the Americas, and several of these species have been investigated for culture (Dobkin *et al.,* 1974). Species of *Macrobrachium* other than *M. rosenbergii* have been cultured in Asia. A few American species have been studied, with respect to larval development and growth rates (Goodwin *et al.,* 1977), but none of the handful of species so far examined has been shown to be superior to *M. rosenbergii.* Fecundities, growth rates, temperature tolerances, and social interactions are factors that vary among species. *Macrobrachium ohione* may be the most cold-tolerant of the American species, occurring at least as far north as Cape Fear River, North Carolina (Holthuis, 1952). The African *M. vollenhoveni* is large and can complete its development entirely in freshwater, according to Prah (1982). Pillai and Mohamed (1973) recommended the Indian species *M. idella* for culture in brackish water. The possibility exists of combining desirable characteristics of different species, as at least two successful hybridizations have already been reported (Uno and Fujita, 1972; Sankolli *et al.,* 1982).

Macrobrachium is subject to several bacterial, fungal, and protozoan diseases (Goodwin *et al.*, 1977; Sindermann, 1977), but none of these appears to be a major problem in production. An unidentified larval mortality factor thought to be disease has appeared in some hatcheries, including the Hawaiian State hatchery, but appears to be controllable with improved sanitation.

Larval rearing techniques have advanced considerably since Fujimura's early efforts. Yields in excess of 50 postlarvae per liter have been attained (Mock *et al.*, 1978; Manzi *et al.*, 1977). These values are much higher than those reported even a few years ago. Better tank design, better water quality control, and more precisely controlled feeding procedures are principal factors in this improved production.

The various techniques for rearing of larval prawns have been described by numerous workers since Ling and have been summarized by Hanson and Goodwin (1977), and have been compared by Meneasveta and Piyatirativo-kul (1980). For most species studied, optimum salinity for larval stages appears to be in the range of 12–16 ppt. *Macrobrachium rosenbergii* has been reared at lower and higher salinities. Some species may require freshwater for larval development (*M. lanchesteri, M. amazonicum*), whereas others require brackish water. Temperatures above 26°C and not more than 30°C are most commonly used, the optimum for *M. rosenbergii* being about 28–30°C. Water may be clear, with or without filtration, and changed daily, or it may be kept green with phytoplankton (Wickens, 1972; Maddox and Manzi, 1976). The role of phytoplankton in these cultures is unclear, for although larvae are known to ingest them sometimes and although water quality improvement through absorption of nitrogenous wastes as well as production of antibacterial ectocrines and other functions of algae has been proposed, equally high production of postlarvae has been achieved without algae in the system. Indeed, Cohen *et al.* (1976) found negligible ingestion of *Tetraselmis* and *Phaeodactylum* by larvae of *M. rosenbergii*. They did find that the phytoplankton removed ammonia and hence was of indirect benefit to larvae.

Food provided for larval prawns typically includes *Artemia*, at least in the first 1 to 2 weeks of development; it is supplemented with fish flesh and fish eggs, flake foods, or other materials as the larvae increase in size. Apparently, larvae find their food by random contact early in development, but later, they may pursue individual prey items. Chironomid insect larvae have been substituted for *Artemia* (V. G. Jhingran, unpublished, in Goodwin *et al.*, 1977, p. 227), and rotifers and other substitutes for *Artemia* have been tried with some success. Duration of larval development may be affected by temperature, diet, and geographical origin of the parent stock (Malecha and Polovina, 1978).

In one production-scale hatchery, larval prawns are stocked in 800-liter conical tanks at 100 larvae/liter. Water is changed every day, without handling of the larvae or changing of temperature. Tetracyclene is added as a prophylactic. *Artemia* nauplii at a concentration of five per ml are fed once a day to the postlarvae, with additional feedings of bonito meat and frozen *Artemia*. Strong aeration keeps the food in suspension. Survival of 64% to metamorphosis, which begins at 28 days after hatching, has been achieved, with production of postlarvae therefore being approximately 54/liter.

Today, the principal growout system for *Macrobrachium* is the outdoor pond. General procedures used in Hawaii have been reviewed by Fujimoto *et al.* (in Goodwin *et al.*, 1977, p. 237). The harvesting method used (seining) requires pond characteristics (smooth flat bottom, without macrovegetation) known to be not optimal for the prawns. Nevertheless, with a stocking rate of 15–20 postlarvae per m² and continuous harvesting, production of approximately 3000 kg/ha per year is achievable. Commercial chicken feed or other artificial ration is provided at rates up to approximately 40 kg/ha/day. Much of this feed is thought to be used only indirectly by the prawns, but little work on quantification of requirements in the field has been published.

A few data regarding water and soil characteristics of successful *Macrobrachium* ponds are available (Goodwin *et al.*, 1977, p. 224). In brackish ponds, salinity of up to 7.2 ppt did not prevent some males from growing from 1 g at stocking to more than 100 g 5 months later. Cripps and Nakamura (1977) demonstrated significant (fivefold) reduction of growth of *Macrobrachium rosenbergii* in waters containing 500 ppm of $CaCO_3$ compared with that in waters containing 65 ppm.

The problems of intensive culture of *Macrobrachium* are not essentially different from those of other crustaceans. McSweeny (1977) has reviewed many of the aspects of intensive culture systems as applied to this prawn.

One of the behavioral differences between *Macrobrachium* and *Penaeus* is their response to currents. *Macrobrachium* tends to move against currents, whereas *Penaeus* tends to move with the currents, at least under some conditions. Thus, in harvesting the former, draining the pond does not result in mass movement to the outlet, as occurs with penaeids. The latter may be harvested from ponds or tanks simply by placing a net over the drain pipe or outlet weir, whereas the former tend to remain in the pond and necessitate a different approach to harvesting, as, for example, seining. In tank culture, harvesting is faster, much simpler physically, and much less stressful for the animals.

An important biological consideration in intensive culture is high mortality in heavily stocked ponds or tanks as a result of intraspecific aggression or cannibalism, or both. Various methods have been proposed to alleviate this problem, the most successful of which provides multiple layers of artificial bottom, such as netting, to increase the total available substrate within the tanks (Smith and Sandifer, 1975). Little work has been done on the nature of the behavioral interaction that leads to increased mortality under crowded conditions; but Peebles (1977) found that most corpses showed damage to sensory and locomotor systems and that this injury was associated with intraspecific aggressive behavior occurring during the postmolt period. Thus, molt-related intraspecific aggressive behavior, not cannibalism per se, is the major cause of behaviorally related mortalities.

Culture of *M. rosenbergii* in North America, originally restricted to areas where naturally or artificially warm waters exist year round, that is, southern Florida, Texas, southern California, geothermal waters, and power plant effluents, is now being tried on a seasonal basis in some southern states. Low-temperature tolerance of the North American species *M. ohione*, which ranges north on the Mississippi drainage basin to West Virginia and Illinois and on the Atlantic coast at least as far as northern North Carolina (Holthuis, 1952), has not yet led to extensive utilization of that species.

The most important group of carideans for commercial culture, other than palaemonids, appears to be the pandalids. None of these is being produced in quantities that are sufficient for marketing; but many pandalids form the basis of cold water fisheries, and some species are large. Accordingly, there is interest in developing methods for commercial culture of these alternatives to tropical species. Unfortunately, the species investigated to date seem to grow relatively slowly, though they may have compensating virtues, such as increased social compatability. In addition, this group contains members that change sex with age (Butler, 1964), offering interesting possibilities for selffertilization, if techniques for *in vitro* culture are further developed. Successful artificial insemination has been achieved for the caridean *Macrobrachium,* a fact that suggests possibilities for hybridization (Sandifer and Smith, 1979). *Pandalus platyceros,* even at optimal temperature (9.5°–21°C) grows too slowly for monoculture with currently available food and management techniques, but this species may have potential in polyculture with other species (Kelly et al., 1977; Rensel and Prentice, 1979).

A neglected aspect of shrimp culture has been the market for ornamental species. In fish culture, as well as other types of agricultural activities, species grown for nonfood purposes generally bring much higher prices per unit than the same or similar species grown for food. The tropical fish industry is based on both fishing and culture, and until recently almost all culture was limited to freshwater species.

The total contribution of nonfood crustaceans to the pet or ornamental trade is still small relative to that of fin fish, but in recent years, advances in marine aquarium keeping (especially the growing popularity of invertebrates

in saltwater tanks) preview considerable growth in the importance of ornamental species. Retail prices for some species of marine shrimp already
exceed $10 per specimen, which, even at much-reduced producer prices,
translates into hundreds of dollars per kilogram. These prices will not be
maintained for any extensively cultured species, of course, because part of
the price is attributable to the relative rarity of living specimens in the trade.

Freshwater crustaceans are rare in the aquarium trade, though a few crabs
are marketed; saltwater decapods are increasing in popularity, partly for
their brilliant colors and partly for their interesting behaviors. Among the
shrimp, stenopids, especially *Stenopus hispidus,* are most common in saltwater display tanks, because they are compatable with many fin fishes.
Other small, but colorful carideans are also popular, particularly parasite-
pickers of the families Hippolytidae and Palaemonidae. With increasing
restrictions on fishing of wild stocks and environmental problems of many
kinds, we may expect increasing emphasis on propagation of species used
in the pet and ornamental trade. Some formerly important habitats are now
legally closed to collecting, and stocks in other areas will receive increasing
pressure. It may be some time before the demand cannot be met by wild
stocks, and demonstration of sufficient market potential would seem to be
crucial for economic commercial culture of marine ornamental crustaceans.
However, there is little doubt that the technology exists and that, in the
future, this culture will become the major supply source for some species.

IV. CULTURE OF ANOMURANS

The anomurans, in general, are less important economically than are any
other decapods, but the group contains some members of major significance
to the fishing industry. The most important of these are the cold water king
crabs of the family Lithodidae. This group, which seems to have evolved
from lines within or close to the presently recognized Paguridae and its
members, have larvae similar to those of the pagurids. Many species have
been reared for identification purposes, and serious effort has been made by
the Japanese to culture them on a large scale. Nakanishi (1979) indicated
that, although larvae of the king crab *Paralithodes camtschatica* can be
reared easily to the postlarva stage in large numbers, survival in the juvenile
stages is very low (10%) with current technologies.

In general, the hermit crabs are of little direct economic importance,
although many of them serve as food for commercially important fishes, and
some are taken on an incidental basis for fish bait. Exceptions are several
members of the land hermit crab family Coenobitidae. *Birgus latro,* the
coconut crab, has long been harvested as a food item in the Pacific Islands,

where it was once widespread. Overharvesting has depleted the supply in many areas, but no efforts have yet been made to propagate it for restocking or culture. The apparently slow natural growth rate of the species mitigates against long-term intensive culture, but culture for restocking purposes may be feasible. Other members, *Coenobita* spp., of the family Coenobitidae have become important in the pet trade within the last few years, and millions of these crabs have been harvested and imported to the United States annually for this purpose. Because of the wide distribution of the species, even intensive harvesting may have an impact only on local populations such as those of the Florida Keys. Though the possibility exists for artificial propagation for restocking purposes, the economics and the desirability of the process are not established. Thus, there is at present no aquaculture for any of the species of hermit crabs.

Members of the Galatheidae, particularly species along the west coast of the Americas, are important in fisheries, and, though some galatheids have had their larval histories determined, commercial culture has not been attempted.

V. CULTURE OF CRABS

Perhaps as many as one-sixth of all crustaceans are crabs. The diversity of this group, both morphologically and ecologically, is extreme; hence, there are many potential candidates for culture. Interest in crab culture always starts with species of commercial fishing interest, not with the biologically most-suited species. Many species are fished commercially, and only a few, mainly xanthids, are known to be toxic (Garth, 1971; Holthuis, 1968; Hashimoto et al., 1969). A general overview of the biology of crabs as a group is given by Warner (1977).

Reproductively, most crabs offer little problem for artificial propagation. All species brood their eggs, multiple broods are common, and the number of young is frequently large. Some species, especially freshwater and some terrestrial forms, have abbreviated development with no free swimming larval stages, whereas other species, especially of portunids, may have as many as six larval stages before the megalopa.

Trophic position is diverse, but most species of commercial interest are carnivores or predators (Portunidae, Cancridae, Xanthidae). There are herbivores or detritivores that are of interest for culture. The spider crabs are among the algae eaters; their spoonlike fingers of the chelae indicate their habit of scraping substrates. Members of this group have seldom been considered for commercial culture, perhaps because few obtain a large size, and their abundance in the field is seldom sufficient to support commercial fishing. There are several cold water exceptions, such as those of the North Pacific that support major fisheries. However, Brownell *et al.* (1977) have shown that larval stages of tropical *Mithrax* can be reared rather easily, and some members of the group may be biologically well suited to culture.

Social suitability for intensive culture also varies. Many predaceous species, as would be expected, cannot be kept in high densities, and many socially gregarious species (such as *Uca*) are of no commercial interest.

The number of species of crab that have been reared from the egg for research purposes is relatively large, and the number of life history descriptions published in the last two decades alone greatly exceeds the total number published previously. Although crabs continue to be a major subject of research, the number of species being cultured commercially for food and for which serious culture is being attempted is small. Partly this is due to the relative recency of aquaculture development and to the ability of natural stocks to satisfy demand. Most species of commercial interest that do not live in deep water and hence might be suitable for culture in artificial environments on land or in shallow seas are aggressive predators, a major factor in the failure of most early attempts to rear these in mass cultures. Thus, crabs generally have not been regarded as top candidates for commercial culture. Nevertheless, there are some examples of successful, though low intensity, culture; and development of more sophisticated techniques, as for lobsters, may change the picture in the near future.

The brachyuran families that contain most species of current or probable future interest for food cultivation include the Cancridae, Portunidae, Majidae, and Xanthidae.

The Cancridae contains many species of major commercial interest, and a number of these have already been investigated with respect to larval rearing. *Cancer magister* of the west coast of the United States is a candidate for culture because the species attains a very large size, demand is always in excess of the apparently dwindling supply, and the price per unit is high (Reed, 1969). Its local congeners also have been reared experimentally. At present, however, no cancrid crab is cultured commercially.

The Portunidae contains some of the most gastronomically and economically attractive crabs. For many years, the Japanese have attempted to culture *Neptunus pelagicus* and *Portunus trituberculatus*. Yatsuzuka (1962) obtained 10% survival of larvae to the megalopa. Live storage of *Portunus trituberculatus* for several months in sandy bottomed impoundments is practiced, the animals being fed fresh fish and shell fish (Tamura, 1970). Larvae of both species have been reared, but mortalities are high; and the technical and economic feasibility of large-scale artificial propagation of these, as well as of many other brachyurans, is presently in doubt. According to

Bardach et al. (1972), *Portunus trituberculatus,* which reaches commercial size in about 1 year, is reared only to 2-cm size for stocking bays and inlets.

In the Indo-Pacific, the large mangrove or mud crab, *Scylla serrata,* is an important fishery species and has been cultured for some years in fish ponds (Matilda and Hill, 1980). Taiwan produced 782,000 kg in ponds in 1973 (Chen, 1976). The species can reach a weight of 0.5 kg in 1 year (Wildman, 1974). Ong (1966a) worked out the larval development and also demonstrated the feasibility of completing the life cycle, egg to egg, within 12 months (Ong, 1966b). This crab is a predator, and, in addition to eating oysters and other shellfish and fish in the ponds in which it occurs, it may damage the banks by burrowing. Nevertheless, the value is sufficiently great so that even 200 crabs/ha per year is a welcome addition to income from fish farming (Bardach et al., 1972).

The most important single species of crab in the United States is *Callinectes sapidus,* which ranges from Cape Cod south to northern South America. It is a rapidly growing species of great value, but its culture has never been seriously attempted for hard crab production, despite some early data gathering by Lunz (1957). The larvae have been reared in mass culture by W. T. Yang (personal communication), who obtained marketable adults in 6 months. The soft-shell crab industry, which is dependent upon molting crabs, may be considered a limited and primitive type of culture, but the residence time of crabs in the process is often only a matter of days and never more than a couple of weeks (Lee and Sanford, 1962). The value of the soft-shell crab is several times that of the market value of the same crab in hard-shelled form. Although this species can reach maturity and market size in much less than a year under optimal rearing conditions, the common problems of low survival, cannibalism, and high trophic level prevent commercial culture at present. Reimer and Strawn (1973) have discussed the potential use of warm water for winter production of soft-shell crabs.

Among the Majidae, few species are both large enough and abundant enough to be of commercial interest, but there are exceptions. The largest crab in the world, the Japanese spider crab (*Macrocheira kaempferi*) is known for its 3–4m arm span. Unfortunately, it lives in deep water and thus may not be suitable for culture. The spider crabs have a number of biological advantages for culture. First, many or most of them are omnivores that scrape the substrate of algae, attached sessile invertebrates, or detritus, rather than prey on active or large species of value. Secondly, all spider crabs studied to date have an abbreviated larval history, making the mass rearing to postlarvae a relatively short and simple process. Unfortunately, there is a lack of basic data on growth rates, stocking density potentials, and the like, and at present, no spider crabs are being commercially cultured.

Among the Xanthidae, several species have attracted interest, and others are likely to become candidates for culture in the future. For example, the stone crab, *Menippe mercenaria,* is of economic importance in the southeastern United States and elsewhere. Its larvae have been grown in mass culture (Yang, 1971), with survivals as high as 9% to metamorphosis; but experiments on intensive growout in ponds (Yang and Krantz, 1976) showed burrowing behavior, wide variation in growth, and high mortality to be problems. The authors suggested that extensive culture techniques may be more suitable. This species and others similar to it prey upon mollusks and other large invertebrates and hence may be expensive to raise. The possibility of rearing large numbers of postlarvae for stocking areas for extensive culture remains, since, very often, species that attain a large size grow rapidly. Under the low population density typical of extensive culture, *Menippe mercenaria* may be cultured in numbers sufficient to supply limited markets. It would be interesting to see whether the Tasmanian giant xanthid *Pseudocarcinus gigas,* which may attain a weight of close to 14 kg (Schmitt, 1965, p. 12), has biological features suitable for culture.

VI. CULTURE OF MACRURAN REPTANTIA—THE LOBSTERS AND THEIR KIN

A. General Characteristics of the Group

The group of decapods represented by the lobsters and their allies constitutes one of the most economically significant fractions of world fisheries products, commanding unit value far in excess of most other seafoods. The fisheries for lobsters and their kin are reviewed by Cobb and Wang (Chapter 2, this volume) and were recently discussed in detail by Bowen (1980) and Dow (1980). FAO statistics indicate 108,134 MT landed in 1980. The United States alone consumes more than one-half of total landings. Because the history of every lobster fishery shows overfishing and depletion of stocks and because of the constantly increasing demand for this luxury seafood, efforts have long been underway to attempt culture of these valuable animals. The biology of various lobster groups makes some of them more suited to culture than others. At the present time, despite great interest and much expenditure of funds, there are no commercial culture operations for any of the marine lobsters, although culture of freshwater relatives, the crayfishes, has been very successful.

Since late in the nineteenth century, workers have been interested in propagation of homarid lobsters, first for restocking depleted waters and more recently, for culture to commercial size to supplement the ever decreasing catches. Biologically and technically, it is possible to culture homarid lobsters from egg to market size with control over the complete life cycle, in much less time than in natural habitats. Still, obstacles, such as the carnivorous and cannibalistic tendencies of lobsters, remain.

The spiny lobsters, tropical counterparts of the north Atlantic lobsters, constitute an equally important group in the economic scene. The spiny lobsters differ from homarids in biological features of major significance for artificial culture. They are less aggressive and less cannibalistic, but control over the life cycle has been difficult to achieve because of the protracted larval life.

The freshwater lobsters, or crayfishes, not all of which, incidentally, are restricted to fresh water, are reproductively more similar to homarids than to the spiny lobsters. Their diets are often largely vegetarian, and their temperament has not prevented large scale culture, at least for some species. In the United States at present, this is the primary crustacean group supporting commercially viable culture operations.

B. The Nephropid Lobsters

Though there are numerous species in the family Nephropidae, nearly all effort to achieve commercial culture has been restricted to three species, *Nephrops norvegicus* of minor importance in western Europe and the Mediterranean, and two of the three *Homarus* spp., *H. gammarus* of western European seas, and *H. americanus* of the northwestern Atlantic. Although culture efforts on these species date from many decades ago, most of the commercially significant research has been accomplished in the last 20 years. In the late nineteenth century, hatchery techniques were developed for the purpose of augmenting the natural recruitment of young to the fisheries. Early efforts failed to provide demonstrable benefit to the fisheries (Carlson, 1954), but this work continued in Massachusetts. Renewed interest in lobster hatching and rearing has been focused mainly on the possibility of rearing hatchery-produced larvae all the way to commercial size in captivity (Hughes, 1975), though several studies have been made recently on the feasibility of introducing populations of homarid lobsters to new environments (Ghelardi and Shoop, 1972; Krekorian et al., 1974).

An exhaustive review of current biological and technological knowledge for culture of *Homarus,* along with an extensive bibliography, has been given by Van Olst et al. (1980) Among the biological characteristics of *Homarus* that are favorable for culture are its relative hardiness, short larval development, relatively large fecundity, willingness to accept a wide variety of food materials, and demonstrated ability to breed in captivity. Among the features detracting from suitability are its generally slow growth rate (at least 2 years even at optimal elevated temperatures), primarily carnivorous feeding habits, inconveniently long reproductive cycle, and, most important, extreme aggressiveness that results in cannibalism under crowded conditions unless individuals are physically separated. The hatching of eggs from captive females and the rearing of larvae to metamorphosis are simple processes, practiced for many decades, though technological developments of modern diets for larvae have greatly increased the efficiency and reliability of hatchery production of postlarval stages (Van Olst et al., 1980, p. 341). Only in the last 10–15 years has much been learned about rearing requirements of juveniles and subadults, to the extent that systems have been designed for growout phases of culture, and economic models have been worked out for evaluating the commercial feasibility of using these systems (Botsford, 1977).

One simple but highly significant advance affecting the technical feasibility of culture of Homarus has been the realization that, like other organisms, lobsters will grow fastest at temperatures that may not occur in the natural habitat. Since McLeese (1956) showed the extraordinary tolerance of Homarus americanus for temperatures ranging from 2° to 32°C, growth of lobsters in culture has been accelerated by high temperature, with the result that commercial size (454 g) can be reached in approximately 2 years from metamorphosis, as opposed to 5–8 years in the field (Hughes et al., 1972). As for many other aquatic organisms requiring year-round elevated temperatures for maximum growth, inexpensive or "waste" heat sources, such as power plant effluents, have been evaluated as to suitability, and closed system culture has been proposed as a method of conserving heat and thereby limiting costs (Van Olst et al., 1976; Hand, 1977).

Food requirements of Homarus have been investigated in detail by a series of workers, who have shown that although the primary trophic role of this lobster is carnivory (Squires, 1970), natural foods are unsuitable for large-scale culture operations. Conklin (1980) has reviewed the status of nutrition research for decapods with special reference to lobsters. Artificially compounded diets are acceptable and probably economically feasible (Gallagher et al., 1976; Conklin et al., 1977). Cholesterol and unsaturated fatty acids are required (Castell et al., 1975; Castell and Convey, 1976). Conklin et al. (1978) have reevaluated artificial diets for H. americanus. The use of living Artemia evidently produces growth superior to all or most formulated diets, and much effort has been spent on identification of the essential growth factor responsible, including lecithin and other phospholipids. A successful purified diet for lobsters has now been developed (Conklin et al., 1980; D'Abramo et al., 1981). Logan and Epifanio (1978) have determined an energy balance for larvae and juveniles.

Although homarid lobsters have been mated in captivity (Hughes and Matthiessen, 1962; Aiken and Waddy, 1976), problems related to egg production remain. Time from mating to oviposition and hatching is a function of temperature, and, in the field, ranges from 12 to 18 months. By maintaining lobsters in elevated temperatures, this has been reduced to 4–5 months (J. C. Van Olst, unpublished; Branford, 1978); but that is still a lengthy period for a commercial operation that requires large numbers of young hatched on a more-or-less regulated production schedule.

Significant intraspecific variation in growth rate among broods from different pairs has been reported (Hedgecock et al., 1976), a complicating factor in experiments, and interspecific hybridization has been achieved with H. gammarus and H. americanus (Carlberg et al., 1978). Hence, improvement of stock through hybridization and selection appears to be feasible and may result in superior growth and other characteristics, when lobster culture reaches the commercial stages.

Diseases of lobsters have been reviewed by Sindermann (1977), Van Olst et al. (1980), and elsewhere in the present series (Volume 6). Diseases in the high-density cultures, a state that is apparently necessary for economic production of reared lobsters, may be a significant factor in success or failure, but most of those should be controllable with prophylaxis and good systems management.

The principal biological feature of lobsters that dominates system design and economics of production is social behavior, especially intraspecific aggression. Communally reared lobsters exhibit high rates of mortality, due to cannibalism (Sastry and Zeitlin-Hale, 1977) that is associated with vulnerability of molting individuals and made worse by inadequate diets. This problem is common among commercially interesting decapods, and, though population densities in culture may be increased over natural population densities through provision of shelters (Ghelardi and Shoop, 1972; Van Olst et al., 1975; Aiken and Waddy, 1978), use of three dimensions in culture vessels is required to approach satisfactory densities for commercially intensive culture. Even with adequate diets, irregular growth will occur, and some mortality will result as dominant individuals defend larger and larger territories. Many recent investigators have proposed physical isolation of individuals, at least for the later phase of culture (Van Olst et al., 1980), though small containers reduce growth and survivorship (Van Olst and Carlberg, 1978).

Mitchell (1975) and Ryther (1977) have demonstrated good growth of young lobsters in a polyculture system requiring no artificial diet; but population densities in this system were much lower than in other systems, and yields of lobsters alone (2500 kg/ha per yr) are probably inadequate for commercial application.

Water quality data for Homarus, including tolerance and production rates of ammonia, have been summarized by Van Olst et al. (1980). Although

Homarus americanus can tolerate 31°C for short periods, at high temperatures its resistance to stress is reduced, and 22°C is near optimal for growth. Because of varying rates of ammonia production, which is determined by feeding rate, size of lobsters, and varying tolerances to concentration as pH fluctuates, much additional work is required to assess filtration and flow rates required in large-scale systems to minimize toxic effects of this metabolite.

Methods for artificial culture of homarid lobsters from egg to commercial size have been presented in detail and evaluated by Van Olst *et al.* (1980). Commercial-scale production is technically feasible, and, although with present systems designs and current costs, economic feasibility has not yet been demonstrated, it is being more closely approximated each year.

C. The Crayfishes

Freshwater crayfishes of the families Astacidae, Cambaridae, and Parastacidae bear considerable physical resemblance to the commercial lobsters of the family Nephropidae. Some members of the Parastacidae reach sizes comparable to those of commercial-sized lobsters of the genus *Homarus*. Freshwater crayfishes are among the most promising crustaceans for aquaculture because of their simple reproductive processes, their suitability for merely aquatic rather than marine environments, their generally mild disposition, and their largely vegetarian food habits. At present, crayfishes are the most successful commercially cultured crustaceans in North America.

Large-scale culture of crayfishes is a recent development that is stimulated by the common pattern of increasing demand and by depletion of natural supplies by overfishing, pollution, and disease. As with the aquaculture of many other organisms, primitive methods, such as simple transplanting of stock, have gradually been supplemented in some countries by hatchery production of young to increase the supply of seed and to provide disease-free animals for introduction to the field. In the United States, however, the development of farming methods apparently developed from observing the results of habitat manipulation for other purposes and the subsequent discovery of increased natural production of crayfishes (Avault, 1973). Here, hatcheries have not been necessary or critical in the development of a multimillion dollar industry. In 1980, the United States alone produced over 10,000 (MT).

FAO data for 1968 indicate only 2000 MT of commercial fishery landings of crayfish, of which at least one-half was in the Soviet Union (Brodsky, 1975). Throughout Europe, the decline of natural stocks was accelerated by the widespread effects of a disease introduced from America—first to Italy in 1860, and later spreading throughout Europe (Unestam, 1973). Over the past 50 years, other species, mainly from North America, have been introduced into European waters in the hope that their resistance to the pest or greater tolerance for deteriorated water quality, or both, would help compensate for the decline of local species, particularly *Astacus astacus*. Some of these species, such as *Orconectes limosus,* have spread rapidly because of disease resistance, rapid reproduction, and tolerance to pollution, but they have not found favor because of inferior taste and a greater difficulty in harvesting that is attributed to behavioral and ecological characters (Kossakowski and Orzechowski, 1975). Other species, particularly *Pacifastacus leniusculus,* have been more acceptable, and they now are rapidly replacing the former species in the commercial catches. The development of farming methods has proceeded independently in diverse regions, with different goals and with different approaches.

The history of the development of crayfish culture in the United States has been reviewed by Comeaux (1975). Although crayfish fisheries and some culture activity have occurred in the Pacific northwest, as well as in Wisconsin and other areas, approximately 90% of the 1980 U.S. crayfish production existed in Louisiana. A decade ago, production ranged between 5000 and 10,000 MT, of which one-half was cultured in artificial impoundments (Avault, 1973) and culture has expanded since then. Fishing for crayfish in the United States is as old as the country. Yet, although small ponds were stocked with crayfish as long ago as 1770 in Louisiana, modern culture practices date from about 1949. Land area in Louisiana that was devoted to crayfish farming increased from 400 ha in 1959 to 18,400 ha in 1975 (Clark *et al.,* 1975). Texas, Mississippi, and California are adding acreage each year.

A third major area for development of crayfish farming may be Australia, home of the largest crayfish in the world. Two species, *Cherax destructor* and *C. tenuimanus,* are being promoted for farm pond production (Frost, 1975).

A few important biological features contribute to the widespread success of crayfishes as a cultured organism. Crayfishes differ from their marine relatives in having direct development, with no free larval stage. Although in most species fecundity is reduced to perhaps a few hundred young per brood, survival is high. The young hatch as nearly independent individuals that require no elaborate hatchery procedure. Moreover, growth is rapid. Many species reach commercial size in a single year or less, and, unlike homarid lobsters, the crayfishes are predominently vegetarian or detritivores. Crayfishes in young stages appear to eat more animal food than do many adults, and some species feed directly on organic debris. Production in ponds can reach in excess of 1000 kg/ha with supplemental feeding, and more than one-half of that with only planted vegetation or fertilization of the pond (Clark et al., 1975).

The crayfish industry of the state of Louisiana and its methods have been reviewed by de la Bretonne (1977). *Procambarus clarkii* and *P. acutus acutus* constitute 90% and 10% of the harvest, respectively. Management for the crop is simple. Brood crayfishes are stocked in May at 50 kg/ha in ponds not over 0.75-m deep. A few weeks later, the water is slowly drained, the crayfish have burrowed, and the females begin laying eggs while in their burrows. One female may have from 100 to 800 young. There is no free swimming stage. Rice or millet is planted for cover and food, and in September, the ponds are flooded again, at which time the female and her young emerge from the burrow. The young may be harvested from September through the following year. Production ranges from 58 to 1750 kg/ha per year, with a statewide average of about 233 kg/ha. The higher production levels are achieved with supplemental feeding, the lower levels result from poor management. Crayfishes feed on plant and animal matter that is naturally produced in the ponds. Thus, no expensive diet is required. Reproduction is not complicated. Harvesting is done mainly with baited traps, as in the commercial fishery, though it is inefficient and labor intensive.

Somewhat different methods are used elsewhere. In France, crayfish farming was important a century ago, but disease destroyed the industry until the recent revival. Current activities are summarized by Arrignon (1975). A more intensive culture method than that in Louisiana is pursued. Terra-cotta pipes, stacked three to four high, are provided as shelters and offer cover for as many as 40 crayfishes/m². At this density, even with 2 to 3 years required per crop, monetary yields are high (Brown, 1977, pp. 200–202.

The most spectacular of freshwater crayfishes are the giant forms of Australia and Tasmania. *Astacopsis gouldi* of Tasmania, which apparently can reach 3.5 kg, lives in cool running water and has an unfavorable ratio of tail to body length (Frost, 1975). *Euastacus armatus* of southeastern Victoria and eastern south Australia reaches 2.5 kg. It, too, is a cold-water form, but has slow growth, a 4-month hatching period, and poorly known food habits, though it apparently can eat mud. There are several other large species, such as *Cherax tenuimanus,* which is being cultivated in Western Australia. Production of *C. tenuimanus* in ponds may reach levels similar to those for the prawn *Macrobrachium rosenbergii* (2100 kg/ha at end of 1 year). Supplemental feeding of vegetable matter, which serves as substrate for detrital formation, is apparently successful (Morrissey, 1979). Much emphasis also appears to be devoted currently to the widely distributed, warmth-loving, highly adaptable yabbie, *Cherax destructor,* which, through rarely exceeding 100 g individual weight, can rival *Macrobrachium* in pond production (Mills and McCloud, 1983). It has the advantage of a short (6 weeks) hatching period, rapid growth (reaching 10 cm in the second year), ready availability of brood stock, high ratio of meat to total weight, and year round breeding (Frost, 1975). It has a high tolerance for salt water. While the other, larger species are fished mainly for sport, *Cherax destructor* has supported a commercial fishery, which reached 273,000 kg landed in 1973–1974, but which fell to 30,000 kg in 1976–1977.

Prospects are excellent for expansion and intensification of the crayfish rearing industry in the United States. Major constraints, at present, appear to include limited marketability outside of the immediate area of production, but that may be overcome by aggressive marketing development. In other areas of the world, trapping or cultivation for bait purposes, or both, promises to continue to develop moderately as recreational fishing continues to outpace commercial fishing in the United States. Techniques for accelerating the molt process and permitting the marketing of soft-shell crayfish for bait are available (Huner and Avault, 1977).

D. The Palinuridea

The spiny lobsters of tropical and subtropical waters, because of their greater diversity and distribution, are far more important economically than are the nephropid lobsters. Spiny lobsters differ from nephropids in ways significant for aquaculture. They are far more gregarious, less aggressive, and less cannibalistic, and hence they are more suitable for communal rearing. Many of them reach a large size, some grow relatively rapidly. Hence, there has been interest in commercial culture of these species. Perhaps, the greatest difference between these species and nephropid and freshwater lobsters (and the primary obstacle to culture of spiny lobster) is in the larval development. Rather than abbreviated development with a short pelagic development or no free-swimming larva, the spiny and sand lobsters have a prolonged larval life and a delicate larval stage, the phyllosoma.

Many species of palinurid lobsters have been hatched in the laboratory, and some have been reared through a number of early stages. Provenzano (1968) reviewed the status of knowledge of larval rearing. Some species have been reared from hatching through as many as 16 molts to an age of approximately 6 months (Saisho, 1966). These results were obtained primarily with use of *Artemia* nauplii as food. Dexter (1972) reared *Panulirus interruptus* through six stages, but no substantial progress has been made over the last decade in developing better techniques. Early stage larvae apparently can capture adequate quantities of nauplii if concentrations near five nauplii/ml are maintained. As the phyllosoma larvae molt and increase size, nauplii no longer provide sufficient nutrition, perhaps because the phyllosoma cannot capture sufficient quantities or perhaps because of inadequate nutritional composition of *Artemia* for later stage larvae (Inoue, 1965).

Larvae of *Jasus edwardsii* rejected *Artemia* nauplii and a variety of natural foods (Batham, 1967; Silverbauer, 1971). Thomas (1963) speculated that phyllosomas, which have extremely small mouthparts relative to their size and relative to those of other decapod larvae, may be adapted to feeding upon large soft objects, such as jellyfishes, with which they have often been associated in the plankton. Batham (1967) and others have fed phyllosomas on polychaete worms and fish flesh. Mitchell (1971) found that *P. interruptus* preferred large prey, such as fish larvae, hydromedusae, and ctenophores.

Though scyllarid (sand or slipper) lobsters are less important in commercial fisheries than are their palinurid relatives, their phyllosoma larvae may be more amenable to artificial culture. *Scyllarus americanus* is the only species with a phyllosoma to be reared from hatching to metamorphosis in the laboratory, and this was accomplished on a diet of *Artemia* nauplii exclusively (Robertson, 1968). A number of other scyllarids were hatched but were not reared to metamorphosis. Perhaps the scyllarids, in general, have somewhat shorter larval periods than have palinurids, or possibly *S. americanus* is atypical in reaching metamorphosis in only 32 to 40 days. The larval development of this rather small species seems to be shorter than that of most panulirids, but knowledge of phyllosoma ecology and larval requirements is so poor and the technology of rearing so primitive that it is not feasible to project progress. Chittleborough and Thomas (1969) have shown that a Western Australian species of palinurid has a pelagic development of 11 months, and the Californian *P. interruptus* also is believed to have a very extended development in the field (Johnson, 1960, 1971). This does not mean that, under artificial conditions, the development time could not be shortened, only that, under the best of circumstances, it is likely that larval period will be inconveniently long for commercial culture. On the other hand, if a large and reliable source of postlarvae were obtainable, some species of spiny lobsters could be grown to commercial size in a moderate time (Serfling and Ford, 1975a).

Numerous examples exist of large-scale aquaculture industries based on wild seed. Although there are certainly disadvantages to such lack of control over the source of seed, it is possible to produce commercial crops if wild seed is abundantly and legally available. Ingle and Witham (1969) and later Chittleborough (1974) and Serfling and Ford (1975b) suggested possible means of spiny lobster culture via the collection of postphyllosoma stages in areas of natural recruitment or of unusual abundance, as in the neuston.

Technically, rearing from the puerulus stage is feasible and has been accomplished numerous times for a variety of species. Though some species may require as much as 5 years of captive rearing to reach reproductive size (Chittleborough, 1974), others can be marketed in as little as 2 years from metamorphosis (Phillips et al., 1977). A detailed summary of current understanding of larval ecology of phyllosomas has been published by Phillips and Sastry (1980). Ecology of juveniles of the Palinuridae has been reviewed by Kanciruk (1980).

VII. SUMMARY AND PERSPECTIVES

The long expected worldwide shortage of food, especially protein, is with us. As late as the early 1960s, cultured fish and shellfish constituted less than 5% of the world's commercial fisheries landings, but now more than 6 million MT of cultured aquatic and marine products amount to at least 10% of world fish production (Glude, 1978). With marine fisheries production nearly stable in the last decade and with the rapid advancement of technology and practice of fish culture, projections for the year 2000 indicate that 15–20% of the world's fish supply will be the products of culture. In terms of the world market, present crustacean culture is relatively small in scale. The combination of abundant natural stocks for fishing and the lack of scientific knowledge delayed serious efforts at developing a variety of crustacean culture techniques. The high unit value of crustaceans and the resulting economic demand has greatly accelerated the pace of this development. The combination of economic demand, decreasing supply, and the availability of necessary technical information has led to the application of that knowledge on an unprecedented scale during the last 2 decades. This rapid development of commercial exploitation of crustacean culture technology has in turn stimulated a wide variety of investigations into basic biology, especially of the decapods, ranging from nutritional requirements to artificial insemination techniques. Just 20 years ago, the quantity of cultured crustaceans as a fraction of total crustacean fishery landings was near zero. Today, it is a large and increasing fraction. For example, yields of pond-cultured penaeid shrimp now equal or exceed fishery yields in Ecuador, and the absolute quantity produced is a large fraction of total U.S. imports. At least one large shrimp fishing company has sold its fishing fleet because of the belief that it is now cheaper to grow shrimp than to hunt them (Rosenberry, 1982, p. 6). Most rapid expansion of large-scale crustacean culture has occurred during the last 5 years, and there is no sign as yet of a decrease in the rate of development. In Taiwan alone, production of both Penaeus monodon and Macrobrachium rosenbergii doubled between 1979 and 1980 from 3200 to 6500 MT and from 65 to 150 MT, respectively

(Sandifer et al., 1982). Prospects for substantial further increases in commercial production of cultured crustaceans in the decades ahead are encouraging. Continued rapid development of aquaculture will produce an even greater impetus to scientific investigations in one of the most exciting marriages of science and industry of our time. This increased level of activity will continue to spin off additional benefits in terms of our understanding of the needs and potentials of many species now of commercial importance.

REFERENCES

Adelung, D., and Ponat, A. (1977). Studies to establish an optimal diet for the decapod crab Carcinus maenas under culture conditions. Mar. Biol. (Berlin) 44, 287–292.

Aiken, D. E., and Waddy, S. L. (1976). Controlling growth and reproduction. The American lobster. Proc. World Maricult. Soc. 7, 415–430.

Aiken, D. E. and Waddy, S. L. (1978). Space, density, and growth of the lobster (Homarus americanus). Proc. World Maricult. Soc. 9, 461–467.

Andrews, J. W., Sick, L. V., and Baptist, G. P. (1972). The influence of dietary protein and energy levels on growth and survival of penaeid shrimp. Aquaculture 1, 341–347.

Anonymous (1971). Shrimp harvest at marifarms. Am. Fish Farmer World Aquacult. News 3, 5–7.

Aquacop (1975). Maturation and spawning in captivity of penaeid shrimp: Penaeus merguiensis De Man, Penaeus japonicus Bate, Penaeus aztecus Ives, Metapenaeus ensis de Haan and Penaeus semisulcatus de Haan. Proc. World Maricult. Soc. 6, 123–132.

Aquacop (1976). Reproduction in captivity and growth of Penaeus monodon Fabr. in Polynesia. Proc. World Maricult. Soc. 8, 927–937.

Arnstein, D. R., and Beard, T. W. (1975). Induced maturation of the prawn Penaeus orientalis Kishinouye in the laboratory by means of eyestalk removal. Aquaculture 5, 411–412.

Arrignon, J. (1975). Crayfish farming in France. In "Freshwater Crayfish, Papers from the Second International Symposium on Freshwater Crayfish. Baton Rouge, LA; 1974" (J. W. Avault, ed.), pp. 105–116. Louisiana State Univ., Baton Rouge.

Avault, J. W. (1973). Crayfish farming in the United States. In "Freshwater Crayfish. Papers from the First International Symposium Freshwater Crayfish, Austria 1972" (S. Abrahamsson, ed.), pp. 239–250. Student Literature, Lund.

Bardach, J. E., Ryther, J. H., and McLarney, W. O. (1972). "Aquaculture: The Farming and Husbandry of Freshwater and Marine Organisms." Wiley (Interscience), New York.

Batham, E. J. (1967). The first three larval stages and feeding behavior of phyllosoma of the New Zealand palinurid crayfish, Jasus edwardsii (Hutton, 1875). Trans. Soc. N. Z., Zool. 9, 53–64, 1 pl.

Beard, T. W., and Wickins, J. F. (1980). Breeding of Penaeus monodon Fabricius in laboratory recirculation systems. Aquaculture 20, 79–89.

Biddle, G. N. (1977). The nutrition of Macrobrachium species. In "Shrimp and Prawn Farming in the Western Hemisphere" (J. A. Hanson and H. L. Goodwin, eds.), pp. 272–290. Dowden, Hutchinson and Ross, Stroudsburg, Pennsylvania.

Boschi, E. E., and Scelzo, M. A. (1974). Rearing the penaeid shrimp, Artemesia longinaris, from egg to juvenile in the laboratory. Proc. World Maricult Soc. 5, 443–444.

Botsford, L. W. (1977). Current economic status of lobster culture research. Proc. World Maricult. Soc. 8, 723–740.

Bowen, B. K. (1980). Spiny lobster fishery management. In "Biology and Management of Lobsters" (J. S. Cobb and B. F. Philips, eds.), pp. 243–264. Academic Press, New York.

Branford, J. R. (1978). Incubation period for the lobster, Homarus gammarus at various temperatures. Mar. Biol. (Berlin) 47, 363–368.

Brodsky, S. Ya (1975). The crayfish situation in Ukraine. In "Freshwater Crayfish, Papers from the Second International Symposium on Freshwater Crayfish, Baton Rouge, LA, U.S.A. 1974" (J. W. Avault, Jr., ed.), pp. 27–29. Louisiana State Univ., Baton Rouge.

Brown, E. E. (1977). "World Fish Farming: Cultivation and Economics." Avi Publ. Co., Westport, Connecticut.

Brownell, W. N., Provenzano, A. J., and Martinez, M. (1977). Culture of the West Indian spider crab (Mithrax spinosissimus) at Los Roques, Venezuela. Proc. World Maricult. Soc. 8, 157–168.

Butler, T. H. (1964). Growth, reproduction and distribution of pandalid shrimps in British Columbia. J. Fish Res. Board Can. 21, 1403–1452.

Carlberg, J. M., Van Olst, J. C., and Ford, R. F. (1978). A comparison of larval and juvenile stages of the lobsters Homarus americanus, Homarus gammarus, and their hybrid. Proc. World Maricult. Soc. 9, 109–122.

Carlson, F. T. (1954). The American lobster fishery and possible applications of artificial propagation. Yale Conser. Stud. 3, 3–7.

Castell, J. D., and Budson, S. D. (1974). Lobster nutrition: The effect on Homarus americanus of dietary protein levels. J. Fish. Res. Board Can. 31, 1363–1370.

Castell, J. D., and Covey, J. F. (1976). Dietary lipid requirements of adult lobsters, Homarus americanus (M.E.), J. Nutr. 106, 1159–1165.

Castell, J. D., Mason, E. G., and Covey, J. F. (1975). Cholesterol requirements of juvenile American lobster (Homarus americanus). J. Res. Board Can. 32, 1431–1435.

Chen, T. P. (1976). "Aquaculture Practices in Taiwan." Page Bros, Horwick (Ltd.),

Chittleborough, R. G. (1974). Review of prospects for rearing rock lobsters. CSIRO, Div. Fish. Oceanogr. Repr. No. 812.

Chittleborough, R., and Thomas, L. (1969). Larval ecology of the Western Australian marine crayfish, with notes upon other panulirid larvae from the eastern Indian Ocean. Aust. J. Mar. Freshwater. Res. 20, 199–223.

Clark, W. H., Jr., Talbot, P., Neal, R. A., Mock, C. R., and Salser, B. R. (1973). In vitro fertilization with the non-motile spermatozoa of the brown shrimp Penaeus aztecus. Mar. Biol. (Berlin) 22, 353–354.

Clark, D. F., Avault, J. W., and Meyers, S. P. (1975). Effects of feeding, fertilization, and vegetation on production of red swamp crayfish, Procambarus clarkii. In "Freshwater Crayfish. Papers from the Second International Symposium on Freshwater Crayfish held at Baton Rouge, LA, U.S.A. 1974" (J. W. Avault, ed.), pp. 125–138. Louisianna State Univ., Baton Rouge.

Clifford, H. C., III, and Brick, R. W. (1978). Protein utilization in the freshwater shrimp, Macrobrachium rosenbergii. Proc. World Maricult. Soc. 9, 195–208.

Cohen, D., Finkel, H., and Sussman, M. (1976). On the role of algae in larviculture of Macrobrachium rosenbergii. Aquaculture 8, 199–208.

Colvin, L. B., and Brand, C. W. (1977). The protein requirements of penaeid shrimp at various life-cycle stages in controlled environment systems. Proc. World Maricult. Soc. 8, 821–827.

Comeaux, M. L. (1975). Historical development of the crayfish industry in the United States. In "Freshwater Crayfish: Papers from the Second International Symposium on Freshwater Crayfish; Baton Rouge, LA., U.S.A. 1974" (J. W. Avault, ed.), pp. 609–619. Louisiana State Univ., Baton Rouge.

Conklin, D. E. (1976). Nutritional studies of lobsters (Homarus americanus). Proc. Int. Conf. Aquacult. Nutr., 1st, 1975 pp. 287–296.

Conklin, D. E. (1980). Nutrition. In "Biology and Management of Lobsters" (J. S. Cobb and B. F. Phillips, eds.), Vol. 1, pp. 277–300. Academic Press, New York.

Conklin, D. E., Devers, K., and Bordner, C. E. (1977). Development of artificial diets for the lobster, Homarus americanus. Proc. World Maricult. Soc. 8, 841–852.

Conklin, D. E., Goldblatt, M. J., Bordner, C. E., Baum, N. A., and McCormick, T. B. (1978). Artificial diets for the lobster, Homarus americanus: A reevaluation. Proc. World Maricult. Soc. 9, 243–250.

Conklin, D. E., D'Abramo, L. R., Bordner, E. E., and Baum, N. A. (1980). A successful purified diet of the culture of juvenile lobsters: The effect of lecithin. Aquaculture 21, 243–249.

Costlow, J. D., and Bookhout, C. G. (1959). The larval development of Callinectes spidus Rathbun, reared in the laboratory. Biol. Bull. (Woods Hole, Mass.) 116, 373–396.

Couch, J. A. (1978). Diseases, parasites and toxic responses of commercial penaeid shrimps of the Gulf of Mexico and South Atlantic coast of North America. Fish. Bull. 76, 1–44.

Couch, J. A., and Nimmo, D. (1974). Detection of interactions between natural pathogens and pollutants in aquatic animals. Proc. Gulf Coast Reg. Symp. Dis. Aquat. Anim., La. State Univ. Center Wetland Resour. Publ. No. LSU-SG-74-05, pp. 261–268.

Couturier-Bhaud, Y. (1974). Cycle biologique de Lysmata seticaudata Risso (Crustacea, Decapode). II. Sexualité et reproduction. Vie Millieu, Ser. A 24 fasc. 3, 423–430.

Cripps, M. C., and Nakamura, R. M. (1977). The effect of $CaCO_3$ in the water on growth of Macrobrachium rosenbergii. In "Shrimp and Prawn Farming in the Western Hemisphere" (J. A. Hanson and H. L. Goodwin, eds.), pp. 245–246. Dowden, Hutchinson and Ross, Stroudsburg, Pennsylvania.

D'Abramo, L. R., Bordner, C. E., Conklin, D. E., and Baum, N. A. (1981). Essentiality of dietary phosphatidylcholine for the survival of juvenile lobsters. J. Nutr. 111, 425–431.

Darnell, R. M. (1958). Food habits of fishes and large invertebrates of Lake Pontchartrain, Louisiana, an estaurine community. Publ. Inst. Mar. Sci., Univ. Texas 5, 353–416.

De la Bretonne, L. (1977). A review of crawfish culture in Louisiana. Proc. World Maricult. Soc. 8, 265–270.

Deshimaru, O., and Kuroki, K. (1976). Studies on a purified diet for prawn. VII. Adequate levels of ascorbic acid and inositol. Bull. Jpn. Soc. Fish. 42, 571–576.

Deshimaru, O., and Shigeno, K. (1972). Introduction to the artificial diet for prawn Penaeus japonicus. Aquaculture 1, 115–133.

Dexter, D. M. (1972). Molting and growth in laboratory reared phyllosomas of the California spiny lobster, Panulirus interruptus. Calif. Fish. Game 58, 107–115.

Dobkin, S. (1969). Abbreviated larval development in caridean shrimps and its significance in the artificial culture of these animals. FAO Fish. Rep. No. 57, pp. 935–946.

Dobkin, S., Azzinaro, W. P., and Van Montfrans, J. (1974). Culture of Macrobrachium acanthurus and M. carcinus with notes on the selective breeding and hybridization of these shrimps. Proc. World Maricult. Soc. 5, 51–62.

Dow, R. L. (1980). The clawed lobster fisheries. In "Biology and Management of Lobsters" (J. S. Cobb and B. F. Phillips, eds.), Vol. 2, pp. 265–316. Academic Press, New York.

Eddie, G. O. (1977). "The Southern Ocean, the Harvesting of Krill." FAO, UNIPUB, New York.

Fenucci, J. L., Lawrence, A. L., and Zein-Eldin, Z. P. (1981). The effects of fatty acid and shrimp meal composition of prepared diets on growth of juvenile shrimp, Penaeus stylirostris. J. World Maricult. Soc. 12, 315–324.

Forster, J. R. M. (1976). Studies on the development of compounded diets for prawns. Proc. Int. Conf. Aquacult. Nutr., 1st, 1975 pp. 229–248.

Forster, J. R. M., and Beard, T. W. (1973). Growth experiments with the prawn *Palaemon serratus* Pennant fed with fresh and compounded foods. *G. B. Minist. Agric. Fish., Food. Fish. Invest., Ser. II* **27,** 1–16.

Forster, J. R. M., and Gabbott, P. A. (1971). The assimilation of nutrients from compounded diets by the prawns *Palaemon serratus* and *Pandalus platyceros. J. Mar. Biol. Assoc. U. K.* **51,** 943–961.

Frost, J. V. (1975). Australia crayfish. *In* "Freshwater Crayfish: Papers from the Second International Symposium on Freshwater Crayfish, Baton Rouge, LA. 1974" (J. W. Avault, ed.), pp. 87–96. Louisiana State Univ., Baton Rouge.

Fryer, G. (1977). Studies on the functional morphology and ecology of the atyid prawns of Dominica. *Philos. Trans. R. Soc. London, Ser. B* **277,** 57–129.

Fujimoto, M., Fujimura, T., and Kato, K. (1977). An idiot's guide to prawn ponds. *In* "Shrimp and Prawn Farming in the Western Hemisphere" (J. A. Hanson and H. L. Goodwin, eds.), pp. 237–242. Dowden, Hutchinson and Ross, Stroudsburg, Pennsylvania.

Fujimura, T. (1966). Notes on the development of a practical mass culturing technique of the giant prawn *Macrobrachium rosenbergii. Proc. Indo-Pac. Fish. Counc., 12th Sess., Hono lulu, Hawaii* (IPEC/C66/WP47) 3 pp.

Fujimura, T. (1972). Notes on progress made in developing a mass culturing technique for *Macrobrachium rosenbergii* in Hawaii. *In* "Coastal Aquaculture in the Indo-Pacific Region" (T. V. R. Pillay, ed.), pp. 313–327. Fishing News (Books) Ltd., Surrey, England.

Fujimura, T., and Okamoto, H. (1970). Notes on progress made in developing a mass culturing technique for *Macrobrachium rosenbergii* in Hawaii. *Proc. Indo-Pac. Fish. Counc., 14th Sess., Bangkok, Thailand, Symp. No. 53,* 17 pp.

Fujinaga, M. (1969). Kuruma shrimp (*Penaeus japonicus*) cultivation in Japan. *FAO Fish Rep. No. 57, 3,* pp. 811–832.

Gallagher, M. L., Conklin, D. E., and Brown, W. D. (1976). The effects of pelletized protein diets on growth, molting and survival of juvenile lobsters. *Proc. World Maricult. Soc.* **7,** 363–390.

Garth, J. S. (1971). *Demania toxica,* a new species of poisonous crab from the Phillipines. *Micronesica* **7,** 179–183.

George, M. J., Mohammed, K. H., and Pillai, N. N. (1968). Observations on the paddy-field prawn filtration of Kerala, India. *FAO Fish. Rep. No. 57, 2,* pp. 427–442.

Ghelardi, R. J., and Shoop, C. T. (1972). Culturing lobsters (*Homarus americanus*) in British Columbia. *Tech. Rep. Fish. Res. Board Can. No. 301,* pp. 32–42.

Glude, J. (1978). The contribution of fisheries and aquaculture to world and U.S. food supplies. *In* "Drugs and Food from the Sea, Myth or Reality" (P. M. Kaul and C. J. Sindermann, eds.), pp. 235–247. Univ. of Oklahoma, Norman.

Goldblath, M. J., Conklin, D. E., and Brown, W. D. (1980). Nutrient leaching from coated crustacean rations. *Aquaculture* **19,** 383–388.

Goodwin, H. L., Hanson, J. A., Trimble, W. C., and Sandifer, P. A. (1977). Book II. Freshwater prawn farming (genus *Macrobrochium*) in the Western Hemisphere. A state of the art review and status assessment. *In* "Shrimp and Prawn Farming in the Western Hemisphere" (J. A. Hanson and H. L. Goodwin, eds.), pp. 193–439. Dowden, Hutchinson and Ross, Stroudsburg, Pennsylvania.

Goyert, J. C., and Avault, J. W., Jr. (1979). Effects of stocking density and substrate on growth and survival of crawfish (*Procambarus clarkii*) grown in a recirculating culture system. *Proc. World Maricult. Soc.* **9,** 731–736.

Guary, M., Kanazawa, A., Tanaka, N., and Ceccaldi, H. J. (1976). Nutrition requirements of prawn. VI. Requirement for ascorbic acid. *Mem. Fac. Fish., Kagoshima Univ.* **25,** 53–57.

Hand, C. (1977). "Development of Aquaculture Systems." Univ. Calif. Sea Grant Publ. No. 58 (IMR 77-105).

Hanson, J. A., and Goodwin, H. L., (1977). "Shrimp and Prawn Farming in the Western Hemisphere." Dowden, Hutchinson and Ross, Stroudsburg, Pennsylvania.

Hartnoll, R. I. G. (1969). Mating in the Brachyura. Crustaceana 16, 161–181.

Hashimoto, Y., Konosu, S., Inoue, A., Saisho, T., and Miyaka, S. (1969). Screening of toxic crabs in the Ryuku and Amami islands. Bull. Jpn. Soc. Sci. Fish. 35, 83–87.

Hedgecock, D., Nelson, K., and Shleser, R. A. (1976). Growth differences among families of the lobster, Homarus americanus. Proc. World Maricult. Soc. 7, 347–361.

Heinen, J. M. (1976). An introduction to culture methods for larval and post-larval penaeid shrimp. Proc. World Maricult. Soc. 7, 333–343.

Helfrich, P. (1973). The feasibility of brine shrimp production on Christmas Island. Univ. Hawaii Sea Grant Tech. Rep., UNIHI-Sea Grant TR-73-02. 173 pp.

Hirata, H., Mori, Y., and Watanabe, M. (1975). Rearing of prawn larvae, Penaeus japonicus, fed soy-cake particles and diatoms. Mar. Biol. (Berlin) 29, 9-13.

Holthuis, L. B. (1952). A general revision of the Palaemonidae (Crustacea Decapoda Natantia) of the Americas. II. The superfamily Palaemoninae. Allan Hancock Found. Publ., Occas., Pap. No. 12, pp. 1–396.

Holthuis, L. B. (1968). Are there poisonous crabs? Crustaceana 15, 215–222.

Hudinaga, M. (1942). Reproduction, development and rearing of Penaeus japonicus Bate. Jpn. J. Zool., 10, 305–493 and 31 pp. of plates (in English).

Hughes, J. T. (1975). Lobster culture. In "Culture of Marine Invertebrate Animals" (W. L. Smith and M. H. Chanley, eds.), pp. 221–227. Plenum, New York.

Hughes, J. T., and Matthiessen, G. C. (1962). Observations on the biology of the American lobster Homarus americanus. Limnol. Oceanog. 7, 414–421.

Hughes, J. T., Sullivan, J. J., and Shleser, R., (1972). Enhancement of lobster growth. Science 177, 1110–1111.

Huner, J. V., and Avault, J. W. (1977). Investigation of methods to shorten the intermolt period in a crawfish. Proc. World Maricult. Soc. 8, 883–893.

Hunte, W. (1977). Laboratory rearing of the atyid shrimps Atya innocuous Herbst and Micratya poeyi Guerin-Meneville (Decapoda, Atyidae). Aquaculture 11, 373–378.

Ibrahim, K. H. (1962). Observations on the fishery and biology of the freshwater prawn, Macrobrachium malcolmsonii Milne-Edwards of River Godavary. Indian J. Fish. 9, 433–487.

Ingle, R. M., and Witham, R. (1969). Biological considerations in spiny lobster culture. Proc. Gulf Carib. Fish. Inst. 21, 158–162, 2 pls.

Inoue, M. (1965). On the relation of amount of food taken to the density and size of food and water temperature in rearing the phyllosoma of the Japanese spiny lobster Panulirus japonicus (Von Siebold). Bull. Jpn. Soc. Sci. Fish 31, 902–906.

Iwai,T. Y. (1976). Energy transformation and nutrient assimilation by the freshwater prawn Macrobrachium rosenbergii under controlled laboratory conditions. M.S. Thesis, Dept. Anim. Sci., Univ. of Hawaii, Honolulu.

Jahnig, C. E. (1977). Artemia culture on a commercial pilot scale. Proc. World Maricult. Soc. 8, 169–172.

Johnson, M. W. (1960). The offshore drift of larvae of the California spiny lobster Panulirus interruptus. Dept. Calif. Coop. Oceanic Fish. Invest. 7(1958–1959), 147–161.

Johnson, M. W. (1971). The palinurid and scyllarid lobster larvae of the tropical eastern Pacific and their distribution as related to the prevailing hydrography. Bull. Scripps Inst. Oceanogr. 19, 1–36.

Johnson, S. K. (1975). "Handbook of Shrimp Diseases." Texas A. and M. Univ. Sea Grant College Publ. #TAMU-SA-77-603, 23 pp. incl. 2 of color illustrations, College Station, Texas (revised 1978).

Johnson, S. K. (1977). "Crawfish and Freshwater Shrimp Diseases." Texas A. and M. Univ. Sea Grant College Publ. #TAMU-SA-77-605, 18 pp. incl. 2 of color illustrations, College Station, Texas.

Johnson, W. C. (1979). Culture of freshwater prawns. (Macrobrachium rosenbergii) using geothermal waste water. Proc. World Maricult. Soc. 10, 385–391.

Joseph, J. D., and Williams, J. E. (1975). Shrimp head oil: A potential feed additive for mariculture. Proc. World Maricult. Soc. 6, 147–155.

Kanazawa, A., Shimaya, M., Kawasaki, M., and Kashiwada, K. (1970). Nutritional requirements of prawn. I. Feeding on artificial diet. Bull. Jpn. Soc. Sci. Fish. 36, 949–954.

Kanazawa, A., Tanaka, N., Teshima, S., and Kashiwada, K. (1971). Nutritional requirements of prawn. II. Requirements for sterols. III. Utilization of the dietary sterols. Bull. Jpn. Soc. Sci. Fish. 37, 211–215 and 1015–1019.

Kanciruk, P. (1980). Ecology of juvenile and adult Palinuridae (spiny lobsters). In "Biology and Management of Lobsters" (J. S. Cobb and B. F. Phillips, eds.), Vol. II, pp. 59–96. Academic Press, New York.

Kato, S. (1974). Development of the pelagic red crab (Galatheidae, Pleuroncodes planipes) fishery in the eastern Pacific ocean. Mar. Fish. Rev. 36, 1–9.

Kelly, R. O., Haseltine, A. W., and Ebert, E. E. W. (1977). Mariculture potential of the spot prawn, Pandalus platyceros Brandt. Aquaculture 10, 1–16.

Kitabayashi, K., Kurata, H., Shudo, K., Nakamura, K., and Ishikawa, S., (1971). Studies on formula feed for Kuruma prawn. I–V. Bull. Tokai Reg. Fish. Res. Lab. No. 65, pp. 91–147.

Kittaka, J. (1976). Food and growth of penaeid shrimp. Proc. Int. Conf. Aquacult. Nutr. 1st, 1975 pp. 249–288.

Kittaka, J. (1977). Technique of prawn culture. In "Aquaculture in Shallow Seas; Progress in Shallow Sea Culture" (T. Imai, ed.), pp. 475–561. Amerind Publ. Co., New Delhi. (Engl. translation available from U.S. Dept. of Commerce, National Technical Information Service, Springfield, Virginia, 22161, U.S.A.)

Korringa, P. (1976). "Farming Marine Fishes and Shrimps." Elsevier Sci. Publ. Co., New York.

Kossakowski, J., and Orzechowski, B. (1975). Crayfish Orconectes limosus in Poland. In "Freshwater Crayfish: Papers from the Second International Symposium on Freshwater Crayfish, Baton Rouge, LA, U.S.A. 1974." (J. W. Avault, ed.), pp. 31–47. Louisiana State Univ., Baton Rouge.

Krekorian, C. O., Summerville, D. C., and Ford, R. F., (1974). Laboratory study of behavioral interactions between the American lobster, Homarus americanus, and the California spiny lobster, Panulirus interruptus, with comparative observations on the rock crab, Cancer antennarius. Fish. Bull. 72, 1146–1159.

Laubier-Bonichon, A., and Laubier, L. (1979). Reproduction controlé chez la crevette Penaeus japonicus. In "Advances in Aquaculture" (T. V. R. Pillay and W. A. Dill, eds.), pp. 273–277. Fishing News (Books) Ltd., Farnham, England.

Lasser, G. W., and Allen, W. V. (1976). The essential amino acid requirements of the Dungeness crab, Cancer magister. Aquaculture 7, 235–244.

Lee, G. F., and Sanford, F. B. (1962). Soft crab industry. Commer. Fish. Rev. 24, 10–12.

Lightner, D. V., Colvin, L. B., Brand, C., and Donald, D. A. (1977). Black death, a disease syndrome of penaeid shrimp related to a dietary deficiency of ascorbic acid. Proc. World Maricult. Soc. 8, 611–623.

Ling, S. W. (1969). Methods of rearing and culturing *Macrobrachium rosenbergii* (de Man). *FAO Fish. Rep. No. 57*, pp. 607–619.

Ling, S. W. (1977). "Aquaculture in Southeast Asia, a Historical Review." Univ. Washington Press, Seattle.

Ling, S. W., and Merican, A. B. O. (1961). Notes on the life and habits of the adults and larval stages of *Macrobrachium rosenbergii* (de Man). *Proc. Indo-Pac. Fish. Counc.* **9**, 55–67.

Little, G. (1968). Induced winter breeding and larval development in the shrimp, *Palaemonetes pugio* Holthius (Caridea, Palaemonidae). *Crustaceana*, Suppl. **2**, 19–26.

Logan, D. T., and Epifanio, C. E. (1978). A laboratory energy balance for the larvae and juveniles of the American lobster *Homarus americanus*. *Mar. Biol. (Berlin)* **47**, 381–389.

Lunz, G. R. (1957). Pond cultivation of shrimp in South Carolina. *Proc. Gulf Carib. Fish. Inst.* **10**, 44–48.

Lyubimova, T. G., Naumov, A. G., and Lagunov, L. L. (1973). Prospects of the utilization of krill and other non-conventional resources of the world ocean. *FAO. Tech. III, Tech. Conf. Fish. Manag. Dev., Vancouver, Can., 13–23 Feb. 1973*. 9 pp.

McLeese, D. W. (1956). Effects of temperature, salinity and oxygen on the survival of the American lobster. *J. Fish. Res. Board Can.* **13**, 247–272.

McSweeny, E. S. (1977). Intensive culture systems. *In* "Shrimp and Prawn Farming in the Western Hemisphere" (J. A. Hanson and H. L. Goodwin, eds.), pp. 255–266. Dowden, Hutchinson and Ross, Stroudsburg, Pennsylvania.

Maddox, M. B., and Manzi, J. J. (1976). The effects of algal supplements in static system culture of *Macrobrachium rosenbergii* (de Man) larvae. *Proc. World. Maricult. Soc.* **7**, 677–698.

Mahler, L. E., Groh, J. E., and Hodges, C. N. (1974). Controlled environment aquaculture. *Proc. World Maricult. Soc.* **5**, 379–386.

Malecha, S. R., and Polovina, J. (1978). Genetic studies in the freshwater prawn (*Macrobrachium rosenbergii*): the influence of parentage and temperature on larvae development rates in Anuenue stocks. Abstr. 9th Annu. Meeting, World Maricult. Soc., 1978.

Manzi, J. J., Maddox, M. B., and Sandifer, P. A. (1977). Algal supplement enhancement of *Macrobrachium rosenbergii* (de Man) larviculture. *Proc. World Maricult. Soc.* **8**, 207–223.

Margarelli, P. C., Jr., and Colvin, L. B. (1978). Depletion/repletion of ascorbic acid in two species of penaeid: *Penaeus californiensis* and *Penaeus stylirostris*. *Proc. World Maricult. Soc.* **9**, 235–242.

Matilda, C. E., and Hill, B. J. (1980). Annotated bibliography of the portunid crab, *Scylla serrata* (Forskal). *Queensland Fish. Serv. Tech Rep. No. 3.* 17 pp.

Meneasveta, P., and Piyatiravitivokul, S. (1980). A comparative study for larviculture techniques for the giant freshwater prawn, *Macrobrachium rosenbergii* (de Man). *Aquaculture* **20**, 239–249.

Meyers, S. P., and Brand, C. W. (1975). Experimental flake diets for fish and Crustacea. *Proc. Fish Cult.* **37**, 67–72.

Meyers, S. P., and Zein-Eldin, Z. P. (1973). Binders aid pellet stability in development of crustacean diets. *Proc. World Maricult. Soc.* **3**, 351–364.

Middleditch, B. S., Missler, S. R., Hines, H. B., Cheng, E. S., McVey, J. P., Brown, A., and Lawrence, A. L. (1980). Maturation of penaeid shrimp: Lipids in the marine food web. *Proc. World Maricult. Soc.* **11**, 463–470.

Mills, B. J., and McCloud, P. I. (1983). Effects of stocking and feeding rates on experimental pond production of the crayfish *Cherex destructor* Clark (Decapoda: Parastacidae). *Aquaculture* **34**, 51–72.

Mitchell, J. R. (1971). Food preferences, feeding mechanisms and related behavior in phyllosoma larvae of the California spiny lobster, *Panulirus interruptus* (Randall). M.S. Thesis, San Diego State Univ., California.

Mitchell, J. R. (1975). A polyculture system for commercially important marine species with special reference to the lobster (*Homarus americanus*) *Proc. World Maricult. Soc.* **6**, 249–259.

Mock, R. C. (1972). Larval culture of penaeid shrimp at the Galveston Biological Laboratory. *NOAA Tech. Rep. Natl. Mar. Fish. Ser. No. 388*, pp. 33–40.

Mock, C. R., and Murphy, M. A. (1970). Techniques for raising penaeid shrimp from the egg to postlarva. *Proc. World Maricult Soc.* **1**, 143–156.

Mock, C. R., Neal, R., and Salser, B. R. (1974). A closed raceway for the culture of shrimp. *Proc. World Maricult. Soc.* **4**, 247–259.

Mock, C. R., Kelly, G. G., and Kelly, C. T. (1978). A hatchery system for freshwater shrimp. *9th Annu. Meet., World Maricult. Soc.* (Abstr.).

Morrissey, N. M. (1979). Experimental pond production of marron, *Cherax tenuimanus* (Smith) (Decapoda:Parastacidae). *Aquaculture* **16**, 319–344.

Nakanishi, T. (1979). Rearing larvae and post-larvae of the king crab (*Paralithodes camtschatica*. *In* "Advances in Aquaculture" (T. V. R. Pillay and W. A. Dill, eds.), pp. 319–321. Fishing News (Books) Ltd., Farnham, England.

New, M. B. (1976a). A review of dietary studies with shrimp and prawns. *Aquaculture* **9**, 101–144.

New, M. B. (1976b). A review of shrimp and prawn nutrition. *Proc. World Maricult. Soc.* **7**, 277–287.

New, M. B. (1980). A bibliography of shrimp and prawn nutrition. *Aquaculture* **21**, 101–128.

New, M. B., ed. (1982). "Giant Prawn Farming." Elsevier, Amsterdam.

Ong, K. S. (1966a). The early developmental stages of *Scylla serrata* Forskål (Crustacea, Portunidae), reared in the laboratory. *Proc. Indo-Pac. Fish. Counc., 11th Session, Sect. 2*, pp. 135–146.

Ong, K. S. (1966b) Observations on the post-larval history of *Scylla serrata* Forskål reared in the laboratory. *Malay Agric. J.* **45**, 429–443.

Parker, J. S., Conte, F. S., MacGrath, W. S., and Miller, B. W. (1974). An intensive culture system for penaeid shrimp. *Proc. World Maricult. Soc.* **5**, 65–79.

Peebles, J. B. (1977). Notes on mortality, cannibalism, and competition in *M. rosenbergii*. *In* "Shrimp and Prawn Farming in the Western Hemisphere" (J. D. Hanson and H. L. Goodwin, eds.), pp. 266–271. Dowden, Hutchinson and Ross, Stroudsburg, Pennsylvania.

Perez Farfante, I. (1978). Intersex anomalies in shrimp of the genus *Penaeopsis* (Crustacea: Penaeidae). *Fish. Bull.* **76**, 687–691.

Persoone, G., Sorgeloos, P., Roels, O., and Jaspers, E. (1980). The brine shrimp *Artemia*. *Proc. Int. Symp. Brine Shrimp, Artemia salina, Corpus Christi, Texas, U.S.A., Aug. 20–23, 1979* 428 pp.

Persyn, H. O. (1977). Artificial insemination of shrimp. United States Patent 4,031,855, June 28, 2977, 4 pp.

Phillips, B. F., and Sastry, A. N. (1980). Larval ecology. *In* "Biology and Management of Lobsters" (J. S. Cobb and B. F. Phillips, eds.), Vol. 2, pp. 11–57. Academic Press, New York.

Phillips, B. F., Campbell, N. A., and Rea, W. A., (1977). Laboratory growth of early juveniles of the western rock lobster, *Panulirus longipes cygnus*. *Mar. Biol. (Berlin)* **39**, 31–39.

Pillai, N. N., and Mohamed, K. H. (1973). Larval history of *Macrobrachium idella* (Hilgendorf) reared in the laboratory. *J. Mar. Biol. Assoc. India* **15**, 359–385.

Prah, S. K. (1982). Possibilities of pond culture of freshwater prawns in Ghana, West Africa. *In* "Giant Prawn Farming" (M. New, ed.), pp. 403–409. Elsevier, Amsterdam.

Primavera, J. H. (1978). Induced maturation of spawning in five-month-old *Penaeus monodon* Fabr. by eyestalk ablation. *Aquaculture* **13**, 351–354.

Provasoli, L. (1976). Nutritional aspects of crustacean aquaculture. *Proc. Int. Conf. Aquacult. Nutr.,* 1st, *1975* pp. 13–24.

Provenzano, A. J. (1968). Recent experiments on the laboratory rearing of tropical lobster larvae. *Gulf Caribb. Fish. Inst. Univ. Miami, Proc. 21st Annu. Sess., Nov. 1969,* pp. 152–157.

Reed, P. H. (1969). Culture methods and effects of temperature and salinity on survival and growth of dungeness crab (*Cancer magister*) larvae in the laboratory. *J. Fish. Res. Board Can.* **26,** 389–397.

Reimer, R. D., and Strawn, K. (1973). The use of heated effluents from power plants for the winter production of soft shell crabs. *Proc. World Maricult. Soc.* **4,** 87–96.

Rensel, J. E., and Prentice, E. F. (1979). Growth of juvenile spot prawn *Pandalus platyceros,* in the laboratory and in net pens, using different diets. *Fish. Bull.* **76,** 886–890.

Robertson, P. B. (1968). The complete larval development of the sand lobster, *Scyllarus americanus* (Smith) (Decapoda, Scyllaridae) in the laboratory, with notes on larvae from the plankton. *Bull. Mar. Sci.* **18,** 294–342.

Rosemark, R. (1978). Growth of *Homarus americanus* on *Artemia salina* diets with and without supplementation. *Proc. World Maricult. Soc.* **9,** 251–258.

Rosenberry, R. (1982). Business developments. *Aquacult. Dig.* **7,** 4–6.

Ryther, J. H. (1977). Preliminary results with a pilot-plant waste recycling marine aquaculture system. *In* "Wastewater Renovation and Reuse" (F. M. D'Itri, ed.), pp. 90–132. Dekker, New York.

Saisho, T. (1966). A note on the phyllosoma stages of spiny lobster. *Inf. Bull. Planktol. Jpn.* **13,** 69–71.

Salser, B., Mahler, L., Lightner, D., Ure, J., Donald, D., Brand, C., Stamp, N., Moore, D., and Colvin, B. (1978). Controlled environment aquaculture of penaeids. *In* "Drugs and Food from the Sea: Myth or Reality?" (P. N. Kaul and C. J. Sindermann, eds.), pp. 345–355. Univ. Oklahoma, Norman.

Sandifer, P. A. Recent advances in the culture of crustaceans. Proc. European Maricult. Soc./World Maricult. Soc. World Conference on Aquaculture, Venice, Italy, 21–25. Sept. 1981 (in press).

Sandifer, P. A., and Joseph, J. C. (1976). Growth responses and fatty acid composition of juvenile prawns (*Macrobrachium rosenbergii*) fed a prepared ration augmented with shrimp head oil. *Aquaculture* **8,** 129–138.

Sandifer, P. A., and Lynn, J. W. (1980). Artificial insemination of caridean shrimp. *In* "Recent Advances in Invertebrate Reproduction." (W. H. Clarke, Jr. and T. S. Adams, eds.), pp. 271–288. Elsevier, Amsterdam.

Sandifer, P. A., and Smith, T. (1979). A method for artificial insemination of *Macrobrachium* prawns and its potential use in inheritance and hybridization studies. *Proc. World Maricult. Soc.* **10,** 403–418.

Sandifer, P. A., and Williams, J. E. (1980). Comparisons of *Artemia* nauplii and non-living diets as food for larval grass shrimp, *Palaemonetes* spp.: Screening experiments. *In* "The Brine Shrimp *Artemia.* Vol. 3. Ecology, Culturing, Use in Aquaculture" (G. Persoone, P. Sorgeloos, O. Roels, and E. Jaspers, eds.), pp. 353–384. Universa Press, Wetteren, Belgium.

Sandifer, P. A., Smith, T. I. J., Stokes, A. D., and Jenkins, W. E. (1982). Semi-intensive growout of prawns (*Macrobrachium rosenbergii*): Preliminary results and prospects. *In* "Giant Prawn Farming" (M. New, ed.), pp. 161–172. Elsevier, Amsterdam.

Sankolli, K. N., Shenoy, S., Jalihal, D. R., and Almekar, G. B. (1982). Crossbreeding of the giant freshwater prawns *Macrobrachium rosenbergii* (de Man) and *M. malcolmsonii* (H. Milne-Edwards). *In* "Giant Prawn Farming" (M. New, ed.), pp. 91–98. Elsevier, Amsterdam.

Santiago, A. C., Jr. (1977). Successful spawning of cultured *Penaeus monodon* Fabricius after eyestalk ablation. *Aquaculture* **11,** 185–196.

Sastry, A. N., and Zeitlin-Hale, L. (1977). Survival of communally reared larval and juvenile lobsters, *Homarus americanus*. *Mar. Biol. (Berlin)* **39,** 297–303.

Schmitt, W. L. (1965). "Crustaceans." Univ. of Michigan Press, Ann Arbor, Michigan.

Serfling, S. A., and Ford, R. F. (1975a). Laboratory culture of juvenile stages of the California spiny lobster *Panulirus interruptus* Randall at elevated temperature. *Aquaculture* **6,** 377–387.

Serfling, S. A., and Ford, R. F. (1975b). Ecological studies of the puerulus stage of the California spiny lobster, *Panulirus interruptus*. *Fish. Bull.* **73,** 360–377.

Shleser, R. A., and Gallagher, M. L. (1974). Formulation of rations for the American lobster, *Homarus americanus*. *Proc. World Maricult. Soc.* **5,** 157–164.

Sick, L. V. (1976). Selected studies of protein and amino acid requirements for *Marcobrachium rosenbergii* larvae fed neutral density formula diets. *Proc. Int. Conf. Aquacult. Nutr., 1st, 1975* pp. 215–228.

Sick, L. V., and Andrews, J. W. (1973). The effect of selected dietary lipids, carbohydrates and proteins on the growth, survival and body composition of *Penaeus duorarum*. *Proc. World Maricult. Soc.* **4,** 263–276.

Sick, L. V., and Beaty, H. (1974). Culture techniques and nutrition studies for larval stages of the giant prawn, *Macrobrachium rosenbergii, Ga. Mar. Sci. Center, Univ. Ga. Tech. Rep. Ser. No. 74-75*.

Sick, L. V., and Beaty, H. (1975). Development of formula foods designed for *Macrobrachium rosenbergii* larval and juvenile shrimp. *Proc. World Maricult. Soc.* **6,** 89–102.

Silverbauer, B. I. (1971). The biology of the South African rock lobster *Jasus lalandii* (H. Milne Edwards). I. Development. *Oceanogr. Res. Inst. (Durban), Invest. Rep.* **92.**

Sindermann, C. J., ed. (1977). "Disease Diagnosis and Control in North American Marine Aquaculture." Elsevier, Scientific Publ. Co., New York.

Smith, T. I. J., and Sandifer, P. A. (1975). Increased production of tankreared *Macrobrachium rosenbergii* through use of artificial substrates. *Proc. World Maricult. Soc.* **6,** 55–66.

Squires, H. J. (1970). Lobster (*Homarus americanus*) fishery and ecology in Port au Port Bay, Newfoundland, 1960–65. *Proc. Natl. Shellfish. Assoc.* **60,** 22–39.

Stephenson, M., and Simmons, M. A. (1976). Section V. Respiration and growth of larval *Macrobrachium rosenbergii*. *In* "Laboratory Studies on Selected Nutritional, Physical, and Chemical Factors Affecting the Growth, Survival, Respiration, and Bioenergetics of the Giant Prawn *Macrobrachium rosenbergii*" (A. W. Knight, principal investigator), *Water Sci. Eng. Pap. 4501,* pp. 38–46. Univ. of California, Davis.

Stewart, J. E. (1980). Diseases. *In* "The Biology and Management of Lobsters" (J. S. Cobb and B. F. Phillips, eds.), Vol. 1, pp. 301–342. Academic Press., New York.

Subramoniam, T. (1981). Protandric hermaphroditism in a mole crab, *Emerita asiatica* (Decapoda anomura). *Biol. Bull. (Woods Hole, Mass.)* **160,** 161–174.

Tamura, T. (1970). "Marine Aquaculture," Translated by National Science Foundation, Washington, D.C. NITS Doc. PB 194-05IT.

Thomas, L. R. (1963). Phyllosoma larvae associated with medusae. *Nature (London)* **198,** 208.

Tobias, W. J., Sorgeloos, P., Bossuyt, E., and Roels, O. A. (1979). The technical feasibility of mass culturing *Artemia salina* in the St. Croix "artificial upwelling" mariculture system. *Proc. World Maricult. Soc.* **10**, 203–214.

Unestam, T. (1973). Significance of disease in freshwater crayfish. *In* "Freshwater Crayfish, Papers from the First International Symposium on Freshwater Crayfish, Austria, 1972" (S. Abrahamsson, ed.), pp 135–150.

United Nations Food and Agriculture Organization, Rome. (1980). "Yearbook of Fishery Statistics." Vol. 50, No. 57, pp. 39–44.

Uno, Y., and Fujita, M. (1972). Studies on the experimental hybridization of freshwater shrimps, *Macrobrachium nipponense* and *M. formosense. Int. Ocean. Dev. Conf. Exhib. Keidanren Kaikan, Tokyo,* **2**, (Abstr.).

Van Engel, W. A. (1958). The blue crab and its fishery in Chesapeake Bay. Part 1. Reproduction, early development, growth, and migration. *Commer. Fish. Rev.* **20**, 6–17.

Van Olst, J. C., and Carlberg, J. M. (1978). The effects of container size and transparency on growth and survival of lobsters cultured individually. *Proc. Annu. Meet. World. Maricult. Soc.* **9**, 469–479.

Van Olst, J. C., Ford, R. F., Carlberg, J. M., and Dorband, W. R. (1976). Use of thermal effluent in culturing the American lobster. *In* "Power Plant Waste Heat Utilization in Aquaculture, 100, Workshop 1" pp. 77–100. Public Service Electric and Gas Co., Newark, New Jersey.

Van Olst, J. C., Carlberg, J. M., and Ford, R. F. (1975). Effects of substrate type and other factors on the growth, survival, and cannibalism of juvenile *Homarus americanus* in mass rearing systems. *Proc. World Maricult. Soc.* **6**, 61–74.

Van Olst, J. C., Carlberg, J., and Ford, R. (1977). A description of intensive culture systems for the American lobster (*Homarus americanus*) and other cannibalistic crustaceans. *Proc. World Maricult. Soc.* **8**, 271–292.

Van Olst, J. C., Carlberg, J. M., and Hughes, J. (1980). Aquaculture. *In* "The Biology and Management of Lobsters" (J. S. Cobb and B. F. Phillips, eds.), Vol. 2, pp. 333–384. Academic Press, New York.

Villegas, C. T., Li, T. L., and Kanazawa, A. (1980). The effects of feeds and feeding levels on the survival of a prawn, *Penaeus monodon,* larvae. *Mem. Kagoshima Univ. Res. Center S. Pac.* **1**, 51–55.

Warner, G. F. (1977). "The Biology of Crabs." Van Nostrand-Reinhold, Princeton, New Jersey.

Watanabe, T. (1980). Studies on the improvement of feeding techniques for rearing the larvae of *Penaeus semisulcatus.* Kuwait Inst. Sci. Res. KISR/PP 1012/FRM-RT-R-8001.

Wickins, J. F. (1972). Experiments on the culture of the spot prawn *Pandalus platyceros* Brandt and the giant freshwater prawn Macrobrachium rosenbergii (de Man). *Fish. Invest. Ministry Agric. Fish. Food (G.B.), Ser. II,* **27**, Vii and 23 pp.

Wickins, J. F. (1982). Opportunities for farming crustaceans in western temperate regions. *In* "Recent Advances in Aquaculture" (J. F. Muir and R. J. Roberts, eds.), pp. 89–177. Westview Press, Boulder, Colorado.

Wigley, R. L., Theroux, R. B., and Murray, H. E. (1975). Deep sea Red Crab, *Geryon quinquidens,* survey off northeastern United States. *Mar. Fish. Rev.* **37**, 1–21.

Wildman, R. D. (1974). Aquaculture and the national Sea Grant Program. *NOAA Tech. Rep. Natl. Mar. Fish. Ser. Circ. No. 388.*

Williamson, D. I. (1968). The type of development of prawns as a factor determining suitability for farming. United Nations food and Agriculture Organization, Fish. Rpt. No. 57, Vol. 2, pp. 77–84.

Yaldwyn, J. C. (1960). Crustacea Decapoda Natantia from the Chatham Rise: A deep water bottom fauna from New Zealand. *N. Z. Dept. Sci. Ind. Res. Bull.* (1), 24.

Yaldwyn, J. C. (1966). Protandrous hemaphroditism in decapod prawns of the families Hippo-
    lytidae and Campylonotidae. *Nature (London)* **209,** 1366.
Yang, W. T. (1971). Preliminary report of the culture of the stone crab. *Proc. World Maricult.
    Soc.* **2,** 53–54.
Yang, W. T., and Krantz, G. E. (1976). Intensive culture of the stone crab. *Menippe mer-
    cenaria. Univ. Miami Tech. Bull.* **35,** 1–15.
Yatsuzuka, K. (1962). Studies on the artificial rearing of the larval Brachyura especially of the
    larval blue crab, *Neptunus pelagicus* Linn. *Rep. Usa Mar. Biol. St.* **9,** 1–88.
Yee, W. C. (1971). Thermal aquaculture design. *Proc. World Maricult. Soc.* **2,** 55–65.
Yee, W. C. (1972). Thermal aquaculture: Engineering and economics. *Environ. Sci. Technol.* **6,**
    232–237.
Zein-Eldin, Z. P., and Meyers, S. P. (1973). General consideration of problems in shrimp
    nutrition. *Proc. World Maricult. Soc.* **4,** 299–317.

Systematic Index*

*Note: Names that have been superseded appear in brackets. Parentheses around name of author of scientific name indicate that currently assigned genus is not the original one.

†For current taxonomic status of *Artemia*, see G. Persoone, P. Sorgeloos, O. Roels, and E. Jaspers, eds. (1980). "The Brine Shrimp *Artemia*," Vol. 1, Morphology, Genetics, Radiobiology, Toxicology. Universa Press, Wetteren, Belgium.

*Editor's note: The spiny lobster of western Australia, Panulirus cygnus George, 1962, has also been referred to as P. longipes and P. longipes cygnus. In some early literature it was misidentified as P. penicillatus. A detailed review and analysis of the situation is given in B. F. Phillips, G. R. Morgan, and C. M. Austin, 1980. Synopsis of biological data on the western rock lobster Panulirus cygnus George, 1962. FAD Fish. Synop. FIR/S **128**, 1–64.

Subject Index

Von Bertalanffy growth equation, 26, 200–
    201, 212, 228

W

Waste utilization
    by-catch, 3, 13, 33, 40, 78–79, 130, 220
    chitin/chitosan, 76–77, 137
    crude meal, 76–77
    organic, 76–77
    proteinaceous, 76–77, 137
    whole shell, 137

Y

Yield-per-recruit model, 35, 52, 201, 214–
    217, 224, 227–228

Z

Zooplankton fisheries
    copepod, 57
    general principles, 54–55
    krill, 57–67
    mysid, 55
    sergestid shrimp, 55–57